The Biology of African Savannahs

Bryan Shorrocks

Environment Department
University of York

OXFORD
UNIVERSITY PRESS

Great Clarendon Street, Oxford OX2 6DP

Oxford University Press is a department of the University of Oxford.
It furthers the University's objective of excellence in research, scholarship, and education by publishing worldwide in

Oxford New York

Auckland Cape Town Dar es Salaam Hong Kong Karachi
Kuala Lumpur Madrid Melbourne Mexico City Nairobi
New Delhi Shanghai Taipei Toronto

With offices in

Argentina Austria Brazil Chile Czech Republic France Greece
Guatemala Hungary Italy Japan Poland Portugal Singapore
South Korea Switzerland Thailand Turkey Ukraine Vietnam

Published in the United States
by Oxford University Press Inc., New York

First published 2007

British Library Cataloguing in Publication Data

Data available

Library of Congress Cataloging in Publication Data

Data available

Typeset by Newgen Imaging Systems (P) Ltd., Chennai, India
Printed in Great Britain
on acid-free paper by
Biddles Ltd., King's Lynn

ISBN 978–0–19–857065–3 978–0–19–857066–0 (pbk)

10 9 8 7 6 5 4 3 2 1

For Jo, who also loves Africa

Preface

This book is an introduction to African savannahs and their biology, concentrating on large mammal ecology, behaviour and conservation. Although it is a book on savannahs I hope that it sets out its ideas within a general framework of ecological and behavioural ideas.

In my 40 years as an ecologist I have worked on both temperate and tropical terrestrial ecosystems. I have carried out field surveys, field experiments and built mathematical and computer models. I believe all these elements are essential for understanding Nature, and they all appear in this book. You cannot simply watch wildlife populations and communities in order to understand how they function. You have to employ field manipulations, either natural or artificial, to tease out the possibilities. You have to describe what you see not just in words but in mathematical models, to allow your less precise verbal suspicions to be tested. And savannah ecology must fit comfortably within the framework of general ecological ideas. There can be no special pleading or special mechanisms that other ecologists just don't understand. I have tried to present such an encompassing view of savannah ecology and behaviour.

African savannahs are magical places. The brightness of the light, the intensity of the colours, and the beauty and excitement of the wildlife, and of course the people. These savannahs are worth preserving for future generations and I hope, in a small way, this book will generate an interest that helps them survive.

I would like to thank all the people that have helped in my savannah work in East Africa, in particular the people at the Mpala Research Centre, in Laikipia, Kenya. Their help and insights have been invaluable during my frequent stays there. Finally my thanks to Paul Ward who, over several beers, over several nights, while running an undergraduate field course in Derbyshire reawakened my latent interest in Africa and its wildlife.

Bryan Shorrocks
June 2007

Contents

1 **Savannahs** 1

Distribution world-wide 2
African savannahs 10

2 **The vegetation** 29

Rainfall, plant biomass, and the grass–tree mixture 29
Morphology and life history 32
Grasses 40
Trees 43
Local vegetation patterns: the Serengeti–Mara ecosystem 57

3 **The animals** 64

The insects 64
The birds 66
The mammals 68

4 **Single species populations** 113

Estimating numbers 114
What changes numbers? 127
Population models 136
Other species, in other areas 141

5 **Species interactions** 155

Predator–prey type interaction (+ −) 156
Competitive interactions (− −) 180
Mutualistic interactions (+ +) 196

6 **The savannah community and its conservation** 205

Energy flow and food webs 206
Assembly rules 219

Island biogeography 224
Conserving savannah ecosystems 229

REFERENCES 240
INDEX 255

1 Savannahs

Savannahs constitute one of the largest biomes of the world, comprising about 20 per cent of the land surface. Stated simply, they are tropical and subtropical grasslands, with scattered bushes and trees. Most savannah occurs in Africa, with a smaller amount in South America, India, and Australia. In Africa the trees tend to be deciduous, while in South America and Australia they tend to be evergreen. Savannahs occur around the equator (between the Tropic of Capricorn and the Tropic of Cancer), where it is warm, but relatively dry. Most experience seasonal drought and the vegetation is influenced by rainfall, soil type, grazing, browsing, and fire. The word savannah, or savanna, is probably derived from a sixteenth century Spanish word *zavanna*, meaning 'treeless plain'. It was recorded in 1535, by the Spanish historian Gonzalo Fernandez de Oviedo, as coming originally from Carib, a language spoken in northern South America and the Caribbean. Savannah therefore shares its origin with other well known words such as barbecue, cannibal and papaya.

Like many plants found in warmer and dryer regions, the dominant grasses in savannahs tend to be C_4 plants. Only in some very wet environments do C_3 grass species become abundant. The terms C_3 and C_4 refer to the type of CO_2 trapping mechanism (photosynthesis) used by the plant. C_4 plants have evolved a secondary carbon fixation pathway. CO_2 is first combined into a 4-carbon compound, in the mesophyll cells of the leaf, and then passed to the cells around the leaf veins where the CO_2 is released at high concentrations. It then enters the usual photosynthetic carbon reduction (PCR) pathway or Calvin-Benson Cycle, used by C_3 plants. C_4 plants are capable of utilizing higher light intensities than C_3 plants, have greater maximum photosynthesis, and use less water in the process. Because of these C_4 grasses, the photosynthetic efficiency of many savannahs is very high. However, C_4 plants are extremely poor quality food for most herbivores, vertebrate or invertebrate, (Caswell *et al.* 1973) unless the animal can break down cellulose. Intriguingly, both ungulates and termites, which are common in many savannah systems, have a symbiotic gut flora that produces an enzyme, cellulase, that can digest this plant cell-wall constituent.

Distribution world-wide

Figure 1.1 shows how global temperatures and precipitation have an influence on the major biomes, and Figure 1.2 shows a map of the world with the major areas of savannah indicated. In total they occupy some 23 million km^2 (Cole 1986). Mean annual rainfall typically varies between about 20 and 150 cm with 60 to 90 per cent of the year's rain falling in a short period of a few months. Although climatic factors, such as the annual temperature and the annual amount of rain, are not the only determinants of the savannah biome, they do potentially, have a major effect. The computer model, BIOME 3, predicts a distribution of savannahs, based solely on climatic factors, almost identical to that of Figure 1.2 (Haxeltine and Prenrice 1996).

On a world scale therefore, savannahs tend to occupy a climatic region between deserts and tropical forests, a picture that is seen very clearly in Africa. Here the huge savannah area surrounds the tropical forests of the Congo basin and to the north is bordered by the Sahara desert and to the south by the Kalahari desert. This climatic position is seen clearly in Figure 1.1. Because savannahs are defined as grasslands with varying amounts of tree cover, they also sit between grasslands and tropical seasonal forests in Figure 1.1. However, the boundaries between these biomes are never very clear cut, and grassland and seasonal forest frequently merge into savannah.

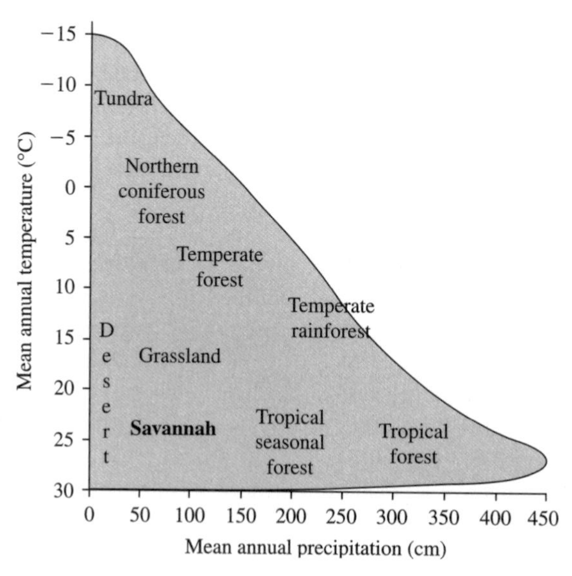

Fig. 1.1 Distribution of the major terrestrial biomes of the World with respect to mean annual temperature and mean annual precipitation (after Whittaker 1975).

Savannah areas usually have a positive water regime (rainfall greater than evaporation) during the wet season and a negative balance during the dry season. When the water regime becomes positive during the dry season, savannahs are replaced by forest.

South America

In South America there are several areas of savannah (Fig. 1.3). South of the equator is the extensive *cerrado* (1), comprising an area of 1.55 million km^2, about 20 per cent of the area of Brazil. In addition to savannah, this cerrado also contains patches of gallery forest. In terms of plant species it is the richest neotropical savannah, with about 430 species of trees and shrubs, about 300 herbaceous species and over a 100 grass species (Sarmiento 1996). Large mammals such as the giant anteater (*Myrmecophaga tridactyla*), yellow armadillo (*Euphractus sexcinctus*), jaguar (*Panthera onca*), and maned wolf (*Chryocyon brachyurus*) compete with the rapidly expanding Brazilian agricultural industry, which focuses primarily on soybean, maize and rice. The cerrado illustrates very well the gradation between typical savannah and other biomes (Fig. 1.1). It contains a gradient of habitats from grassland (*campo limpo*), through grassland dotted with shrubs and small trees (*campo sujo*), to low open woodland (*cerrado typica*). The trees of the cerrado have a characteristic contorted appearance. The bark is usually thick, and the leaves leathery, features that are thought to provide resistance to fire. The soil of the cerrado contains a hard layer, formed by the accumulation of iron oxides. Grasses can grow in the soil above this layer but trees can only

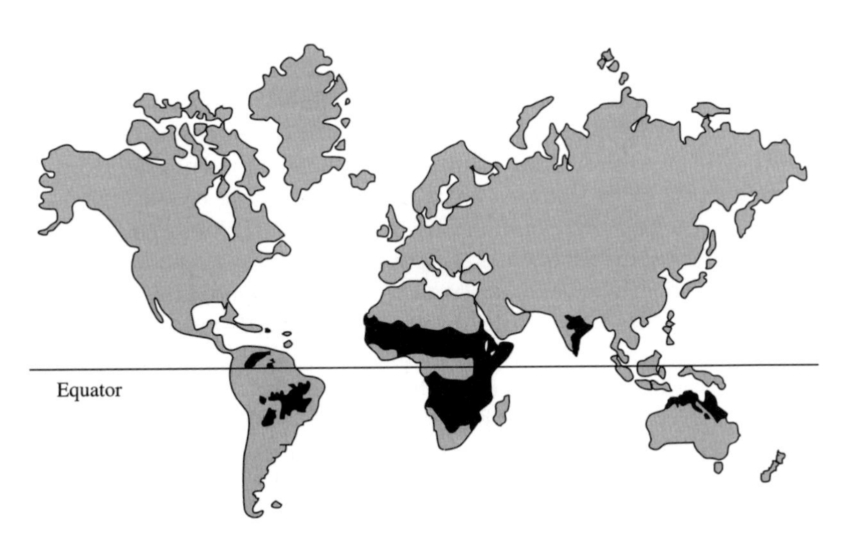

Fig. 1.2 World distribution of the savannah biome. It comprises about 20% of the land surface. On this familiar world projection it appears less, because those land areas near the poles (North America, Europe, and Asia) are artificially enlarged.

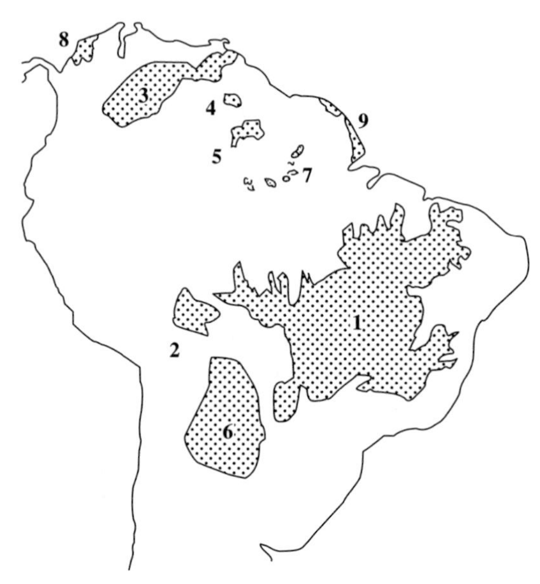

Fig. 1.3 The major South American savannahs. 1. cerrado, 2. llanos de Moxos, 3. llanos del Orinoco, 4. Gran Sabana, 5. savannahs of the Rio Branco-Rupununi, 6. chaco, 7. Amazonian campos, 8. llanos of the Magdalena, 9. coastal savannah of the Guayanas (after Solbrig 1996, Biodiversity and Savanna ecosystems processes. *Ecological Studies*, vol. 121, p. 5. with kind permission of Springer Science and Business Media.)

establish themselves where cracks in this layer allow their roots to reach deeper groundwater.

In the southwest of Brazil is a giant flooded area known as the *pantanal.* Here the vegetation is a mosaic of forest, and grassland and savannah vegetation more typical of the cerrado. The pantanal is home to many aquatic birds such as the jaburú (*Jabiru mycteria*), jacanas (*Jacana jacana*), and the anhinga (*Anhinga anhinga*), along with the endangered hyacinth macaw (*Anodorhynchus hyacinthinus*). Capybara (*Hydrochaeris hydrochaeris*), several species of monkeys (the howler monkey, *Alouatta caraya* and the brown capuchin, *Cebus apella* are the most common) also thrive in this mosaic of wetland and savannah, along with the occasional maned wolf, jaguar, several small forest cats (of the genus *Felis*), giant anteaters, and giant river otters (*Pteronura brasiliensis*).

Another area of wet and frequently flooded savannah is the *llanos de Moxos* (2). This savannah contains a mosaic of vegetation that varies from grassland to evergreen forest, depending upon the degree of flooding. South of this area is the *chaco* (6), a region of woodland and savannah. It occupies an area of approximately 1,000,000 km^2. The southern part of the chaco (and certain savannahs in southern Africa) are unique in that they can have winter frost. The chaco has many of the same mammals as the cerrado, including the giant anteater, jaguar, and maned wolf. As in the cerrado they are threatened by the rapid expansion of soya bean plantations.

North of the equator is an area called the *llanos del Orinoco* (3), situated around the Apure-Orinoco river system in Colombia and Venezuela and dominating an area of approximately 500,000 km^2. This is grassland with scattered trees, mostly belonging to three species, *Curatella americana*, *Bowdichia virgilioides*, and *Casearia sylvestris*. Trees are of low stature with narrow and tortuously branched stems. Most of the 12 million cattle in Venezuela are bred here, on 30 per cent of the country's land area, but with only 13 per cent of the population. Parts of the llanos del Orinoco is flooded in winter (May to October) by the Orinoco river and here, many plants have adapted to growing for long periods in standing water. However this standing water can also inhibit the growth of most trees. In these flooded areas the capybara and marsh deer (*Blastocerus dichotomus*) have adapted themselves to a semi-aquatic life, wading in the silt-bearing floodwater by day and sleeping on higher ground at night.

Other smaller areas of savannah north of the equator are the *Gran Sabana* (4) in Venezuela characterized by a very unique flora, the coastal savannahs of the Guayanas (9), and the savannahs of the Rio Branco-Rupununi (5) in Brazil (Figure 1.3).

In South America a distinct savannah fauna is not well developed; most South American mammals and birds are not restricted to savannah habitats. Most large herbivores of grasslands are ungulates (mammals with hooves) and although the continent has 21 species of ungulates, only three are savannah species. The small, delicately built, pampas deer (*Ozotoceros bezoarticus*) is restricted to dry open areas of the Brazilian cerrado. It shares part of its range with the much larger marsh deer (*Blastocerus dichotomus*) which tends to frequent wetter habitats. Its hoofs are wide and connected with a membrane as an adaptation to walking on soft, often waterlogged, ground. It feeds on grasses and swamp plants. The white-tailed deer (*Odocoileus virginianus*) has a more northern distribution, part of which overlaps with the llanos del Orinoco. However it is not restricted to the llanos savannah, being found also in the high Andes from Venezuela to Peru and lowland deciduous forests. In the Venezuelan llanos it appears to feed on leaves and twigs of *Mimosa* plus various grasses and forbs such as *Caperonia* and *Desmodium*. Fruits of savannah trees such as *Copernicus tectorum* and *Genipa caruto* are also eaten (Ojasti 1983). All three species of savannah deer live in small herds of between two and twenty individuals. As already noted, there is a fourth neotropical herbivore, the capybara, associated with savannah, particularly the flooded llanos. It is the largest living rodent (average weight 49 kg, although a maximum of 91 kg has been recorded). When the first European naturalists visited South America they called these giant rodents 'water pigs' or 'Orinoco hogs'. Capybaras are exclusively herbivorous, feeding mainly on grasses that grow in or near water. Unfortunately, a major constituent of these grasses is cellulose, which no mammal can break down. However, the capybara has a fermentation chamber, the caecum, containing symbiotic microrganisms which can digest cellulose. However, because the caecum (our appendix) is located after the small intestine, most of the products cannot be absorbed. The capybara therefore resorts to coprophagy (reingestion of faeces) in order to take full advantage of the symbionts efforts (see also Chapter 3).

Africa

In Africa, savannahs are very widespread (Fig. 1.2 and Fig. 1.9). They extend in a broad semicircle, sandwiched between rain forest and desert, from the West Coast to East Africa and then round to Angola and Namibia. However, this more or less continuous band of savannah vegetation differs considerably in its detailed composition. In the north, the band of savannah changes from more wooded to more open grassland as you move from the wetter Congo basin to the dryer Sahara. South of the Congo basin the same gradation occurs, within the miombo woodland savannah, as you move south towards the dryer regions of Angola, Namibia and the South African veldt. Joining these two regions are the classical savannahs of Kenya and Tanzania. The second half of this chapter looks at these African savannahs in more detail.

Australia

Savannahs are widespread in the north of Australia (Fig. 1.4). However, these savannahs are not uniform. In the east, temperatures are lower, rainfall higher and the dry season shorter and in the north, the climate is warmer with less rainfall and a longer dry season. In addition, there is a

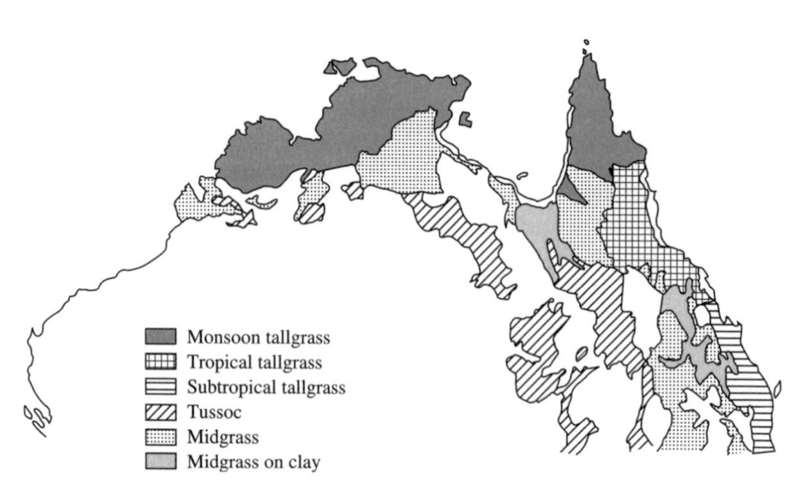

Fig. 1.4 Savannahs of northern Australia (after McKeon *et al.* 1991 and Mott *et al.* 1985).

gradient of rainfall from the wetter coast to the dryer interior. These combinations of temperature and rainfall produce a variety of savannahs from the so called 'tallgrass savannahs' of the north and east coast, to the 'midgrass savannahs' of the interior.

The 'monsoon tallgrass' savannahs, occurring across the northern part of the continent, are open low woodlands dominated by tree species such as *Eucalyptus tetradonta*, *E. dichromophloia* and *Melaleuca* species. The grass layer is composed of species such as *Themeda triandra*, *Heteropogon* species, *Sorghum* species, *Schizachyrium fragilis* and *Chrysopogon fallax*. The tropical and subtropical tallgrass savannahs also contain open woodland with several *Eucalypus* species and an under story of *Heterpogon* and *Themeda* species. More to the interior are the 'midgrass savannahs', that are a mixture of grasslands and open woodlands of *Eucalyptus populnea* and *E. microneura*, and tussock grasslands. The fertility of Australian soils is generally low (Nix 1981) and many native trees such as *Eucalyptus* have an array of nutrient-conserving mechanisms, such as the ability to extract almost all nutrients from dying leaves. Consequently there is a low decomposition rate of litter.

These Australian savannahs are dominated by invertebrates, particularly by species of grasshoppers, ants and termites (Table 1.1). These invertebrates are better able to cope with the infertile soils and harsh conditions. Termites can fix nitrogen using symbiotic microrganisms in their hindgut and are able to store grass in their large mounds or terminaria. These grass-eating, or harvester, termite mounds are a conspicuous feature of many Australian savannahs, with densities very similar to those in African savannahs. The giant mounds of *Nasuititermes triodiae* in Australia (and *Macrotermes* species in Africa) often exceed a height of 5 m, but occur at low densities ($<5\,ha^{-1}$). However, the smaller mounds of *Tumulitermes* species (Australia) and *Trinervitermes* species (Africa) can occur at densities of many hundreds per hectare. Consequently, termites are a major force in nutrient cycling in these Australian savannahs, particularly during the dry season when microbial decomposition virtually stops. In a study in the Townsville region of north Queensland Holt (1987) found that two species, *Amitermes laurensis* and *Nasutitermes longipennis* were responsible for the decomposition of approximately $250\,kg\ ha^{-1}\,yr^{-1}$ of organic matter. This

Table 1.1 Numbers of various animal species found, in savannah habitat, in Kakadu National Park, in the north of the Northern Territory, Australia.

Taxon	Birds	Bats	Other mammals	Snakes	Lizards	Frogs	Termites	Grasshoppers
Number of species	93	14	16	10	36	18	36	47

Data from Braithwaite 1991.

amounts to approximately 10 per cent of the annual carbon turnover at the site. If the unquantified population of other, subterranean detritivorous, termite species were also included, Holt believes, that this could rise to 20 per cent of organic matter decomposition. Australia has a diverse grasshopper fauna, with over 800 species. Unlike termites, which harvest mainly dead grass, grasshoppers eat almost only live plant material, mainly leaf. Surprisingly, the ecology of Australian grasshoppers is not well known and there are no available estimates of density. However they are very common, and comparison with African studies, such as Gander (1982a, b) and Sinclair (1975), would suggest that grasshoppers are probably the most important grazers in these Australian savannahs.

Because of the low fertility of Australian tropical soils, and the consequent low nutritional value of their grasses, large grazing mammals occur at relatively low densities in these Australian savannahs. In fact the major grazing mammals are often feral ungulates, such as water buffalo, donkeys, cattle and horses (Freeland 1990). Interestingly, these introduced feral species are frequently found at a higher density than in their native habitat (Freeland 1991) a fact that is probably due to a combination of reduced competition, lack of predators and pathogens, and reduced plant defences. The indigenous grazing mammals are macropods, of which there are some 19 species in the northern savannah regions. These include five species of kangaroo, including the large red kangaroo (*Macropus rufus*) weighing 95 kg, eight species of wallaby and six species of rock wallaby. Macropod digestion is aided by a fermentation chamber in the enlarged fore stomach, a modification similar to that found in many eutherian ruminants (Chapter 3). Other, smaller marsupials, found in these Australian savannahs include the northern quoll (*Dasyurus hallucatus*) an endangered marsupial carnivore looking superficially like a mongoose, the broad-footed marsupial mouse (*Antechinus bellus*), the northern brown bandicoot (*Isodon macrourus*), and the common brushtail possum (*Trichosourus vulpecula*) (Braithwaite 1998).

The Australian savannahs are similar, in many ways, to those of Africa but with *Eucalyptus* species and kangaroos replacing *Acacia* species and wildebeest. However, as mentioned above, the main animal biomass is insect rather than large mammal. Of course the large numbers of insect herbivores in the Australian savannahs means that there is lots of food for insectivores, such as lizards. This may account for the fact that Australia has the highest diversity of lizard species in the world (Table 1.1).

India

Savannahs in India are widespread in the north and east (Fig. 1.2), but are thought to be derived from woodland systems through deforestation, abandoned cultivation and burning (Misra 1983; Gadgil and Meher Homji 1985). These savannahs are prevented from returning to woodland by repeated grazing and burning, two factors that are not unimportant in other savannah systems. However, some of the anthropic savannahs of the north and central region of Rajasthan may have been derived from natural savannah. Although several types of Indian savannah have been described by Dabadghao and Shankarnarayan (1973), they are all rather similar in form and appearance and are defined by their major grass species. By far the largest of these, covering the whole of the eastern half of peninsular India is a *Sehima-Dichantium* savannah with bushes of *Acacia catechu*, *Mimosa rubicaulis*, *Zizyphus* species, and sometimes fleshy *Euphorbia*. In addition, there are low trees of *Anogeissus latifolia* and *Soymida febrifuga*. In the north east, is a *Phragmites-Saccharum-Imperata* type, west of this a savannah with *Themeda-Arundinella* grass cover and further west a *Dichanthium-Cenchrus-Lasiurus* type. Shrubby trees include *Acacia* species, *Calotropis gigantea*, *Anogeissus latifolia*, and *Zizyphus nummularia*.

These derived Indian savannahs contain several species of mammal, although ranges are frequently restricted. There are Bengal tigers (*Panthera tigris tigris*), although these are not only found in savannah habitats. Other mammals include the Indian rhinoceros (*Rhinoceros unicornis*), Asian elephant (*Elephas maximus indicus*), wild water buffalo (*Bubalus arnee*), Manipur brow antler deer (*Cervus eldii eldii*), Reeve's muntjac or barking deer (*Muntiacus reevesi*), sambar (*Cervus unicolor*), and chital or spotted deer (*Axis axis*). The IUCN conservation status of the Bengal tiger, wild water buffalo, Manipur brow antler deer, Indian rhinoceros, and Asian elephant, is 'endangered'. Interestingly, the water buffalo, sambar, and chital have been introduced to Australia. Biomass data for these Indian species is unfortunately lacking, but it is known that some areas of Indian grass savannah can support 3.5 cattle ha^{-1} (Yadava 1991).

A comparison of the major savannahs

The savannahs of South America, Africa, and Australia are produced by similar climatic conditions of temperature and rainfall distribution. They are all grasslands with trees. However, the savannahs of Africa could be portrayed as vast fertile grasslands with *Acacia* trees and large mammals, the Australian savannahs as eucalypt open woodlands with marsupials, and the South American savannahs as attenuated rainforests with large rodent herbivores. Of course this is a simplistic picture and we shall see in the next section, on Africa, that savannah structure within a continent, or even local area, can be quite different. Nonetheless, a rather generalized view of some important differences between the savannahs of the three continents are summarized in table 1.2.

The fauna and flora of savannahs on different continents share very few of their species. The introduction of water buffalo, sambar and chital into Australia and the invasion of South America and Australia by African grasses is a recent event, for which humans are responsible. In fact

Table 1.2 A generalized comparison of the major ecological features of savannahs on three continents.

Ecological feature	Africa	South America	Australia
Primary production	High	Intermediate	Low
Nutrient acquisition by plants	Easy	Intermediate	Difficult
Decomposition rate of litter	Rapid	Intermediate	Slow
Plant dispersal by	Large vertebrates	Smaller vertebrates	Invertebrates
Fire tolerance of vegetation	Medium	Low	High
Type of pollination	Specialized	Very specialized	Unspecialized
Main herbivores	Large vertebrates	Smaller vertebrates	Invertebrates
Main predation by	Large vertebrates	Medium vertebrates	Small vertebrates
Mutualism	Uncommon	Common	Common
Animal migration	Major	Intermediate	Minor
Animal productivity	High	Medium	Low
Competition	Widespread	Patchy	Patchy

Modified from Braithwaite 1991.

savannahs from different continents usually show more similarities with adjacent, different, biomes than with each other. For example, the flora of the Brazilian cerrado shows more affinities with the flora of Amazonia than with savannahs in Africa and Australia. Savannahs on different continents often 'look similar', but the individual species are quite different. The grasses are different species, the trees are different species and the herbivores and carnivores are different species. The rodent capybara in the llanos, the ungulate wildebeest in the Serengeti and the marsupial kangaroo in Northern Australia are all large, herbivorous, mammals but they are taxonomically quite different. The similarities are due to the fact that the biotas of savannahs, in different areas of the world, are the product of convergent evolution. Different species function in similar ways when they have to deal with similar environments.

African savannahs

Climatic patterns

As we saw in Figure 1.1, temperature and rainfall are major determinants of savannahs world-wide. The same is true in Africa, although in most of tropical Africa temperature does not have the same limiting effect on plant growth as it does, for example, in temperate Europe or North America. However, the average air temperature falls by about 0.6°C per 100 m increase in altitude and therefore African montane areas have a temperature regime not suitable for savannah vegetation to develop. Rainfall is very variable across the continent, and its amount largely determines the location of the three major African biomes; tropical forest, savannah and

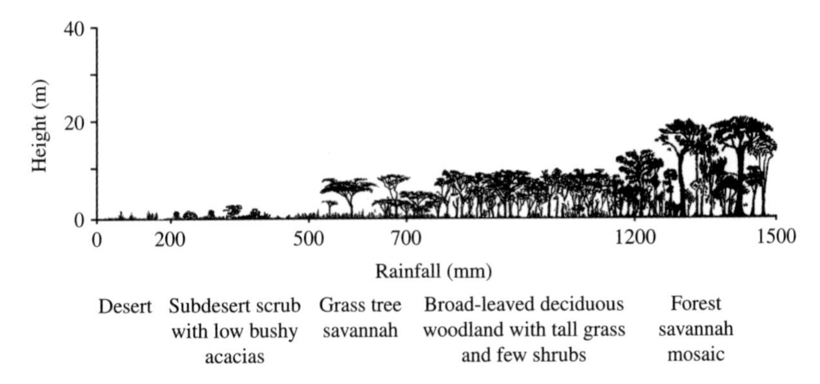

Fig. 1.5 Idealized relationship between rainfall and African savannah vegetation.

desert and the different types of African savannah (Figure 1.5). It is therefore important to understand the climate patterns of Africa, particularly for rainfall. It is not just the amount of rain falling in an area that determines the vegetation, but also its seasonal distribution throughout the year. Whether there is one or two rainy seasons, and the length and severity of the dry season are also important to the structure of the vegetation, particularly the development and establishment of trees.

During the course of a year, as the earth orbits the sun, with its axis tilted relative to the plane of its orbit, the sun changes its position in the sky and produces the changing seasons. The limits of this annual movement of the sun occur in June, when the sun is overhead at the Tropic of Cancer (23.45°N) and December when the sun is overhead at the Tropic of Capricorn (23.45°S). During March (spring or vernal equinox) and September (autumnal equinox) the sun is overhead at the Equator. In effect, the sun's position moves north and south across the Equator twice a year. A consequence of this seasonal change in the sun's position is a change in the pattern of air movement over the continent. To illustrate this, a simple climate model, that influences Africa, is shown in Figure 1.6.

The atmosphere above Africa can be seen as two circulating air systems, rising in the centre, and moving outwards towards the poles. During March, air is heated over the equator (Figure 1.6a) and therefore rises, causing a low pressure area called the Intertropical Convergence Zone (ITCZ). The ascending air is cooled by expansion resulting from reduced pressure, causing saturation, condensation, cloud formation, and rain. This 'dryer' air moves towards the poles, cools, and descends, forming a subtropical high pressure region. This descending 'dry' air is warmed by compression, reducing its relative humidity even further. This results in two arid belts, the Sahara desert in the north and the Kalahari and Namib deserts in the south. Winds from the northern and southern high pressure zones blow (converge) towards the low pressure ITCZ. These winds always blow from the

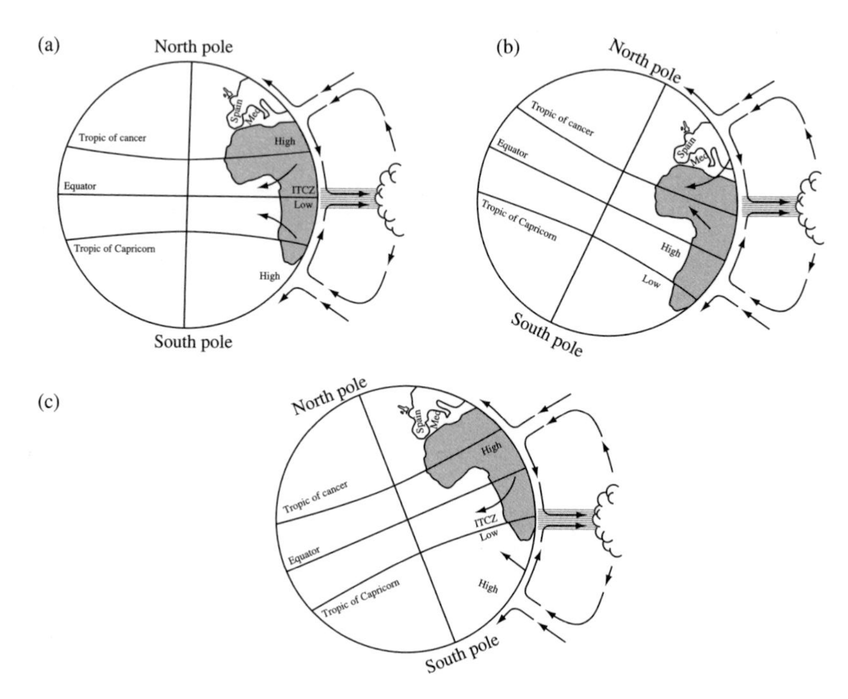

Fig. 1.6 Simple climate model showing how the Intertropical Convergence Zone (ITCZ) moves with the seasons. (a) Vernal and autumnal equinox, rain near the equator to about 4°N–S. (b) June–August, sun over Tropic of Cancer, rain in northern tropics, dry in southern tropics (c) November–February, sun over Tropic of Capricorn, rain in southern tropics. ← winds converging at ITCZ.

east because of the direction of the earth's spin. Following the sun's movement by about one month, this system of circulating air, with its attendant low pressure, wet, centre (ITCZ) and high pressure, arid, margins, moves north and by June–August is centred over the Tropic of Cancer (Figure 1.6b), it then moves south again and by September is centred over the equator once again (Figure 1.6a). By November–February it has moved further south and is centred over the Tropic of Capricorn (Figure 1.6c), eventually returning north to be centred over the equator again by March. In the days of sailing ships the winds that converge on the ITCZ were called the 'trade winds', a term that comes not from any reference to commerce but from the expression 'to blow trade', meaning to blow regularly. Under the ITCZ there was no wind, an area known as the 'doldrums'. Because of these age-old wind patterns, ancient mariners leaving Europe to sail round the African Cape had to sail across the Atlantic on the northeast Trades, down the South American coast and then back across the Atlantic on the southeast Trades.

As already noted, one consequence of this circulation of air above Africa is that there is a central 'wet' zone around the equator, and two 'arid' zones

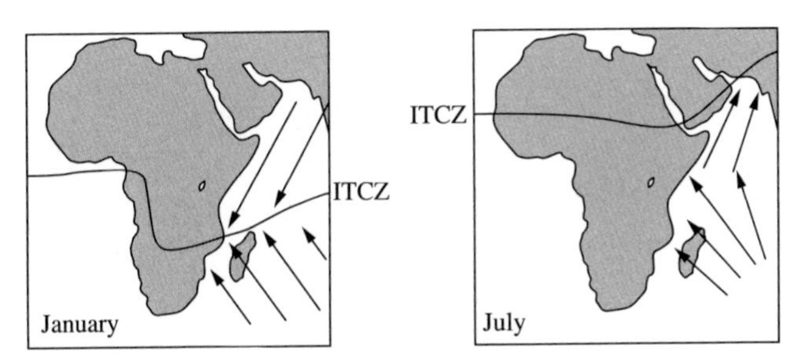

Fig. 1.7 Seasonal movement of the ITCZ and its associated trade winds (indicated by the arrows).

to the north and south. However because the circulation of air moves north and south during the year there is, in general, a gradation from wet, to moist, to dry to arid as we move away from the equator. However, the ITCZ does not move north and south, over Africa, as a straight-line zone, running east–west. It bends (Fig. 1.7).

This bending of the ITCZ distorts the moist-arid, north–south, gradient so that East Africa is relatively dry. This is because in January, when the ITCZ is 'south', the northeast trade winds, blowing across east Africa, come across relatively dry land areas. At the same time the ITCZ is still over parts of central and west Africa. In July, when the ITCZ is 'north' the southeast trade winds come across the wet Indian Ocean. This creates the pattern of rainfall seen in Figure 1.8 and the consequent pattern of biomes seen in Figure 1.9. In the area with most rainfall, around the equator, there is tropical forest. This grades into forest savannah mosaic, woodland savannah, tree and bush savannah, grass and shrub savannah, and finally desert (Fig. 1.5). However in the east, the dry savannah extends down from Sudan, through Kenya, into northern Tanzania. Also, because the ITCZ passes over the central equatorial area twice, there are two rainy seasons here (bimodal rains), one from late March to mid-June and one from late October to mid-December. These short rainy seasons are separated by short dry seasons. As you move away from the equator one of the rainy seasons tends to be longer and more reliable (the 'long rains'), and the other shorter and less reliable (the 'short rains'). As you go further north, or south, the rain pattern becomes unimodal. In northern areas, rain falls in one long wet season from April to October, with a long dry season from November to March. In the drier, northeastern parts, the rainy season tends to be shorter, and unreliable, and the intervening period extremely dry. In southern areas the unimodal rainfall regime is the opposite of that in the north with rain falling mainly from November to April. However, these latitudinal rainfall patterns can be moderated by closeness to oceans and large water bodies such as Lake Victoria. There are also large rainfall variations due to the modifying influences of montane areas. For example, the Laikipia plateau and Amboseli, in East Africa, receive lower rainfall, being in the shadow of

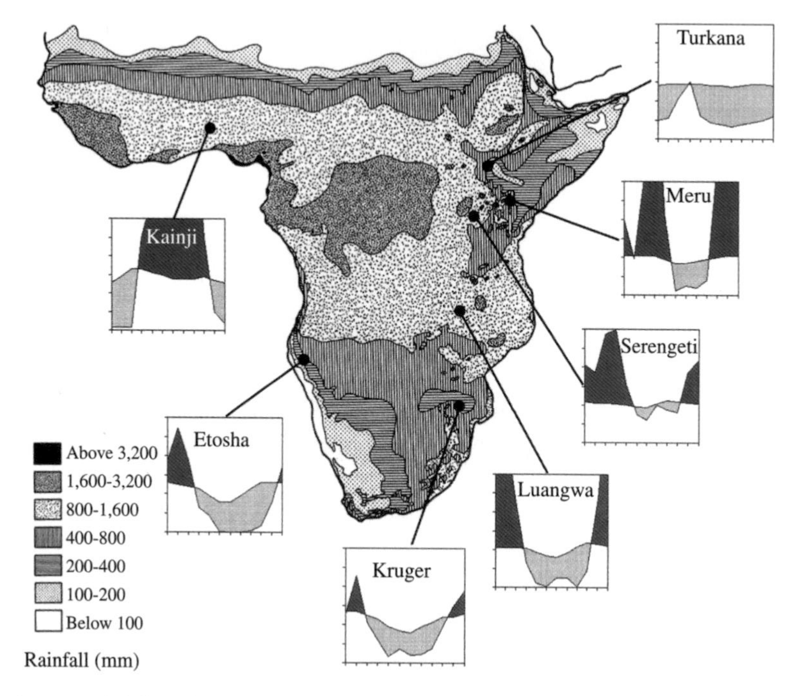

Fig. 1.8 Average rainfall in Africa with 7 climograms. Both temperature (the rather level line) and rainfall are plotted on the climograms. The ordinate shows both temperature and rainfall, with one division = 20 mm of rain and 10°C. The abscissa shows months from January to December. A relative dry period occurs when the rainfall curve falls below the temperature curve (stippled area) and a relative humid period occurs when the rainfall curve is above the temperature curve (dark grey area) (map from Kingdon 1989).

Mt Kenya and Mt Kilimanjaro, respectively. Figure 1.8 shows a selection of climate diagrams, or climograms, for different savannah sites in Africa. These climograms are superimposed upon a map of Africa showing mean annual rainfall. They are a convenient way to allow an immediate visual assessment of climates within savannah regions.

Of course, the amount of water available to a plant is a product not only of the amount of rainfall but how that rainfall is converted into available soil moisture. In practice, the idealized relationship between vegetation and total rainfall , seen in Figure 1.5, is affected by the seasonality of that rainfall. An area with two short rainy seasons and two short dry seasons may produce woodland savannah, while another area with the same amount of rain, but restricted to one short wet season followed by a long dry season, might have no trees, only low *Acacia* scrub. Additionally, the long dry season might make the second area more prone to fires, that would also reduce tree cover. The influence of total annual rainfall can also be modified by topography, drainage and soil type. For example, sandy or

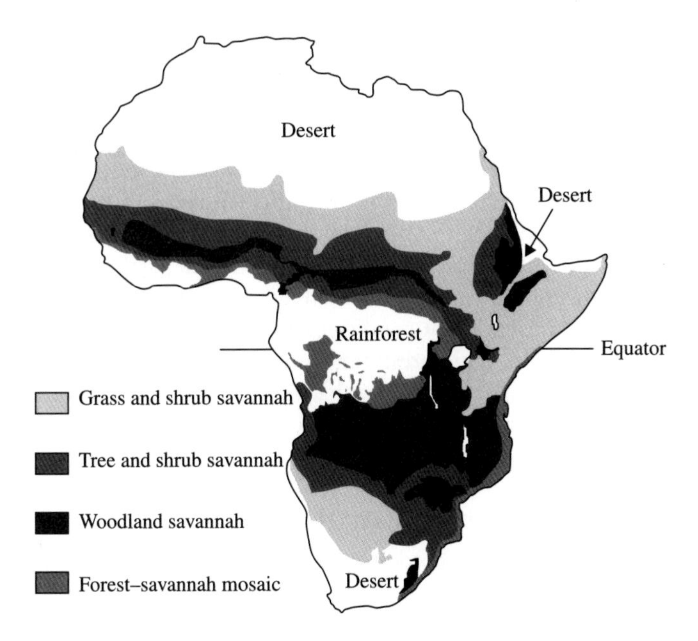

Fig. 1.9 The principal African savannahs. The black areas are montane.

free-draining loamy soils readily absorb rainfall, whereas the surface of heavy clay soils quickly becomes sealed and further rain simply runs off and is lost to the vegetation.

Vegetation patterns

There have been several attempts to classify the vegetation of Africa. At one extreme some 'Safari Guides' talk only about tropical forest, savannah, desert, montane and perhaps 'Mediterranean' vegetation. The latter is a type of dwarf shrubland found on the extreme north, Mediterranean coast (called *maquis*) and southern Cape (called *fynbos*). At the other extreme we have the detailed subdivisions of Frank White (1983) who recognized 17 major vegetation types, based on the cover and height of the vegetation, rather than the species involved. For example, forests were defined as all vegetation comprising a continuous stand of trees at least 10 m tall with interlocking crowns. Woodland was defined as open stands of trees at least 8 m tall, with a canopy cover of 40 per cent or more, and with a ground layer usually dominated by grasses and other herbs. This physionomic approach to vegetation classification (looking at form or appearance) has largely been replaced by the phytosociological method which looks at species associations. However, in the context of African it is still frequently used because the 'physionomy' of a plant is a useful, indirect, measure of the environment, particularly temperature, rainfall and soil. I will therefore adopt a simplified version of White's 1983 classification, concentrating on savannahs, but making mention of species associations within the broad physionomic classification.

In Africa, different types of savannahs make up approximately 50 per cent of the land area (Figure 1.9). Of course the sharp demarcation between the vegetation zones shown on such maps of Africa is artificial, since the environmental conditions that produce the biomes (e.g. rainfall) follow gradients (Figure 1.5). This gradation is particularly true of the 'grass and shrub' (sometimes called tree and shrub steppes) and 'tree and shrub' savannahs. Local conditions can also blur the edges of these biomes. For example, rain forest can penetrate deeply into the savannah biome, as gallery forest along river banks. For this reason the 'lines' between the areas on the map have been shown blurred. A range of savannah profiles is shown in Figure 1.10 and some photographs are shown in Figure 1.11.

Grass and shrub savannah

The northern border of this type of savannah is called the Sahel and the predominant tree genus is *Acacia*. It stretches across Africa from northern Senegal and Mauritania on the Atlantic coast, to Sudan on the Red Sea. The

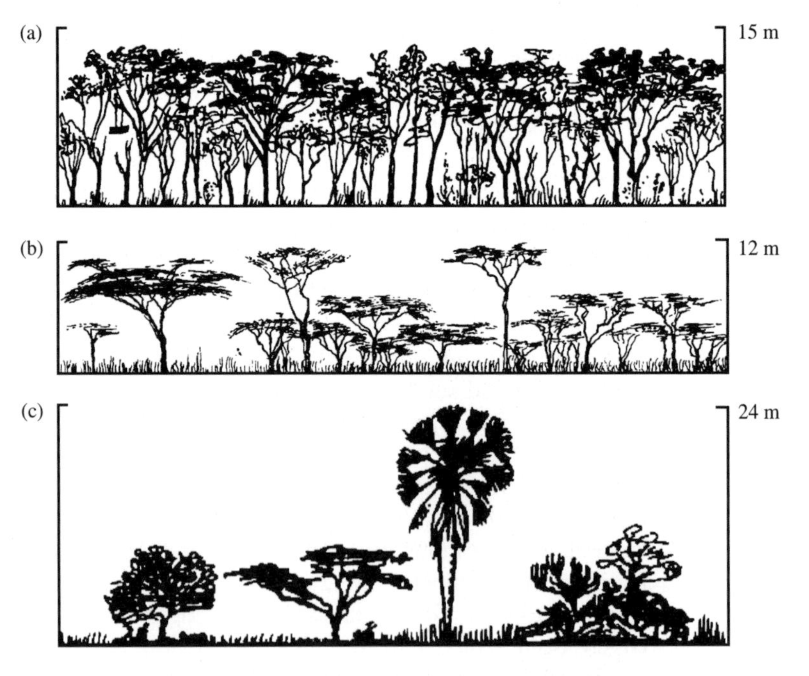

Fig. 1.10 Savannah profiles. (a) *Brachystegia*, *Terminalia* woodland savannah, (b) *Acacia* tree and shrub savannah, (c) *Combretum*, *Acacia*, *Borossus* tree and shrub savannah (modified from Kingdon 1997).

word 'sahel' means 'shore' in Arabic and is a reference to this region being a transition zone between the more wooded savannahs of the south and the Sahara Desert to the north. This arid, 'grass and shrub' savannah continues into the *Acacia-Commiphora* savannah of the Horn of Africa, to the east of the Ethiopian highlands, and down into East Africa as the *Somali-Masai* dry savannah.

In the Sahel, Horn of Africa and northern Kenya, mean maximum temperatures vary from 30° to 36°C and mean minimum temperatures between 15° and 21°C. Further south, on the Kenyan Tanzanian border, temperatures are more moderate with mean maximum temperatures of 30°C at lower elevations and only 24°C at the higher elevations. Mean minimum temperatures are between 9° and 18°C, and normally between 13° and 16°C. Annual rainfall varies from 600 mm to 100 mm in the Sahel and rain falls mainly in the summer months of May to September, followed by a 6 to 8 month dry season. In central Kenya rainfall starts to show bimodality, but with most precipitation occurring in the long rains, from March to June, and less in the short rains of October to December. However, the timing and amount of rainfall varies greatly from year to year, and frequently one, or even both, rainy seasons fail. Further south still, in northern Tanzania, the rainfall is clearly bimodal. The long rains occur from March to May and the short rains from November to December. Mean, annual, rainfall is 600 to 800 mm through most of the region. Rainfall is still variable and the short rains may fail in a given year, or dry season rain may join the two wet seasons.

In the northern areas, typical woody species include *Acacia tortilis*, *A. laeta*, *Commiphora africana*, *Balanites aegyptiaca* and *Boscia senegalensis*. Grass cover is continuous, with annual species such as *Cenchrus biflorus*, *Schoenefeldia gracilis*, and *Aristida stipoides*. In some areas *Acacia* and *Commiphora* species are joined by *Euphorbia* and *Aloe* species, as well as grasses such as *Dactyloctenium aegyptium* and *Panicum turgidum*. Two species of tree in this region are characterized by aromatic resins used in Biblical times for incense and perfumes. Frankincense comes from the resin of *Boswellia carteri*, found in Somalia, and myrrh is extracted from *Commiphora abyssinica* and other species found in Ethiopia. In the southern part of this region predominant plants again include species of *Acacia* and *Commiphora*, along with *Crotalaria* and the grasses *Themeda triandra*, *Setaria incrassata*, *Panicum coloratum*, *Aristida adscencionis*, *Andropogon* species, and *Eragrostis* species. In northern Tanzania, the region is bisected by patches of grassland on volcanic soil (southern Serengeti) and patches of montane forest. The volcanic grassland has no trees because of the nature of the soil but is non-the-less an integral part of the Serengeti savannah ecosystem.

The Sahel region is not especially rich in mammal species although it does possess at least four species of endemic gerbil (*Gerbillus bottai*, *G. muriculus*, *G. nancillus* and *G. stigmonyx*). The scimitar oryx (*Oryx dammah*)(now regarded as extinct in the wild), dama gazelle (*Gazella dama*), dorcas gazelle (*G. dorcas*) and red-fronted gazelle (*G. rufifrons*), all now rare, were formerly abundant and widespread (East 1999). Predators such as wild dog (*Lycaon pictus*), cheetah (*Acinonyx jubatus*) and lion (*Panthera leo*) were once common, but have now been exterminated over most of the region. The Horn of Africa has a number of unique antelopes, such as the dibatag (*Ammodorcas clarkei*), beira (*Dorcatragus megalotis*), and Speke's gazelle (*G. spekei*). The African ass (*Equus africanus*), desert warthog (*Phacochoerus aethiopicus*) and Ethiopian wolf (*Canis simensis*) are also unique to this region. Several species are found in reduced numbers in the Horn and adjacent regions, but are more abundant further south. These include Grevy's zebra (*Equus grevyi*), beisa oryx (*Oryx gazella beisa*), gerenuk (*Litocranius walleri*), greater kudu (*Tragelaphus strepsiceros*), lesser kudu (*T. imberbis*), elephant (*Loxodonta africana*) and African buffalo (*Syncerus caffer*). Lion, leopard (*Panthera pardus*), cheetah, striped hyaena (*Hyaena hyaena*), and spotted hyaena (*Crocuta crocuta*) are the main large carnivores in this region. The endangered wild dog is found in Ethiopia, in Mago and Omo National Parks (Woodroffe *et al.* 1997). The central Laikipia region of Kenya, between Mount Kenya and the Rift Valley, probably hosts the only intact savannah mammal community outside a Kenyan National Park. At least 75 mammal species and 400 bird species can be found in the region. Elephant, eland (*Taurotragus oryx*), common zebra (*Equus burchellii*), Grevy's zebra, beisa oryx, and reticulated giraffe (*Giraffa camelopardalis reticulata*) are seasonally abundant, migrating long distances across the region, depending on rainfall and forage availability. Defassa waterbuck (*Kobus ellipsiprymnus defassa*), impala (*Aepyceros melampus*), Grant's gazelle (*Gazella granti*), Jackson's hartebeest (*Alcelaphus buselaphus jacksoni*), gerenuk, lesser kudu, and bushbuck (*Tragelaphus scriptus*) are resident. Lion, leopard, cheetah, spotted hyaena, black-backed jackal (*Canis mesomeles*), and bat-eared fox (*Otocyon megalotis*) are regularly seen, and striped hyaena, aardwolf (*Proteles cristatus*), and wild dog are present. Of course it is the southern Kenya–northern Tanzania part of this region that is most well known for its outstanding concentrations of large mammals. The Serengeti–Mara migration of approximately 1.3 million wildebeest (*Connochaetes taurinus*), 200,000 common zebra, and 400,000 Thomson's gazelle (*Gazella thomsoni*) is the most spectacular mass movement of terrestrial mammals anywhere in the world. Not surprisingly the area also supports one of the highest concentrations of large predators, with approximately 7,500 spotted hyaena and 2,800 lion, along with leopard and cheetah. Both Tarangire and Serengeti National Parks have approximately 350 to 400 recorded bird species.

In southern Africa, the equivalent 'grass and shrub' savannah is an arid transition zone between the northern mopane savannah and the southern desert. Annual rainfall varies between about 200 and 500 mm, although in the narrow escarpment belt that lies inland of the Namib Desert

(Namaland), annual rainfall can be as low as 60 mm. Most of the rain falls in thunder storms in the summer months, from October to March. Average temperatures over the whole region are about 21°C, but are quite variable.

Vegetation varies from a dense, short, shrub savannah (bushveld) to an open tree savannah and reflects the diverse topography, soil and microclimate. In the north of the area is *Euphorbia guerichiana*, a shrub or small tree with conspicuous, shiny, brownish-yellow, papery bark growing to a height of 5 m. Also common are *Cyphostemma* species with succulent stems, *Adenolobus* species, the quiver tree (*Aloe dichotoma*), and *Moringa ovalifolia*. Two species of *Acacia* are confined to this vegetation type; these are the Brandberg acacia (*Acacia montis-ustii*) and *A. robynsiana*. *Acacia senegal* and *A. tortilis* are found along ephemeral rivers. To the south, the vegetation becomes more open and is dominated by karoo shrubs (*Rhigozum trichotomum* is characteristic) and grasses. *Parkinsonia africana*, *Acacia nebrownii*, *Boscia foetida*, *B. albitrunca*, and *Catophractes alexandri*, as well as smaller karoo bushes such as *Pentzia* species and *Eriocephalus* species are also typical. Tufted grasses, mainly *Stipagrostis* species, are found scattered between the woody plants. On rocky ridges, the conspicuous quiver tree becomes very abundant.

Endemic and near-endemic mammals are mainly bats, rodents, and small carnivores. The only endemic large mammal is the mountain zebra (*Equus zebra*), which is found to the west of this region, and is the only large mammal endemic to Namibia. This western edge is also well known for its desert-dwelling populations of elephant and black rhinoceros (*Diceros bicornis*). The black rhino population in this area is one of the few unfenced populations of black rhinos in the world, and it is estimated to number more than 100 individuals. Other large mammals found within this southern arid region are greater kudu, springbok (*Antidorcas marsupialis*), gemsbok (*Oryx gazella gazella*), Kirk's dik-dik (*Madoqua kirkii*), and black-faced impala (*Aepyceros melampus petersi*). Predators include lion, leopard, cheetah, bat-eared fox, and Cape fox (*Vulpes chama*).

Protected areas within the 'grass and shrub' savannah region include many of the famous East African safari parks. These include Samburu National Reserve (165 km^2) in central Kenya, Nairobi National Park (117 km^2) in southern Kenya, Serengeti National Park (14,763 km^2) in northern Tanzania, and the adjacent Masai Mara Game Reserve (1,510 km^2) in southern Kenya, Amboseli National Park (392 km^2) and Tsavo, East and West, National Parks (11,747 km^2 and 9,65 km^2) in southern Kenya, Tarangire National Park (2,600 km^2) in northern Tanzania and Mkomazi Game Reserve (1,000 km^2) in northern Tanzania.

Tree and shrub savannah

Like the 'grass and shrub' savannahs, the 'tree and shrub' savannahs form two separated blocks of vegetation, lying north and south of the rainforest and miombo woodland savannahs of central Africa.

In the north the climate is tropical and strongly seasonal and the vegetation is composed mainly of *Combretum* and *Terminalia* shrub and tree species and tall elephant grass (*Pennisetum purpureum*). This region lies south of the Sahel and stretches from the west coast of Africa to Sudan, and also extends south into northwestern Uganda. Mean maximum temperatures range from 30° to 33°C and mean minimum temperatures are between 18°C and 21°C. The annual rainfall is as high as 1,000 mm in the south, but declines to the north with only 600 mm found on the border with the 'grass and shrub' savannah.

Typical trees include *Anogeissus leiocarpus*, *Boswellia papyrifera*, *Balanites aegyptiaca*, *Lannea schimperi*, *Stereospermum kunthianum*, *Kigelia aethiopica*, *Acacia seyal*, *Commiphora africana*, *Prosopis africana*, *Tamarindus indica*, *Ziziphus mucronata*, and, of course, species of *Combretum* and *Terminalia*. Dominant grasses include tall species of *Hyparrhenia*, *Cymbopogon*, *Echinochloa*, *Sorghum*, and *Pennisetum*. Mammals include elephant, African buffalo, oribi (*Ourebia ourebi*), western hartebeest (*Alcelaphus buselaphus major*), wild dog, cheetah, leopard, and lion. The giant eland (*Taurotragus derbianus*) still survives in parts of the eastern section of this savannah. The roan antelope (*Hippotragus equinus*) is widespread throughout this northern region, but not in large numbers. Remnant populations of the western giraffe (*Giraffa camelopardalis peralta*), numbering about 2,000 are still found in Chad and northern Cameroon (East 1999).

To the south of the central African block of miombo 'woodland savannah', described in the next section, are 'tree and shrub' savannahs characterized by the dominance of the mopane tree *Colophospermum mopane*. Mopane is a single-stemmed tree or shrub with distinctive, butterfly-shaped leaves. Two regions of mopane can be identified, the southeast 'Zambezian' region and the southwest 'Angolan' region. The Zambezian region extends into South Africa, Mozambique, Botswana, Zambia, Zimbabwe, Swaziland, Namibia, and Malawi. Here, the mopane tree is frequently the sole canopy species but can be associated with other prominent trees and shrubs. These include *Kirkia acuminata*, African blackwood (*Dalbergia melanoxylon*), baobab (*Adansonia digitata*), *Combretum apiculatum*, *C. imberbe*, *Acacia nigrescens*, *Cissus cornifolia*, and *Commiphora* species. These Zambezian mopane communities show considerable variation in height and density. Trees in dense woodland or in more open savannah woodland may reach heights of 10 m to 15 m on deep alluvial soils, and even attain 25 m in the so-called 'cathedral mopane' of Zambia. In contrast, mopane tends to be stunted and shrubby (1 to 3 m) where it occurs on impermeable alkaline soils. These two structural forms, often occur together in a mosaic depending on micro-climatic factors and soil conditions (White 1983; Smith 1998). The ground layer can also vary markedly. For example, dense grass swards are found beneath gaps in the mopane canopy on favourable soils, while grasses are almost completely absent in shrubby mopane communities on heavy, impermeable alkaline clays. Typical grasses include *Aristida* species, *Eragrotis* species,

Digitaria eriantha, *Brachiaria deflexa*, *Echinochloa colona*, *Cenchrus ciliaris*, *Enneapogon cenchroides*, *Pogonarthria squarrosa*, *Schmidtia pappophoroides*, *Stipagrostis uniplumis*, and *Urochloa* species. The Zambezian mopane woodland-savannah experiences rain largely in the period November to April. Annual average rainfall generally varies between 450 mm to 710 mm, although parts of the region, may receive up to 1,000 mm. Temperatures range between −4° and 46°C, with a mean annual temperature of 18° to 24°C.

The Angolan mopane woodland-savannah is located in Namibia and Angola, completely surrounding the large Etosha salt pan, in northern Namibia. Again, mopane trees dominate the vegetation. They often form a dense, single species stand under which grass is virtually absent. Hence fire damage is minimal, even though the mopane tree itself is resinous and flammable. However, if the canopy is opened up (for example by browsing elephants), grasses invade and the fire frequency and intensity is increased. Browsing elephants frequently push down mopane trees, which then regrow into low (0.3 to 1.6 m), multi-stemmed, shrubs. The 'typical' mopane woodland is then converted into a tall grassland, with the grass frequently as tall as the mopane coppice (White 1983). In Angola, mopane grows over vast areas in a low, thorny bushveld. It is then associated with *Acacia kirkii*, *A. nilotica subalata*, *A. hebeclada tristis*, *A. erubescens*, *Balanites angloensis*, *Combretum apiculatum*, *Commiphora* species, *Dichanthium papillosum*, *Dichrostachys cinerea*, *Grewia villosa*, *Indigofera schimperi*, *Jatropha campestris*, *Melanthera marlothiana*, *Peltophorum africanum*, *Rhigozum brevispinosum*, *R. virgatum*, *Securinega virosa*, *Spirostachys africana*, *Terminalia prunoides*, *T. sericea*, *Ximenia americana*, and *X. caffra*. Rain normally falls in the summer months, between August and April, with most falling in late summer. Mean annual rainfall is between 400 mm to 600 mm, although the annual rainfall total is unpredictable. In the Etosha National Park, for example, the mean annual rainfall in 1946 was 90 mm, but in 1950 it was 975 mm. Mean maximum temperatures range from around 24°C, near the coastal deserts, up to 30°C further inland. The mean minimum temperatures similarly increase inland, from 9°C towards the coast to up to 12°C further inland.

Because the vegetation in the mopane region is more nutritive than in miombo woodland-savannah, coupled with the extensive and well-maintained system of protected areas, the mopane region supports large concentrations of ungulates. These includes elephant, black rhinoceros, white rhinoceros (*Ceratotherium simum*), hippopotamus (*Hippopotamus amphibius*), African buffalo, wildebeest, mountain zebra, nyala (*Tragelaphus angasii*), gemsbok, eland, greater kudu, roan, steenbok (*Raphicerus campestris*), dik-dik, and the near-endemic black-faced impala. Large predators include lion, leopard, cheetah, spotted hyaena, brown hyaena (*Hyaena brunnea)*, and wild dog. Smaller predators include the black-backed jackal and the bat-eared fox. In addition to biomass differences for many mammals, there are differences in the species assemblages between miombo and mopane. Side-striped jackal (*Canis adustus*), sable antelope (*Hippotragus niger*), roan antelope, and Lichtenstein's hartebeest (*Alcelaphus lichtensteinii*) are associated with miombo, while black-backed jackal, greater kudu, and impala (*Aepyceros melampus*) are identified with Mopane (Huntley 1978). This region has a diverse avian fauna, with over 375 species recorded.

On the southern borders of the mopane region are Baikiaea woodlands, a mosaic of dry deciduous *Baikiaea plurijuga* dominated forest, thicket and grassland. It forms a belt along the Angola–Namibia border, and extends southwest into Botswana, Zimbabwe, and the northern province of South Africa. Mean annual rainfall ranges from less than 300 mm, in the drier southwest, to more than 600 mm, in eastern Zimbabwe, and is strongly concentrated from October to April. The mean maximum temperature is between 27° and 30°C and the mean minimum temperature ranges from about 9° to 12°C. In well-developed *Baikiaea* communities, species of *Brachystegia, Julbernardia* and *Colophospermum mopane* (species typical of miombo and mopane woodlands) are totally absent. *Baikiaea plurijuga* forms a fairly dense, dry, semi-deciduous forest with trees up to 20 m in height. There is a dense and shrubby lower story of *Combretum engleri, Pteleopsis anisoptera, Pterocarpus antunesii, Guibourtea coleosperma, Dialium engleranum, Strychnos* species, *Parinari curatellifolia, Ochna pulchra, Baphia massaiensis obovata, Diplorhynchus condylocarpon, Terminalia brachystemma, Burkea africana, Copaifera baumiana*, and *Bauhinia petersiana serpae*. This type of savannah has a diverse birdlife, with 468 species recorded.

Protected areas within the 'tree and shrub' savannah region include Kruger National Park (19,624 km^2) in northern South Africa with its wild dogs, Chobe National Park (10,570 km^2) in Botswana with its elephants, and Etosha National Park (22,270 km^2) in northern Namibia with its gigantic salt pan. This salt pan covers an area of 6,133 km^2, and is believed to have once been a great inland lake fed by a large river (probably the Kunene), which over time changed its route and caused the lake to dry out and form a salt desert. Etosha has over 340 bird species recorded and the park is particularly rich in raptors, with 46 species recorded, including all the vultures found in Namibia.

Woodland savannah

There are two areas of 'woodland savannah'. A huge area, called *miombo*, in central/south Africa, and a reduced area called *doka* in the north.

Miombo covers an estimated 3 million km^2 of Zimbabwe, Zambia, Mozambique, Angola, Malawi, Katanga province of Zaire, and southern Tanzania. It is the largest vegetation unit in the Zambezian centre of endemism, or Zambezian phytochorian. A phytochorian is a plant-geographic area, recognized by its species composition rather than its structure or physionomy. Miombo woodland savannah takes its name from the miombo

(Muuyombo) tree, a species of *Brachystegia*, (*B. boehmii*), the dominant tree genus over this entire area. In the centre, the miombo area experiences a seasonal tropical climate. Most of its rainfall is concentrated during the hot summer months of November to March/April and this is followed by a pronounced winter drought, which can last up to seven months in some areas. In general, mean maximum temperatures range from around 27°C to 30°C, although in areas of higher elevation, such as central Zambia, it can be around 24°C. The hottest temperatures are in the lowland areas. In the south, the miombo region has a climate with three distinct seasons: hot and dry from mid-August to October; hot and wet from November to March; and warm and dry from April to early August. Mean maximum temperatures here range from 18°C to 27°C, but are typically around 24°C. Mean minimum temperatures, over the whole miombo region, range from 9° to 21°C. Throughout the central miombo region rainfall varies from about 800 to 1,200 mm annually, but with figures as high as 1,400 mm recorded at some higher elevations. However, in parts of Zimbabwe, in the south of the region, it can be reduced to around 600 to 800 mm.

Typically, mature miombo trees are 15 to 20 m tall, with a shrub and grass understory. In Angola, the canopy height is lower, from 5 to 10 m, with little or no shrub layer. Dominant tree species include *Brachystegia spiciformis*, *B. boehmii*, *B. allenii*, *B. glaberrima*, *B. taxifolia*, *B. utilis*, *Marquesia macroura*, *Julbernardia globiflora*, *J. paniculata*, and *Copaifera baumiana*. *Brachystegia floribunda*, *B. gossweilerii*, *B. wangermeeana*, *B. longifolia*, *B. bakerana*, *Guibourtea coleosperma*, and *Isoberlinia angolensis* are locally dominant. In the southern part of the miombo region, other common tree species include *Uapaca kirkiana*, *Monotes glaber*, *Faurea saligna*, *F. speciosa*, *Combretum molle*, *Albizia antunesiana*, *Strychnos spinosa*, *S. cocculoides*, *Flacourtia indica*, and *Vangueria infausta*. Most of the miombo tree and shrub species shed their leaves in the late dry season, and the miombo woodland is bare for two or three months. A few weeks before the rainy season starts, the trees produce their new, predominantly bright reddish new foliage. Fire is an important ecological factor in miombo woodland. The strong seasonality in rainfall leaves the vegetation dry for several months of the year, and fires, either natural or man-made, can be frequent.

Miombo does not support large mammals in high densities, although due to the vast size of the region its overall importance for mammal species is quite high. The low, large mammal, density is probably caused by the poor soils, which generally support vegetation of low nutritional value. Conditions are made worse by the harsh dry season and long droughts. Of course many of the 'miombo large mammals' are found in other savannah regions and include elephant, black rhinoceros, and African buffalo. These are able to survive on poor quality forage by consuming it in large quantities. Specialized grazers are also common. They selectively feed on grass, and include sable antelope, Lichtenstein's hartebeest, and southern reedbuck (*Redunca arundinum*), all species largely restricted to the miombo belt, as well as the more widespread roan antelope. Browsers such as eland, and mixed feeders such as greater kudu are also present. Many species make use of the wooded margins or open areas of the numerous grassy flood plains scattered through the miombo region. These include lechwe (*Kobus leche*), puku (*K. vardoni*), tsessebe (*Damaliscus lunatus*), oribi (*Ourebia ourebi*), wildebeest (*Connochaetes taurinus*) and sitatunga (*Tragelaphus spekii*). Common waterbuck (*K. ellipsiprymnus ellipsiprymnus*), and bushbuck are mostly found in more wooded areas close to permanent water. Other large ungulates include common zebra (*Equus burchelli*) and the restricted, but abundant, Thornicroft's giraffe (*Giraffa camelopardalis thornicrofti*). Hippopotamus are relatively common near water. Due to annual droughts and frequent fires, many species are seasonally dependent on non-miombo vegetation, within or adjacent to the region, to provide food, water, or shelter. For example, sable antelope, are largely confined to the miombo belt but move onto more open grassy areas during the dry season (Kingdon 1997).

There are many carnivores, although most are not confined to miombo woodland savannah. These include lion, leopard, cheetah, spotted hyaena, striped hyaena, African wild dog, side-striped jackal, wild cat (*Felis sylvestris*), serval (*F. serval*), caracal (*Caracal caracal*), miombo genet (*Genetta angolensis*), Selous's mongoose (*Paracynictis selousi*), and bushy-tailed mongoose (*Bdeogale crassicauda*). Two large insectivores, distributed over much of the African savannah, the ground pangolin (*Smutsia temminckii*), and the aardvark (*Orycteropus afer*) feed on the numerous ants and termites found in the miombo.

The miombo area supports a number of primate species, mostly on its northwestern borders, next to areas of 'forest–savannah mosaic' and rainforest, and where it grades into sub-montane forest habitats in Uganda. The Gombe Game Controlled Area, although largely covered by evergreen forests, includes substantial miombo habitat on its lower slopes. The Controlled Area is well known as the site of Jane Goodall's long-term study of the chimpanzee (*Pan troglodytes*) (Goodall 1988). This area also supports several species of red colobus (*Procolobus speciesi*), black and white colobus (*Colobus angolensis*), blue monkey (*Cercopithecus mitis*), and red-tailed monkey (*C. ascanius*). More widespread primates that are typical of many African savannahs are vervet monkey (*Chlorocebus aethiops*), and savannah baboon (*Papio hamadryas*). Miombo bird life is rich in species. However, most are not restricted to this type of savannah. Birds breeding in miombo woodlands generally have relatively short breeding seasons and start nesting before or during the early rains.

Protected areas within the miombo woodland savannah region include South Luangwa National Park (9,050 km^2) in Zambia, with its Thornicroft's giraffes, Selous Game Reserve (50,000 km^2) in central Tanzania, with a sizeable wild dog population, and Gombe Game Controlled Area (3,000 km^2) in western Tanzania, famous for its well-studied chimpanzees.

Fig. 1.11 African savannahs. (a) Laikipia, central Kenya. (b) Laikipia, central Kenya. (c) Amboseli, southern Kenya. (d) Serengeti, northern Tanzania.

Like miombo and mopane savannahs, the northern doka savannah takes its name from a dominant tree, *Isoberlinia doka*. Here, where human population remains sparse, patches of dense dry forest remain, dominated by *Isoberlinia doka* with *Afzelia africana*, *Burkea africana*, *Anogneissus leiocarpus*, *Terminalia* species and, *Borassus aethiopum*. In many ways these northern doka woodland savannahs are similar to the forest–savannah mosaics described below.

Forest–savannah mosaic

Encircling the tropical rainforest of the Congo Basin, the 'forest–savannah mosaic' forms the edge of the 'true' savannah. This encircling, sometimes narrow, transition zone can be conveniently divided into three regions. In the north, and to the west of the Cameroon Highlands, is the Guinean forest–savannah mosaic of West Africa running through Guinea, Ivory Coast, Ghana, Toga, Benin and Nigeria. The interlacing forest, savannah and grassland habitats are highly dynamic, and the proportion of forest versus other habitat components has varied greatly over time. The protected areas in this region are under funded and only cover two per cent of the area. In the north, and to the east of the Cameroon Highlands, is the northern Congolian forest–savanna mosaic. This narrow transition zone marks an abrupt habitat discontinuity between the extensive southern rain forests and the dryer savannahs to the north and east. It extends east through the Central African Republic, northeastern Democratic Republic of Congo and into southwestern Sudan and a sliver of north-western Uganda. To the south and west of the Congo Basin is the Zambezian forest-savanna mosaic. Because all these mosaics are edge, or ecotonal, regions they frequently have high species richness.

The northern forest–savannah mosaic is characterized by a single wet season and a single dry season but forested areas exhibit high relative humidity even in the dry season. Mean annual precipitation ranges locally from about 1,200 mm to 1,600 mm per year. There are only small seasonal changes in average temperature, with rainy season average maximum temperatures of 31° to 34°C and dry season average minimum temperatures of 13° to 18°C.

The vegetation of this region is an interesting mosaic of rainforest and savannah elements. Gallery forests, extending along rivers, interdigitate with drier, semi-evergreen, rainforest, which in turn grades into grassland and wooded grassland. Widespread gallery species include *Berlinia grandiflora*, *Cola laurifolia*, *Cynometra vogelii*, *Diospyros elliotii*, *Parinari congensis*, and *Pterocarpus santalinoides*. Species restricted to the drier forests, and widespread across this region, include *Afzelia africana*, *Aningeria altissima*, *Chrysophyllum perpulchrum*, *Cola gigantea*, *Combretum collinum*, *Morus mesozygia*, and *Khaya grandifoliola*. Trees found in the wooded grasslands include *Annona senegalensis*, *Afzelia africana*, *Burkea africana*, *Butyrospermum paradoxum*, *Daniellia oliveri*, *Hymenocardia acida*, *Maranthes polyandra*, *Pariniari curatelifolia*, *Parkia biglobosa*, *Piliostigma thonningii*, *Psuedocedrela kotschyi*, *Pterocarpus erinaceus*, *Stereospermum kunthianun*, *Strychnos* species, *Terminalia* species, and *Vitex* species. Common grasses, many growing taller than two metres, include *Andropogon* species, *Hyparrhenia* species, and *Loudetia* species.

The southern forest–savannah mosaics are similar in structure to the northern mosaics except that the drier southeast boundary merges more gradually into the adjacent miombo woodland savannah already described. Covering a broad area of the southern Democratic Republic of Congo, these southern forest–savanna mosaics are a blend of forest, woodland, shrubland and grassland habitats.

Some mammals of these forest–savannah mosaics include the forest subspecies of elephant (*Loxodonta africana cyclotis*), giant eland, and in the eastern sector, bongo (*Tragelaphus eurycerus*). Predators are lion, leopard, and the Nile crocodile (*Crocodylus niloticus*), which is present in northern waterways across the region.

Protected areas within the forest–savannah mosaic include Queen Elizabeth National Park (1,978 km^2) and Murchison Falls National Park (4,000 km^2), both in Uganda. Both still have good numbers of Uganda kob (*Kobus kob thomasi*), and a very wide range of bird life.

Species richness patterns

So far we have looked at climatic patterns in Africa and seen how these influence and determine patterns in the structure (physionomy) of savannah vegetation. But do these patterns of climate and vegetation physionomy influence species distributions and biodiversity? A detailed look at these two topics will be delayed until later chapters, but it is appropriate to end this chapter with a sneak preview.

Species richness varies with area (Gaston 1996) and to make meaningful comparisons it is necessary to compare areas of a similar size. For vegetation, Cailleux (1953) considered a reference area of 10,000 km^2 to be adequate and this is the data (areal plant richness) shown in Figure 1.12. The floristic richness of the African savannahs stands out clearly. The average areal plant richness of African savannahs is about 1750 species, not much lower than that of rain forests (c. 2020 species). The miombo savannahs of East Africa have an even greater (>3000) areal plant richness than the rain forest. Notice also that Figure 1.12 is remarkably similar to Figure 1.8 and Figure 1.9. Rainfall (Figure 1.8) appears to be a major determinant of savannah physionomic types (Figure 1.9), and floristic richness (Figure 1.12). There is therefore, a great similarity between the distribution of the areas of comparable floristic richness and that of the major physionomic categories of African vegetation.

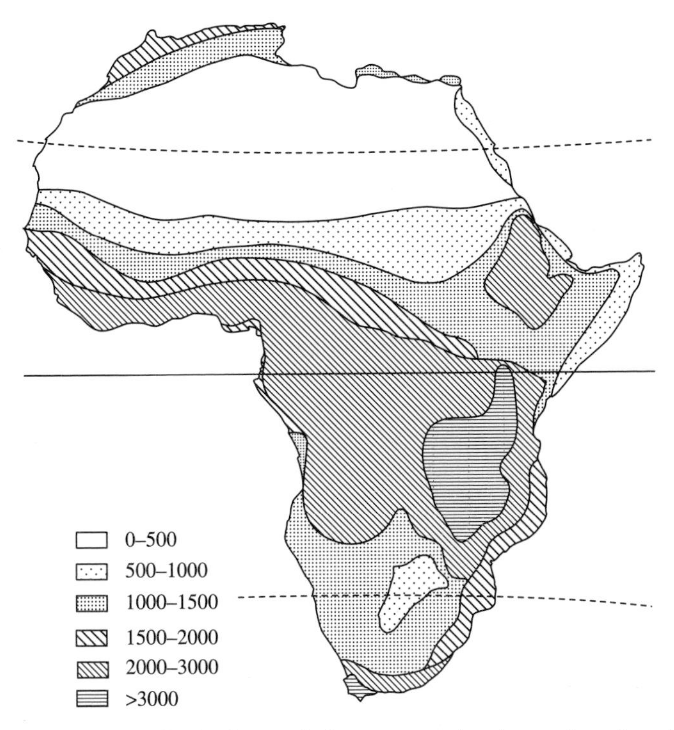

Fig. 1.12 Spatial variation in the floral richness of Africa, expressed as the number of species per 10,000 km^2 (from Menaut 1983; redrawn from Lebrum 1960).

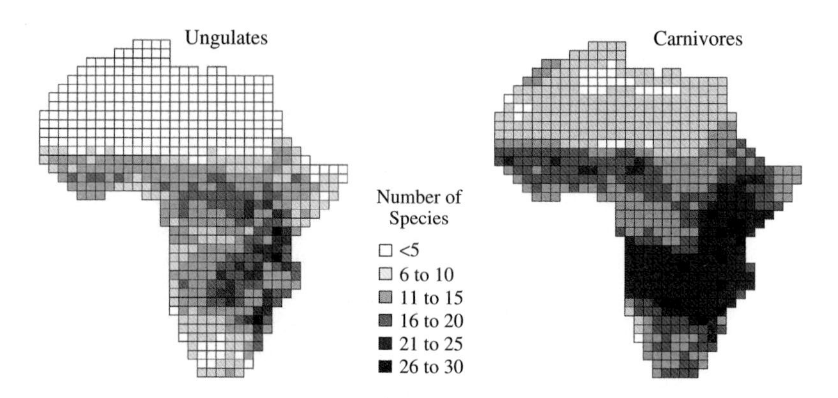

Fig. 1.13 Spatial variation in the mammal richness of Africa.

Figure 1.13 shows an equivalent pair of maps for two groups of African mammals. These maps have been constructed from the individual species distribution maps in Stewart and Stewart (1997). Each of these species maps was enlarged and the information transferred onto the grided maps of Africa shown in Figure 1.13. These grided maps therefore accumulated the number of species present in each square (equivalent to about 46,500 km^2). Two maps were produced. One for 94 ungulate species and one for 67 carnivore species. Like the areal vegetation map (Fig. 1.12), these mammal maps show high species richness in savannah areas, particularly down the eastern side of Africa, from Kenya to South Africa. Much of this area of high species richness is miombo woodland savannah, but northern areas are more arid grass and shrub savannah.

Savannah biomes are therefore extremely rich in both plant and animal species, and consequently have an extremely rich and fascinating network of species interactions. In Chapter 2 we will explore the plants, and in Chapter 3 the animals, in more detail. In the later chapters we will then explore the ecological interactions that take place between them.

2 The vegetation

This chapter will examine the biology and ecology of the plants found in African savannahs. Of course, not all species of plants can be mentioned individually. I will examine two groups of plants—grasses and trees—and describe some of the more important savannah species. These two types of plant dominate African savannah vegetation (Chapter 1), and this vegetation is an expression of the interactions of climate, soils, herbivores, fire, and human activities. Climate was examined briefly in Chapter 1, and soils will be mentioned in the case study at the end of this chapter. The important interaction between fire, elephants, and trees will be examined in Chapter 5, and human activities will be dealt with in Chapter 6. To start this chapter, I will briefly look at the effect of rainfall on plant biomass, and the grass/tree mixture. As we shall see in Chapter 4, the relationship between rainfall and plant biomass is important for the animal populations that graze and browse on savannah vegetation. In the final section of this chapter I will examine vegetation at a 'local scale' using the Serengeti–Mara ecosystem of northern Tanzania and southern Kenya as an example.

Rainfall, plant biomass, and the grass–tree mixture

Plant biomass, in African savannahs, is closely associated with the annual amount of rain falling in an area, and this has an important effect on the biomass of herbivores and carnivores (Chapter 4). One African study that demonstrates this relationship clearly is that of Deshmukh (1984), who compiled published information on above ground herbage production and rainfall for several eastern and southern African sites. All these studies harvested the herb layer (grasses, forbs, and dwarf shrubs), away from tree and shrub canopy effects. The resulting data are shown in Figure 2.1, along with the linear regression line describing the relationship between rainfall and biomass. This significant regression ($P < 0.001$) predicts approximately 800 kg ha^{-1} herb layer biomass for every 100 mm of rainfall.

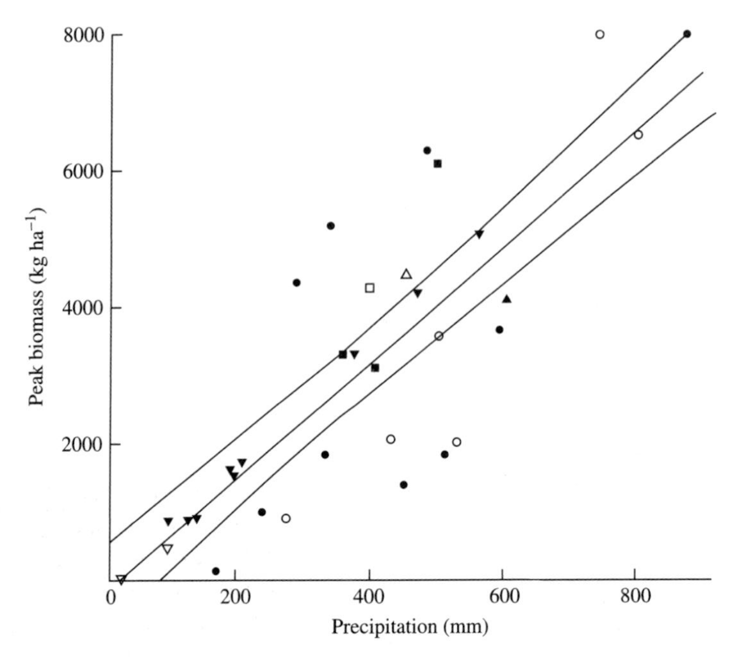

Fig. 2.1 The relationship between rainfall and above-ground herb biomass, for sites in East and southern Africa: ● Marsabit, northern Kenya, △ Kaputei, southern Kenya, □ Nairobi NP, southern Kenya, ▽ Namib Desert, western Namibia, ▼ Namibian grasslands, western Namibia, O Serengeti NP, northern Tanzania, ■ Mkomazi GR, northern Tanzania, ▲ Mweya Peninsula, Queen Elizabeth NP, Uganda. Also shown are the regression line and 95% confidence limits (from Deshmukh 1984).

Several of the studies that produced the data points for Figure 2.1 also calculated regression lines for their local data (Marsabit, Mweya, Namib Desert, and Serengeti). They all fit within the 95 per cent confidence limits of Figure 2.1 and have very similar slopes. This suggests that although herb layer biomass production may be influenced by edaphic factors that affect soil water-holding capacity and water runoff, the major determinant of the amount of grazing material available to herbivores is rainfall. The other potential modifying influence is that of the grass species present in the herb layer. Not all grass species will respond to rain in quite the same way. O'Connor, Haines, and Snyman (2001) looked at this same relationship for three conditions of southern African savannahs. They termed these 'good' (dominated by the tufted perennial grass, *Themeda triandra*), 'medium' (dominated by the tufted perennial grass, *Eragrostis lehmanniana*) and 'poor' (dominated by the stoloniferous perennial grass, *Tragus koeleriodes*), reflecting the quality of forage. Ground cover decreased from 'good' to 'poor', and the medium and poor sites where maintained artificially. Their results are shown in Figure 2.2. Clearly species composition, and perhaps more importantly ground cover, affect the precise nature of this plant

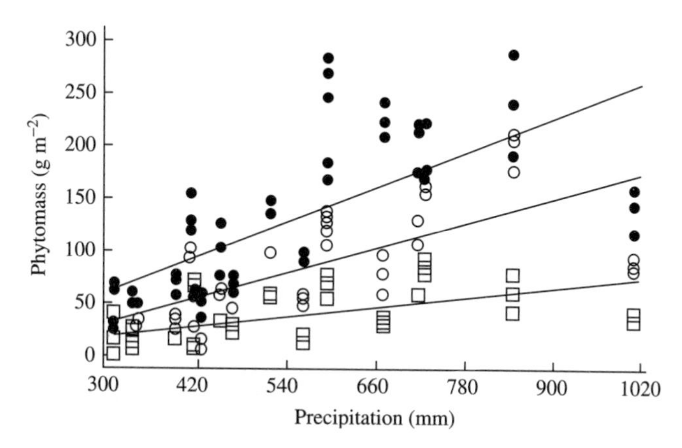

Fig. 2.2 Relationship between plant biomass and rainfall for three 'grass composition' sites in southern Africa ● good, ○ medium, □ poor (from O'Connor *et al.* 2001).

biomass/rainfall relationship. However, the over-riding influence of the amount of rainfall still remains. When we turn to the other major component of savannah vegetation, trees, this type of detailed information, about the effect of rainfall on biomass, is more difficult to find. However, Birket and Stevens-Woods (2005), working at a savannah site in Laikipia, Kenya found that tree growth is also influenced by the amount of rain. When rainfall was high (109 mm per month), *Acacia drepanolobium* grew by 4.8 cm per month, but when rainfall was low (45 mm per month) it grew by only 1.4 cm per month. Perhaps more importantly, rainfall, or mean annual precipitation (MAP), has been shown to limit the amount of tree cover in African savannahs (Sankaren *et al.* 2005). They compiled data from 854 African savannah sites, for which annual rainfall, fire incidence and tree cover were known (Fig. 2.3). Notice that tree cover ranges from 0 to 90 per cent across these sites, and that the 'maximum' tree cover increases linearly with rainfall. However, the actual tree cover, for any one site, can be lower than this maximum. For arid and semi-arid savannah sites (MAP < ~650 mm) the maximum tree cover is constrained by, and increases linearly with, MAP. Their analysis suggests that below a MAP of ~650 mm, tree-grass coexistence is 'stable' to the extent that disturbances such as fire and herbivory, although they occur, are not needed for coexistence. These could be referred to as 'climatically determined savannahs'. Above a MAP of ~650 mm, rainfall is sufficient for the tree canopy to approach closure, and disturbance (fire and herbivory) is required for the coexistence of trees and grasses. These could be referred to as 'disturbance driven savannahs'. This is a topic that will reappear in Chapter 5, and further comments on the interaction between rainfall, trees and grasses (and fire) will be left until that chapter.

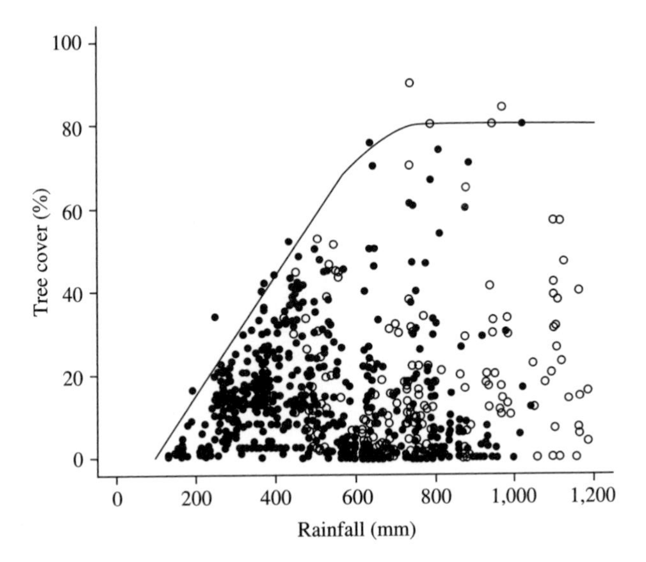

Fig. 2.3 Tree cover as a function of mean annual precipitation (MAP), as related to the fire regimes of 854 African savannah sites. The line representing maximum tree cover was obtained from a regression analysis that used the 99th quantile of y, rather than the mean, as in conventional regression (Cade and Noon 2003). Filled circles represent savannahs with low fire frequencies (average fire return times of >3 years), and open circles represent sites with high fire frequencies (fire return intervals of <= 3 years). Fires typically tend to be more frequent in sites with greater rainfall. For sites receiving <650 mm MAP, the upper bound on tree cover exists despite low fire frequencies (some of the sites had not burned for >50 years) (from Sankaren *et al.* 2005, Determinants of woody cover in African Savannahs. *Nature*. Reprinted by permission of Macmillan Publishers Ltd.).

Morphology and life history

In African savannahs, both the woody and the herbaceous plant species have to overcome two major environmental problems: seasonal drought and/or flood and periodic burning. They do this in various ways. However, rather than describe these morphological and life history adaptations individually, for each plant species, I will examine, and discuss these features within a simple classification. The term morphology, or life form, will be used to encompass all those traits of external morphology and overall organization shown by the plant. The term life history, or phenology, will be used to encompass all those traits of sequential development of plant structures during an annual cycle.

Life forms

The Danish botanist, Christen Raunkier, proposed a system of life forms (Raunkier 1934) which, with various later improvements, has been widely

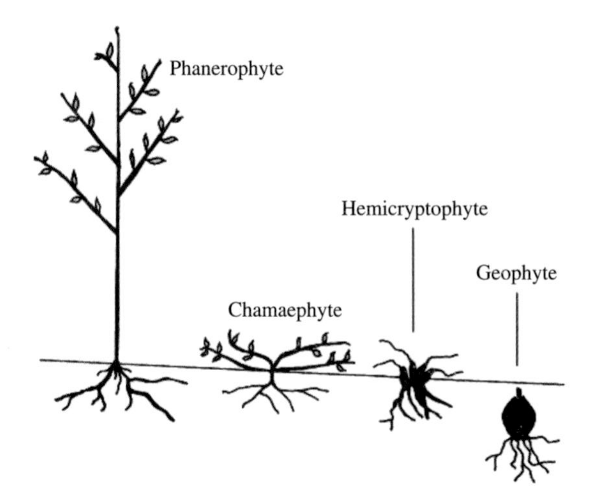

Fig. 2.4 The life forms of Raunkier most relevant to savannahs. Therophytes are not shown.

employed to compare floras from different biomes, in different parts of the World. The system classifies plants into a series of categories (Fig. 2.4) primarily according to the position of their buds during the unfavourable season of the year. The Raunkier categories most encountered in African savannahs are:

1. **phanerophytes**: woody or herbaceous perennials, taller than 20 cm, whose shoots do not die back. In other words, trees and large shrubs.
2. **chamaephytes**: woody or herbaceous perennials from 10 to 20 cm tall, whose shoots die back periodically. These plants are small shrubs.
3. **hemicryptophytes**: perennial (or biennial) herbaceous plants in which the stems die back to a remnant shoot system that lies on the ground. These are herbaceous plants with runners along the ground.
4. **geophytes**: perennial (or biennial) herbaceous plants in which the stems die back to a remnant shoot system with storage organs that are imbedded in the soil. These storage organs are variously called bulbs, corms, rhizomes, and tubers.
5. **therophytes**: annuals, or plants that die after seed production and complete their entire life cycle within one year.

Some authors (Aubréville 1963) consider this Raunkier system to be of limited use for savannah plants, since the unfavourable season was originally assumed to be a low temperature winter. In most African savannahs low winter temperature does not represent a serious limiting factor. Here the main environmental limitations are an extended drought, and periodic burning. None the less, this system of life forms is widely used by African botanists (Menaut 1983) and, as an example, Table 2.1 shows the distribution of the relevant Raunkier life forms across some western African

Table 2.1 Percentages of species in the various Raunkier life forms, in nine west African savannahs. There is a moisture gradient from the dry Sahelian savannahs to the more humid Guinean savannahs.

	Phanerophytes	Chamaephytes	Hemicryptophytes	Geophytes	Therophytes
Guinean savannahs					
Ivory Coast	28	—	42	6	24
Congo	14	25	26	16	19
Nigeria	30	—	23	21	25
Sudan–Zambezian savannahs					
Central African Republic	13	24	8	11	40
Rwanda	29	30	12	7	20
Zaire	38	44	9	4	5
Sahelian savannahs					
Southern Mauritania	24	7	6	2	61
Northern Senegal	19	2	2	2	75
Northern Chad	14	2	12	5	67

For the original references see Menaut 1983.

savannahs of differing humidity. Notice that the proportion of phanerophytes does not change much along the humidity gradient, although it diminishes slightly in the dryer savannahs. More humid savannahs appear to be especially rich in geophytes and hemicryptophytes. These forms are particularly resistant to fierce bush fires fed by an abundant grass layer. Chamaeophytes seem to thrive in mesic savannahs and theophytes are largely dominant in arid savannahs. However, one of the problems with using the Raunkier system in this comparative way is that different botanists frequently assign the same species to different categories. Also the same species may appear to have individuals that fall into different categories depending upon the local conditions of soil, grazing and fire, for example phanerophytes versus chamaephytes. Another problem, which certainly distorts the 'community structure' of savannah plant assemblages, is that botanists have usually compared the percentage of species that fall into the Raunkier categories (Table 2.1). This gives a wrong impression of the 'importance' of annual grasses (therophytes) in savannah systems. For example, using species, the grass layer in one savannah site (Hopkins 1962) gave: hemicryptophytes 32 per cent, geophytes 29 per cent, and therophytes 39 per cent. Using the percentage of individual plants gave: hemicryptophytes 71 per cent, geophytes 14 per cent, and therophytes 15 per cent.

Below, therefore, I combine many of the Raunkier categories and use a simpler classification of life forms (Sarmiento and Monasterio 1983). This has only three categories: perennials with permanent above ground woody structures, perennials whose above ground structures are seasonal, and annuals.

Perennials with woody ground structures (trees and shrubs)

The main 'architectural' type within this group are trees, often looking like shrubs, while palms, woody vines, succulents and so forth are much less frequent. In the Raunkiaer system this group would include mainly phanerophytes and chamaephytes. One characteristic feature of savannah trees is their relatively modest aerial development. In many savannah communities an overwhelming proportion of the trees only have a mean height ranging between 2 and 6 m, although the tallest specimens may reach 12 m. An exception are species such as *Brachystegia* which occur in what I have called the miombo woodland savannah, where the taller trees may reach a height of 20 or 25 m. Another obvious exception are the familiar fever trees (*Acacia xanthophloea*), found along the rivers that penetrate many savannahs, that may reach a height of 30 m. Menaut (1971) sampled the woody populations in the Lamto savannahs of West Africa and found that all trees were lower than 10 m, except the palm *Borassus aethiopum* which could reach 20 m. Associated with this feature, stem girth rarely exceeded 150 cm, with most individuals in the range 20 to 40 cm. Another feature of many savannah trees is that they are low-branched and often ramify from the base. This is frequently a response to fire damage, or some other mechanical injury such as grazing. New shoots sprout from the damaged stump, or sucker from the lateral roots.

The thickness of the bark of many savannah trees has been interpreted as a protection against repeated bush fires. This might be the case in some species, but the lack of a thick bark does not prevent many other species from surviving in savannahs that are regularly burnt. Many savannah trees have spines. For example, *Acacia* species, *Balanites aegyptiaca*, and *Euphorbia* species. These spines may prevent, or interfere with, browsing by some species of herbivore. Many savannah trees, such as *Acacia drepanolobium*, certainly put more 'effort' into producing spines if they are heavily grazed. However, spines do not prevent browsing by some species such as giraffe and black rhinoceros. The leaves of savannah trees are generally moderate or small in size, and except for the palms (e.g. *Borassus aethiopum*), plants with large leaves are rarely found. Many species (e.g. thorn trees or *Acacia*) have moderate sized compound leaves but they are divided into many, quite small, leaflets. The leaves of many savannah species live for about one year, with leaf fall preceding the development of new leaves. They therefore appear to be evergreen. However, some savannah trees are deciduous, or semi-deciduous, with the alternation of leaf and leafless conditions corresponding with the wet and dry seasons. This phenology can vary between species even within the same genus, and also between geographical areas (see the three *Acacia* species in Figure 2.6 below). Leaf fall is one method of adaptation to drought, particularly for those species with 'soft' mesomorphic leaves. Another method is the production of xeromorphic leaves that can survive the dry season and have adaptations that reduce water loss through transpiration. These adaptations include hard leaves with a thick cuticle and cuticular layers, stomata (the leaf's breathing pores) placed at the bottom of deep depressions, and small leaves reduced to scales. Some trees have enhanced water storage facilities in their trunk, such as the cactus-like *Euphorbia* species and the baobab, *Adansonia digitata*. A final characteristic of savannah trees is their extensive root system that allows them to exploit the water and mineral resources of a great volume of soil. Many savannah trees are reputed to have a downward growing tap-root, from which a number of lateral, horizontal, roots develop (Hopkins 1962, Sarmiento and Monasterio 1983). However, Walter (1973) has shown this is often not the case. Particularly in arid areas, root systems tend to flatten out in order to provide the best opportunity to absorb water from the upper soil layers after relatively light rain.

Perennials with above ground seasonal vegetation

In the Raunkiaer system this group would include mainly geophytes and hemicryptophytes. It includes two types: perennial species with woody underground storage organs and perennial species with non-lignified storage organs, such as fleshy rhizomes. Dominant among this latter group are the tussock grasses, a very obvious and dominant part of the savannah flora. The stem of grasses, bearing the leaves and flower-head, is called the culm. It is cylindrical and hollow except at the nodes, or joints, which are of solid tissue. The hollow sections between the nodes are called the internodes, and the basal nodes may be swollen. The nodes are where the leaves are attached, in two rows on opposite sides of the stem. The upper, expanded, part of the leaf is the blade or lamina, and the basal part, surrounding the stem, is the sheath (Fig. 2.5). At the point where the blade forms the sheath there is usually a membranous outgrowth called the ligule.

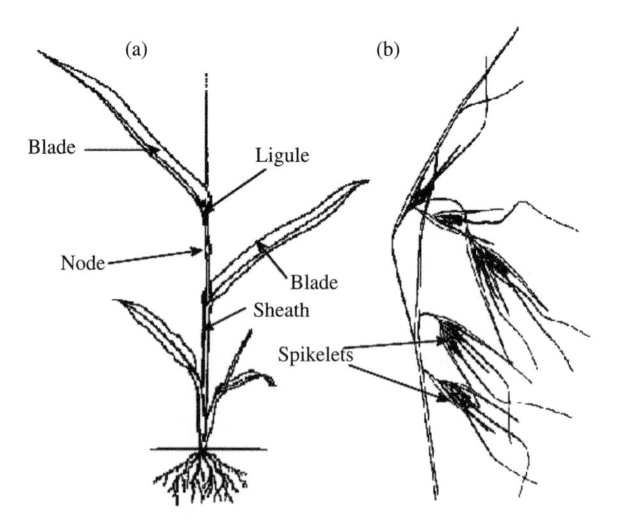

Fig. 2.5 (a) Diagram showing the basic structure of a grass stem and (b) a *Themeda* inflorescence.

The normal method of branching in grasses involves the production of new stems from the axils of older leaves, and is called tillering. If the new stem grows upwards, within the sheath, dense tufts of grass are produced. If the new stem breaks through the sheath near its point of origin then loose or open tufts result. The thin and fibrous roots are adventitious (meaning they arise from 'abnormal' positions, in this case the stem), and arise from the lower node or nodes of the stem. In some perennial species (e.g. *Chloris gayana*) the roots are produced at every node of a surface creeping stem, or stolon. In other perennial grasses (e.g. *Pennisetum purpureum*) they arise from the nodes of underground creeping stems, or rhizomes. The 'flower head', or inflorescence, is usually at the end of the stem and consists of a much-branched structure called the panicle. The panicle branches bear structures called spikelets (Fig. 2.5). These spikelets are composed of one or more flowers (usually called florets in grasses) with enveloping scales that conceal the flowers from view until flowering time. The florets of perennial grasses are often self-sterile and achieve cross-pollination using wind. The fruit, or grain, of grasses is called a caryopsis, although the term 'grass-seed' is often applied to the entire spikelet.

The growth habit of perennial grasses has made them ideal primary producers for grazing ecosystems. The production of new shoots by tillering provides a rapid means of recovery from the grazing pressure imposed by herbivores, and new growth takes place chiefly at the base of the leaves where it is least likely to be damaged by grazing mammals. The root system also improves the 'condition' of the underlying soil. The roots bind the soil particles together, forming a 'sod', and brings to the surface layers nutrients that have been leached into the subsoil by heavy rain. A morphological characteristic of many perennial savannah grasses is the protection of the apical bud from fire and desiccation by a thick tunic formed by the old leaf sheaths. These species, having their buds protected at ground level, are able to regrow rapidly after fire. Some geophytic species may have their buds protected in the soil at depths of 10 cm, or more.

Although the tussock grasses are the dominant growth form, both in species and biomass, among plants with non-woody underground storage organs, there is a rich diversity of other herbaceous, bulbous, geophytes. These are represented by numerous species of Amaryllidaceae (daffodils), Iridaceae (irises), Liliaceae (lilies), and Orchidaceae (orchids).

Perennial plants with woody underground storage organs, but with all shoots annual, form another characteristic feature of African savannahs. The annual shoots dry out completely during the dry season. They have been variously called subshrubs, geofrutescent, geoxyles or hemixyles. Some of these growth forms are permanent geoxyles, conserving this habit under all circumstances, while others are traumatic geoxyles, in which this habit results from external injury. These traumatic geoxyles revert to a normal tree habit when circumstances allow. Many African savannahs have a rich flora of these species. In the Lake Edward plain, subshrubs such as *Vigna friesiorum* and *Cissus mildbraedii* can form up to 35per cent of the 'grassland' flora (Sarmiento and Monasterio 1983). This growth form is also characteristic of the grasslands of the high plateaus covered by the Kalahari sands.

Annuals

The contribution of annuals to the savannah herb layer is generally inconspicuous. In the Raunkiaer system this group would include the therophytes. Although many of these annuals are grasses, they are not very common in savannahs, compared to perennial grasses. Annuals are more characteristic of savannahs that have been modified by overstocking of cattle, at least in mesic and arid savannahs inhabited by pastoral tribes. Bush fires can also contribute to the elimination of perennials by favouring annuals whose seeds ripen before burning takes place, and survive in the ground. When burning is repeated, or is started very late in the dry season, its influence is very much the same as that of overgrazing. The basic above ground structure of annual grasses is similar to that of perennial grasses, however, in contrast to perennial grasses, the flowers of annual grasses are usually self-fertile and do not rely on wind for pollination. They also tend to have thin spreading root systems compared to the deeper root systems of perennial grasses. Sillans (1958) considered that perennial grasses make up the stable, basic component of the savannah grass layer and that annuals only constitute a fleeting component, 'filling the gaps' when the opportunity arises.

Phenology

As I have already said, variations in plant structure and function associated with the annual cycle of day length, temperature, and rainfall are called phenology. These phenological events include the germination of seeds, the appearance of leaves (the time between leaf bud opening and leaf senescence), flowering, fruit maturation, and the growth of vegetative parts such as stems and roots. As I have already alluded, the overriding influence on the phenology of savannah plants is the alternation of wet and dry seasons. The deciduous nature of many trees, as a response to this, has already been mentioned. Three *Acacia* examples are shown in Figure 2.6 and since the phenology varies in different parts of Africa, these three examples are specifically for Tanzania. For example, in Kenya, for *Acacia sieberiana*, the no flowers period would extend from March to October. Leaf expansion in some trees may precede the rains, especially if the previous rainy season was exceptionally wet, or prolonged, leaving residual water in the soil. Early leaves in some species (e.g. red-leaved rock fig *Ficus ingens*) are bright red because anthocyanin pigments predominate over chlorophyll. One important aspect of phenology is the timing of pollination. A lot of work on pollination biology has considered how plants might avoid competition for pollinators. Plants may differ in the pollinators they recruit, but many East African acacia flowers (Stone *et al.* 1998) are visited by a wide diversity of

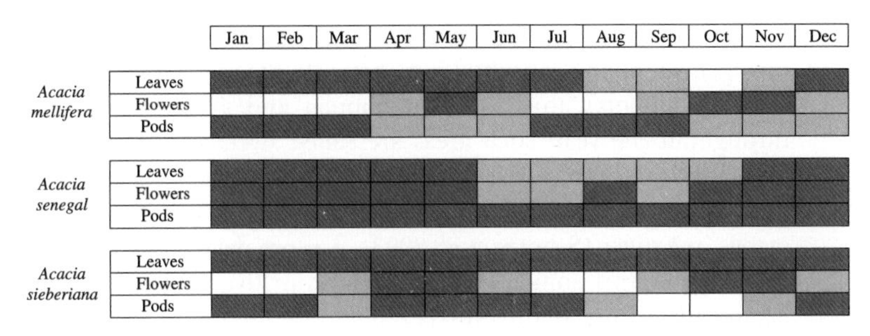

Fig. 2.6 Phenology of three *Acacia* species in Tanzania. The varying degrees of shading indicate different intensities of leaves, flowers, and pods, in different months, with dark grey indicating prolific, and white none (modified from Dharani 2006).

Table 2.2 Seasonal flowering patterns of Mkomazi *Acacia* species. An asterisk (*) indicates mass flowering, and a cross (+) slight, scattered flowering.

Acacia species	Jan.	Feb.	Mar.	Apr.	May	Jun.	Jul.	Aug.	Sep.	Oct.	Nov.	Dec.
A. brevispica	+				*	*	+					+
A. bussei									+	*	*	+
A. etbaica								+	*	*		
A. reficiens	+							+	*	*	+	+
A. thomasii							*	*	+		+	
A. drepanolobium	*	+			+	+	+					*
A. nilotica	*	+				+	+					*
A. senegal	*	+				+	+				+	*
A. tortilis	*	+					+	+				*
A. zanzibarica	*					+	+				+	*

Data from Stone *et al.* 1989.

species, at least some of which are shared by more than one acacia species. Another solution is to use different populations of pollinators over time and certainly in parts of East Africa this appears to be the case. Table 2.2 shows the seasonal flowering patterns of *Acacia* species present in the Mkomazi reserve, Tanzania (Stone *et al.* 1998). There is still some overlap between *Acacia* species, but there is also phenological separation.

One interesting correlate of plant phenology is that the annual cycle of grasses, and to a lesser extent trees, is associated with changes in their palatability (see Table 2.4 later). For example, in many plants the nitrogen content of young tissue is high and the fibre content low. As the plant grows, and the dry season approaches, the ratio of leaf nitrogen to fibre tends to fall. Scholes *et al.* (2003) comment that as a consequence of this, on nitrogen poor soils in Kruger NP (granite uplands and sandstones), the nitrogen content may drop below the threshold for ruminant digestion (see Chapter 3).

These areas are called sourveld in southern Africa. On soils with a high nitrogen supplying capacity (granitic bottomlands and soils derived from basalt sediments), this does not happen and grazers can be sustained throughout the year. Such areas are called sweetveld in southern Africa. This difference in nutritive phenology is thought to underlie the difference in herbivore biomass and composition between the granite and basalt landscapes in Kruger (Scholes *et al.* 2003). A more detailed example, with buffalo, of how this annual change in the nutritive value of vegetation can greatly affect some herbivores, is given in Chapter 4.

Grasses

Savannahs are essentially tropical grasslands with trees. African savannahs are dominated by herbivores, many of which eat grasses, either as grazers or mixed feeders (Chapter 3). It is estimated that there are about 10,000 species belonging to the family of grasses (Poaceae) in the world. However, the number of grass species in any one savannah is usually only between 30 and 60 species, with six to ten dominant species. Some widespread, African savannah species, are listed in Table 2.3. Many of these are species mentioned in other chapters. Some species, such as *Ctenium newtonii*, are representative of a widespread group of very similar species. For example, *Ctenium newtoni* is found in open scrub savannah across western African, from Senegal to Angola, extending eastwards into the Sudan. In East Africa it intergrades with a similar species, *C. somalense*, which extends down to Zambia. A third, southern African species *C. concinnum*, then intergrades with *C. somalense*. This species complex therefore extends across scrub savannahs from western Africa, through East Africa to southern Africa. Many grass genera, for example *Cymbopogon*, are notorious for their considerable variation within species and the weak separation between them. Consequently their taxonomy is still in a very fluid state. Sometimes even different genera, for example *Cymbopogon*, *Andropogon*, and *Hyparrhenia*, are quite difficult to separate.

The grass family is divided into five subfamilies (Bambusoideae, Arundinoideae, Pooideae, Chloridoideae, and Panicoideae). The most primitive grasses are thought to be the bamboos, but Africa is relatively poor in these species. In fact, tropical African grasses are predominantly of the subfamily Panicoideae, with the two tribes Andropogoneae (e.g. *Andropogon*, *Cenchrus*, *Chrysopogon*, *Cymbopogon*, *Heteropogon*, *Hyparrhenia*, *Imperata*, *Monocymbium*, *Pennisetum*, *Schizachyrium*, *Themeda*) and Paniceae (e.g. *Brachiaria*, *Digitaria*, *Echinochloa*, *Panicum*, *Setaria*) accounting for the majority of species. Some examples of African grasses are shown in Figure 2.7. Notice that the inflorescence (the panicle and its component spikelets) is quite variable. In grass taxonomy it tends to take the place of the flower in flowering plant (Angiosperm) taxonomy.

Table 2.3 Some principal grass species of African savannahs (g = open grass and shrub savannah, t = tree and scrub savannah, w = woodland savannah, m = miombo).

Scientific name	Common name	Type	Height (cm)	Savannah type
Aristida adscensionis	common needle grass	Annual or perennial	10–100	g
Aristida stipoides	needle grass	Tufted annual	60–100	t
Andropogon gayanus	bluestem grass	Tufted perennial	up to 300	t w m
Andropogon greenwayi	bluestem grass	Tufted perennial	45–60	g
Brachiaria brizantha	common signal grass	Erect perennial	up to 120	w
Brachiaria deflexa	signal grass	Annual	up to 45	t w
Cenchrus biflorus	African foxtail	Tufted annual	up to 90	t
Cenchrus ciliaris	common African foxtail	Rhizomatous perennial	20–110	g t
Chloris gayana	Rhodes grass	Stoloniferous perennial	30–120	g m
Chrysopogon aucheri	Aucher's grass	Tufted perennial	up to 50	t
Ctenium newtonii	sickle grass	Tufted wiry perennial	up to 100	g
Cymbopogon afronardus	blue citronella	Tufted perennial	200	g w
Cymbopogon plurinodis	citronella	Tufted perennial	40–100	g
Cynodon dactylon	Bermuda grass	Stoloniferous perennial	8–180	g t
Dactyloctenium aegyptium	common crowfoot	Annual	up to 60	g
Digitaria eriantha	Pangola grass	Tufted perennial	50–250	g
Digitaria macroblephora	woolly finger grass	Tufted perennial	40–100	g t
Echinochloa colonum	barnyard grass	Tufted annual	up to 60	g
Echinochloa pyramidalis	antelope grass	Reed like perennial	up to 500	g m
Enneapogon cenchroides	grey head grass	Tufted annual	30–60	g
Enteropogon macrostachyus	bush rye	Tufted perennial	90	g t
Eragrostis lehmanniana	Lehmann's lovegrass	Tufted perennial	45–61	g t
Eragrostis patens	desert lovegrass	Tufted annual	15–30	g t
Eragrostis superba	Masai lovegrass	Tufted perennial	30–90	g t w
Heteropogon contortus	spear grass	Tufted perennial	up to 90	g t w
Hyparrhenia cymbaria	coloured hood grass	Tall perennial	180– 600	g w
Hyparrhenia diplandra	sword hood grass	Tufted perennial	150– 300	g
Hyparrhenia dissoluta	yellow hood grass	Tufted perennial	up to 300	g w m
Hyparrhenia filipendula	fine hood grass	Slender perennial	up to 300	w m
Hyparrhenia rufa	brown hood grass	Variable perennial	up to 300	w m
Imperata cylindrica	cotton grass	Perennial	up to 120	g w
Loudetia kagerensis	sour russet grass	Tufted perennial	up to 110	g
Loudetia simplex	common russet grass	Tufted perennial	up to 150	g m
Monocymbium ceresiiforme	wild oatgrass	Tufted perennial	up to 120	g
Oryza barthii	wild rice	Tall annual	60–200	g w
Panicum coloratum	blue guinea grass	Tufted perennial	8–100	g
Panicum maximum	slender guinea grass	Tufted perennial	70–150	g w
Panicum turgidium	Taman guinea grass	Tufted perennial	up to 100	g
Pennisetum clandestinum	Kikuyu grass	Stoloniferous perennial	up to 46	g
Pennisetum mezianum	bamboo grass	Tufted perennial	40–160	g
Pennisetum purpureum	elephant grass	Tufted perennial	up to 450	g
Pennisetum schimperi	wire grass	Tufted perennial	up to 120	g
Pogonarthria squarrosa	herringbone grass	Tufted perennial	up to 120	w m
Schizachyrium semiberbe	crimson bluestem	Tufted perennial	60–150	g w
Schmidtia bulbosa	Zand Kweek grass	Tufted perennial	up to 80	g t w
Schoenefeldia gracilis	spiky grass	Tufted annual	70–120	g t
Setaria incrassata	setaria	Tufted perennial	60–100	g

Table 2.3 (*Continued*)

Scientific name	Common name	Type	Height (cm)	Savannah type
Setaria sphacelata	common setaria	Tufted perennial	45–180	g w
Sporobolus pellucidus	dropseed	Tufted perennial	up to 60	g
Sporobolus pyramidalis	pyramid dropseed	Tufted perennial	45–150	g t w
Sporobolus robustus	reed dropseed	Tufted perennial	100–200	g
Sporobolus spicatus	spike dropseed	Stoloniferous perennial	9–30	g
Stipagrostis uniplumis	silky bushman grass	Short-lived perennial	up to 75	g t
Themeda triandra	red oat grass	Tufted perennial	30–200	g t w
Tragus koelerioides	creeping carrot grass	Stoloniferous perennial	12–95	g

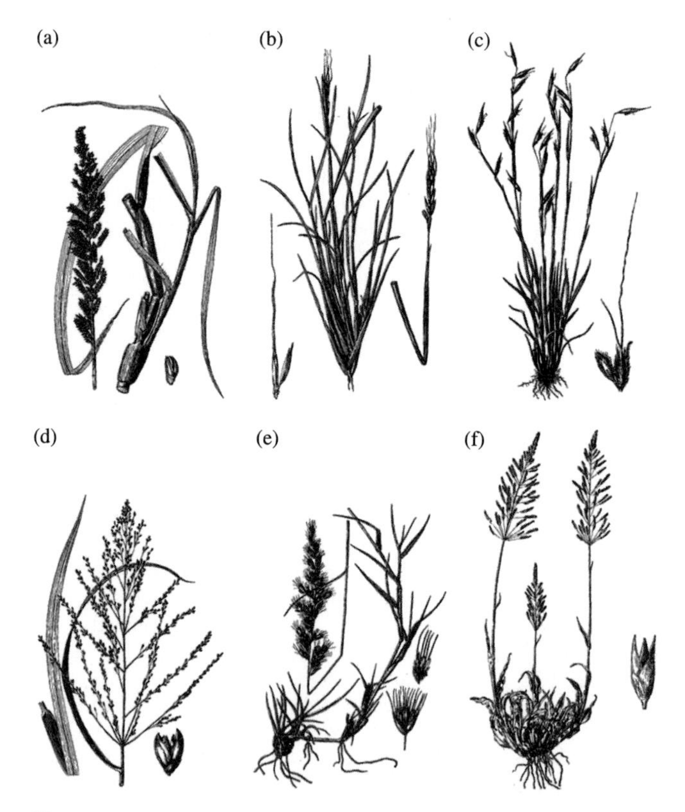

Fig. 2.7 Some African grasses. (a) *Echinochloa pyramidalis*, (b) *Heteropogon contortus*, (c) *Monocymbium ceresiforms*, (d) *Panicum maximus*, (e) *Schmidtia bulbosa*, (f) *Sporobolus nitens* (drawings by Cythna Letty, from Bews 1929).

Trees

Some of the more important families of savannah trees are described below, and within each family some representative and important species are described. The information on trees is taken from Coates Palgrave (1983), Dharani (2006), Hines and Eckman (1993) Noad and Birnie (1989) and my own observations. Most of the families with tree-forms are dicotyledons. However, the moncotyledons (in addition to the grasses) produce a few savannah forms that have tree-like growth forms. The palms and the aloes will be briefly mentioned.

Palm family (Arecaceae)

Palms are widespread throughout the tropics and subtropics and are remarkably constant in their growth pattern. They have a slender, smooth, stem with a crown of large leaves. There are over 2,700 species. The fruits are often rich in food and at least three species have economic importance. These are the **coconut palm** (*Cocos nucifera*), the **date palm** (*Phoenix dactylifera*), and the **oil palm** (*Elaeis guineensis*). The **borassus palm** (*Borassus aethiopum*) is a widespread African species, being found from East Africa to southern Africa.

Lily family (Liliaceae)

The main savannah species within this family are the aloes. These plants are more or less succulent, and some are shrubby to tree-like. The leaves are long and fleshy, the base clasping the stem. There are some 250 species of aloes, the majority of which are indigenous to the drier parts of eastern, central and southern Africa. The medicinal properties of some species has been recognized in Europe, for many centuries. *Aloe vera*, from North Africa, is widely cultivated and an extract from the pulp is used to sooth and heal the skin. The **quiver tree** (*A. dichotoma*) reaches 3 to 5 m, occasionally 7 m, in height. It is a thickset and massive aloe, with a stem up to 1 m at ground level. The bushmen of southern Africa make quivers, for their arrows, from the soft branches of this tree, hence its common name. Birds and locusts are attracted to the plentiful nectar, and baboons tear the flowers apart to get at the sweet liquid.

Grevillea family (Proteaceae)

This is one of the most notable families of the southern hemisphere, with related species in South America, southern Africa and Australasia. The trees are suited to dry conditions with leathery, evergreen, foliage. The **beechwood** (*Faurea saligna*) is a small to medium-sized (7 to 10 m), deciduous, shrub, widespread in many highland savannahs in East Africa, and the 'woodland' (miombo) savannahs of southern Africa. In the miombo it is frequently associated with the **broad-leaved beechwood** (*F. speciosa*), a small (4 to 7 m) leafy tree. Both *Faurea* species have flowers in spikes and fruits that are small nuts. The genus *Proteus* dominates the Cape flora in South Africa but most are montane rather than savannah species. The high-veld protea (*Protea caffra*) is found in highland savannahs.

Sour plum family (Olaceae)

The sour plum family contains the genus *Ximenia*. The **small sour plum** (*X. americana*) and the **large sour plum** (*X. caffra*) are both found in the mopane 'woodland' savannahs of southern Africa. They are small to medium bushes (4 to 6 m in height), often associated with termite mounds. They have grey bark, oblong elliptic leaves (5 to 2.5 cm) and, like many savannah trees, spines. The flowers are small and white, and the fruit red and sour.

Caper family (Capparaceae)

This is a medium-sized, mainly tropical family of shrubby trees and herbs. They are common in dry areas and recognized by the conspicuous spreading stamens of the flowers. The caper, commonly used as a pickle, is the bud of *Capparis spinosa*, a native of the Mediterranean region. In Africa, the genus *Boscia* is frequently encountered in dry 'grass and shrub' savannahs, often on rocky outcrops. *Boscia senegalensis* is common along the Sahel, and in East Africa, and the **sheppard's tree** (*B. albitrunca*) and the **smelly boscia** (*B. foetida*) are common in southern Africa. They are stocky trees (3 to 7 m height), stiffly branched, often associated with termite mounds.

Mobola family (Chrysobalanaceae)

The mobola plum (*Parinari curatellifolia*) is found in the wooded grasslands associated with the forest savannah mosaic, and also the 'tree and shrub' (baikian) savannah and 'wooded' savannahs of central and southern Africa. It is a large, evergreen, spreading, mushroom-shaped, tree reaching up to 13 m in height, often forming a conspicuous feature of the landscape. When David Livingstone died, on 1 May 1873 at Chitambo, in central Zambia, a commemorative inscription was carved on the trunk of a fine specimen of *P. curatellifolia* (Coates Palgrave 1983).

Pod-bearing family (Leguminosae)

The Leguminosae is the third largest family of flowering plants, containing some 17,000 species world-wide. It contains three major sub-families: the *Cassia* sub-family (Caessalpiniodes), the *Acacia* sub-family (Mimosoideae), and the Pea sub-family (Papilionoideae). However, they all share two features. First, the fruit is always a one-chambered pod, and second, they have root nodules containing 'nitrogen fixing' bacteria. The latter enables legumes to grow in relatively poor soil lacking little available nitrogen.

Undoubtedly, one of the most characteristic features of savannahs are the thorn trees of the genus *Acacia*, and as a consequence of this I will describe some of these species in some detail. All African species bear spines (Figure 2.8b) or thorns (hooks)(Figure 2.8a), and get their name from the Greek *akis*: a sharp point. The spine is a hardened, modified, stipule and may be either straight or recurved. They have compound leaves (Figure 2.8a), subdivided into pinnae, which in turn are usually subdivided into many, long, feathery leaflets. Occasionally, the leaflets are fewer and more rounded (e.g. *A. mellifera* and *A. nigrescens*). The flowers are small, but they are grouped into a flower head that is either a spike (bottle-brush or spicate inflorescence) or a sphere (capitate inflorescence) (Figure 2.8b). In the subgenus *Aculeiferum* (such as *A. senegal*) the flowers are usually in spikes and produce nectar, while in the subgenus *Acacia* (such as *A. nilotica*) they are usually in a sphere and lack nectar. The flowers of both subgenera contain both male and female parts, and last for only a day. The flowers are white, cream, or yellow. In many savannah areas the small leaves of these trees provide the only green browse in the dry season. The stem sap is also eaten by vervet monkeys and baboons, and the stringy bark by elephants and giraffe. Bees and butterflies utilize the nectar, and bruchid beetles the seeds. The pods of some *Acacia* species, such as *A. tortilis* and *A. nilotica*, are also utilized extensively by both wild mammalian species and domestic livestock. The pods vary in shape, size, texture, and colour and are important in field identification. They may be thick and woody (e.g. *A. nilotica*) or thin and papery (e.g. *A. mellifera*).

The indigenous African acacias are frequently described as 'taxonomically difficult', and some species can be very variable in their growth form.

Fig. 2.8 The structure of an *Acacia* (a) a leaf and hooks of *A. senegal*, (b) flowers and spines of *A. tortilis*, (c) pods of *A. senegal* subspecies *leiorhachis* (from Dharani 2006).

Following in the wake of the 17th International Botanical Congress all 142 species of African *Acacia* have been controversially re-classified under two newly constituted genera: *Vachellia* (73 species) and *Senegalia* (69 species). This renaming was carried out despite the fact that an African *Acacia*, the Egyptian thorn (*A. nilotica*), was the original type specimen for the genus. Many African botanists, and end-users, feel quite aggrieved about this change and intend to carry on using the name *Acacia* for the African species. I will adopt this strategy in this book.

The **prickly acacia** (*Acacia brevispica*), also called the **wait-a-bit-thorn**, is a shrub or small tree, 1 to 8 m high, with long, thin, rambling branches covered with small scattered thorns. The leaves are fairly large and the flowers are in white balls. The pods are thin, flat and broad and leathery. The young pods are used as fodder for goats and cattle. It is widespread in Africa, being found in the Sudan, Ethiopia, Somalia, Kenya, Zaire, Angola, Natal, and Cape Province.

The **whistling thorn** (*A. drepanolobium*), is a small shrub or tree (1 to 7.5 m), usually either short and robust or tall and slender, with a rounded canopy. It is abundant on poorly drained black-cotton and clay soils in East Africa. Its most noticeable feature are the swollen, purplish or black, galls at the base of the pair of long, whitish, spines (1.5 to 4.5 cm). These galls contain various species of ant (Chapter 5). The pods (4 to 7 by 0.5 cm) are brown and slightly spiralled. Young galls, pods and leaves are all eaten by wildlife, and the pods are much liked by giraffe. Whistling thorn is a rather variable species, with soil conditions, browsing and fire all contributing to its final shape.

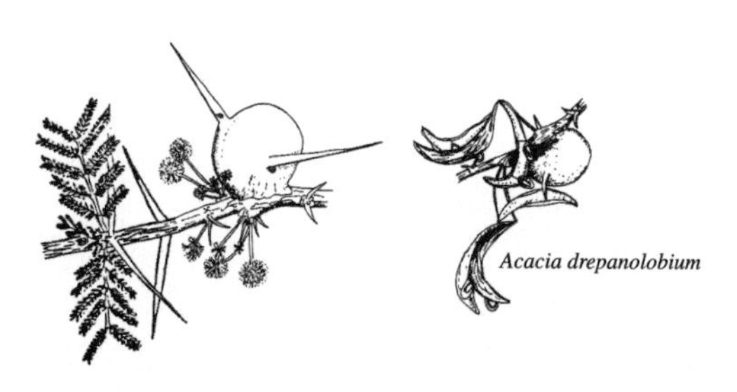
Acacia drepanolobium

The **blue thorn** (*A. erubescens*) can either be a multi-stemmed shrub, or a 10 m high tree. The name *erubescens* is Latin for 'becoming red' and refers to the colour of the young flowers. It is found in dry 'tree and shrub' savannah and is quite characteristic of Angolan mopane woodland (Chapter 1). The bark is yellowish to greyish-brown, and scaly or papery. The thorns are in pairs (no spines), strongly hooked and up to 7 cm long. It has classic *Acacia* compound leaves, with three to seven pairs of pinnae, each with 10 to 25 pairs of leaflets. The flowers are in short, squat, spiked inflorescences, and are, eventually, yellowish white. The pods are light brown to dark brown, leathery, and 3–13 by 1–1.8 cm in size. It is found in Tanzania, and southwards to Namibia, Botswana and South Africa.

The **grey-haired acacia** or **Gerrard's thorn** (*A. gerrardii*) is a widespread species, occurring from Nigeria to Sudan in the north, and southwards to

Botswana and South Africa. Along with *A. drepanolobium*, *A. robusta* and *A. tortilis* it is one of the dominant acacias in the Serengeti ecosystem. The bark is dark grey or reddish and the stipular spines, in pairs, are greyish. In East Africa they can occasionally have galls like *A. drepanolobium*. The leaves have 2 to 12 pairs of pinnae, each with 8 to 23 pairs of leaflets, and the flowers are white, tinged with pink, in a capitate inflorescence. The clustered pods (5–16 by 0.6–1.7 cm) are dark brown, curved, and covered with grey hairs. The tree is browsed, improves soil quality and is extensively used by bees. The bark is used as a traditional remedy for coughs and soar throats.

Acacia gerrardii

The **candle-pod acacia** (*A. hebeclada*) is a shrub or small tree up to 7 m in height, often branching near the ground, often forming thickets. Along with *A. kirkii*, *A. nilotica*, and *A. erubescens* it is a typical species of Angolan mopane savannah woodland (chapter 1). The bark is dark grey, fissured and often flaking. The stipular spines are quite variable, sometimes short and sometimes long (up to 3.5 cm). The leaves have two to nine pairs of pinnae, each with 7 to 16 pairs of leaflets. The flowers are creamy white, in balls, and the pod is hard and woody (4–15 by 1.4–4 cm) covered in yellow-grey hairs.

Kirk's acacia (*A. kirkii*), also known as the **flood-plain acacia**, is a handsome, flat topped, shrub or tree which can reach a height of 2 to 15 m. Like *A. ebeclada* it is a typical species of Angolan mopane savannah woodland (chapter 1). It is found in Uganda, Kenya, and Tanzania, extending southwards to Zambia, Botswana and southwest Africa. The bark is brown, fissured and rough with each leaf composed of 3 to 14 pairs of pinnae, each with 9 to 18 pairs of leaflets. The flowers are cream or white in a capitate inflorescence, and the short pods (2–10 by 0.8–2 cm) are reddish-brown. The leaves and pods are used as fodder and the Masai make 'tea' from the bark.

Acacia kirkii

The **black-hooked thorn** (*A. laeta*) is a shrub or small tree up to 6 m high, with a greyish-green bark looking blackish from a distance. The leaves have two to three pairs of pinnae, each with 3 to 5 pairs of leaflets. The thorns, or hooks, are paired (sometimes in threes), and the flowers are creamish white or pinkish, in 5 cm long spicate inflorescences. The mature pods are pale brown (3.5–8 by 1.7–2.8 cm). It is found in grass and shrub savannah along the whole Sahel, from Niger to Egypt, Sudan, Ethiopia, and Somalia, and south into Tanzania and Kenya. Many butterfly and beetle species feed on the flowers. It may hybridize with *A. mellifera.*

The **wait-a-bit-thorn** or **hook thorn** (*A. mellifera*) is a tall, rounded shrub or small tree with a round crown, reaching occasionally 9 m in height in the southern limit of its distribution. The branches are covered with very sharp recurved thorns which attach themselves to clothing and pull the wearer back. Hence the common name. The bark is smooth grey, with white lenticels on the young branches and shoots. The leaves have 1–2 pairs of pinnae with 1–2 pairs of asymmetrical leaflets. It simultaneously develops leaves and flowers in the early rainy season. There are no spines, just the recurved thorns, in pairs. The fragrant flowers are white to cream-coloured, 3.5 cm long, and gathered in short, dense, hanging spikes. The pods are flat, oblong, 3–8 x 1.5–2.5 cm, with slightly constricted margins between the seeds, papery, and generally containing 3 seeds. The pods, young twigs, leaves and flowers are all very nutritious and eagerly consumed by wildlife. It is a widespread species, occurring in both the Sahel, and eastern and southern Africa.

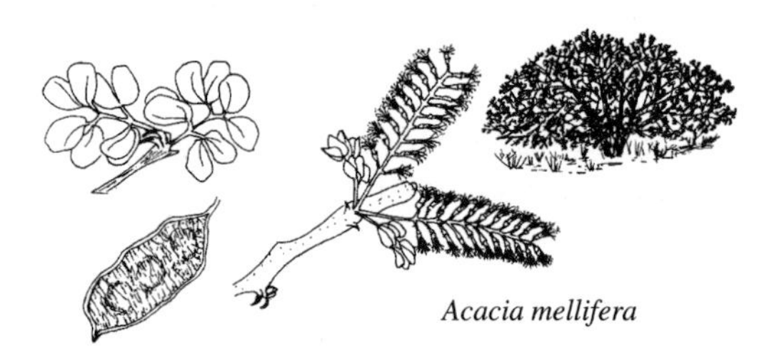

Acacia mellifera

The **nob-thorn** (*A. nigrescens*) is usually a low tree (8 to 10 m in height) but can grow up to 10 m. The bark is dark brown and covered with large, conspicuous, knoby prickles. There are no spines. Each leaf has only two to four pinnae, each with one or two, round, leaflets. The inflorescence is a spike (one to 10 cm long), with many yellowish white flowers. The pods are green when young becoming dark brown when mature. The species is both fire and drought resistant. It is widespread in 'tree and shrub' savannah, being found from Tanzania southwards into Namibia, Botswana and South Africa. The powder from the dry root is used as a traditional treatment for snake bite.

The **Egyptian thorn** (*A. nilotica*) is a small to medium sized tree (5–20 m high), with a dense spherical crown. The stems and branches are usually dark to black coloured, with fissured bark. The thin, straight, light grey spines are in pairs, 5 to 7.5 cm long in young trees. Mature trees are commonly without thorns. The leaves are bipinnate, with 3–6 pairs of pinnae and 10–30 pairs of leaflets each. The flowers are in round heads, 1.2–1.5 cm in diameter, and bright yellow. The pods are green when young becoming darker with age and eventually drying black. They are velvety, straight or slightly curved, 5 to 15 cm long by 0.5 to 1.2 cm wide, with constrictions between the seeds giving a necklace appearance. It is widespread and has been divided into seven subspecies which vary in shape, size and hairiness of the pods.

The **splendid acacia** (*A. robusta*) has a very variable growth form from a short, multi-branched, shrub to a tall (up to 25 m) flat-topped tree. The latter height is attained on wet soils. The bark is often deeply fissured, and the straight spines are in pairs (up to 14 cm long). However, old branches frequently have short spines, up to 1.2 cm, thickened at the base. The leaves are bipinnate, with 1–10 pairs of pinnae and 6–27 pairs of leaflets each. The flowers are white, in a round (capitate) inflorescence, and the flattened, curved, pods (10–19 by 0.7–1.5 cm) are red-brown to black. The pods and leaves are occasionally browsed, but the pods are suspected of causing prussic acid poisoning in livestock.

The **umbrella thorn** (*A. tortilis*) is widespread throughout African savannahs and the quintessential savannah *Acacia*. It is a wide-spreading, flat-topped or umbrella-shaped tree, up to 4 m high with mixed spines, some white, straight, and slender, up to 7.5 cm long, and others grey with black or brown tips, sharply curved, and very small. The pinnae are in three to ten pairs, and leaflets in 7 to 15 pairs. It is almost always evergreen. The flower heads are white to cream, and the pods are yellow-brown. The pods are spirally twisted, sometimes even curled into rings, slightly constricted between the seeds, circular in cross-section, and 7.15 to 15 cm long by 0.6 to 0.8 cm thick. It occurs in the drier areas of Africa, in the Sudan, Kenya, and Tanzania, in southern Africa and Namibia.

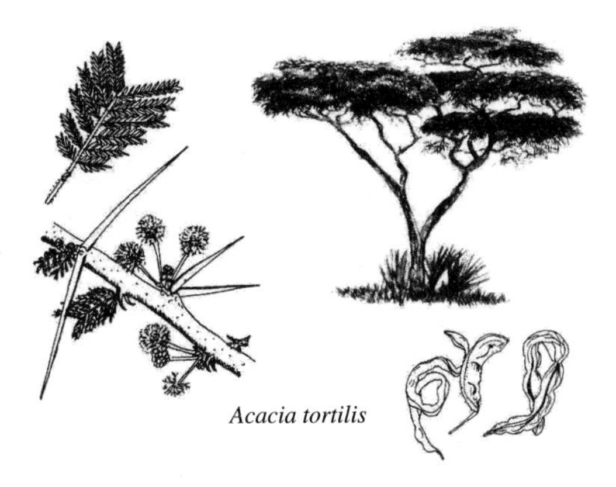

Acacia tortilis

The **three-thorned** or **gum arabic acacia** (*A. senegal*) is a bush or small tree, usually 2 to 8 m high, occasionally reaching 10 m under optimal conditions, frequently forming thickets. It has a short stem, is usually low branched with many upright twigs, with the crown eventually flattened and umbrella-shaped. The prickles are in threes, up to 0.5 cm long, with the centre one sharply curved downwards and the other two curved upwards. The leaves are bipinnate, small, greenish-grey, with 3–6 pairs of pinnulae having 10–20 pairs of leaflets each. The leaflets are grey-green, 3–8 × 1–2 mm. The flowers are very fragrant, creamy white, usually appearing before the leaves in spikes, 3–10 cm long, either solitary or two to three together. The pods are 7–10 cm long by 2 cm wide, flat and thin, papery, attenuated at both ends, yellowish to brown. The leaves and

Acacia senegal

pods are protein-rich and are eaten by wildlife such as giraffe, black rhinoceros and gerenuk. It is found in the Sahel from the Atlantic to Ethiopia, extending southwards into eastern and southern Africa.

The **white thorn** (*Acacia seyal*) is a small, slender tree, reaching 6 to15 m in height, which can develop a characteristic umbrella-shaped canopy when mature. The twigs have paired thorns, up to 7 cm long, which are narrow, straight, sharp-ended, and grey in colour. The leaves are dark green, with 4–12 pairs of pinnae, each with 10–22 pairs of leaflets. The flowers are clusters of 2–3, with bright yellow globose heads about 1.5 cm in diameter, on peduncles about 3 cm long. The pods are hanging, slightly curved, dehiscent, light brown when mature, 10–15 cm long by 1 cm wide, containing 6–10 seeds each. A. *seyal* is found in the Sahel, from the Atlantic Ocean to the Red Sea, and south to East and South Africa. In East Africa there is a variety called *fistula*, with myrmecophilous, swollen-based thorns, or 'ant-galls' similar to those of *A.drepanolobium*.

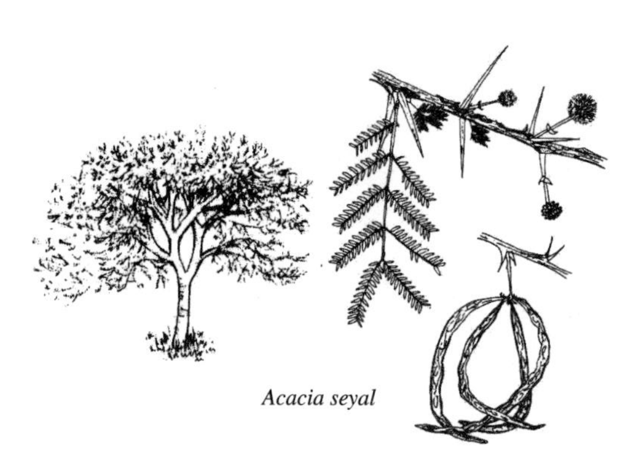
Acacia seyal

The **paperbark acacia** (*A. sieberiana*) is another, very common, flat-topped acacia. Like *A. tortilis*, it is also called the umbrella thorn. It can reach 15–25 m in height, with strongly fissured bark, which is yellow to cream-coloured in young trees and twigs, scaly in old trees. The thorns are greyish white, strong and straight, in pairs up to 6–10 cm long. The leaves are bipinnate, with 10–25 pairs of pinnulae having 15–50 pairs of leaflets each. The flowers are fragrant, cream-coloured or light yellow, in round heads. The pods are straight or slightly curved, woody and without hairs, yellowish to reddish-brown, and varnished and shining when mature. It is widely distributed in Africa from the central Sudan to the southern Sudan, all of Central and Southern Sahel, Kenya, Zambia, Zimbabwe, Angola, Botswana and the Transvaal.

Although most *Acacia* species are utilized to some degree by browsers, the actual frequency of different *Acacia* species in the diet varies with the area (containing different *Acacia* species, in different frequencies), and with

Table 2.4 Seasonal selection of the six principal *Acacia* species browsed by giraffe in the Serengeti N.P. The symbols represent degrees of selection from ++ (very positive), through +, 0, and −, to −− (very negative).

Acacia species	Wet season	Dry season
A. robusta	−−	−−
A. gerrardii	++	0
A. hockii	0	−
A. senegal	+	−
A. tortilis	0	−
A. xanthophloea	−	+

Data from Pellow 1981.

the preference of the browsing species. For example, Pellew (1981) looking at giraffe in the Serengeti found that six principal *Acacia* species were browsed, but that their selection rating varied with the season (Table 2.4).

In the dry season the giraffe positively selected for other tree species such as *Balanites aegyptiaca*, and *Albizia harveii*. The latter species, known as the **sickle-leaved albizia,** is also a tree in the sub-family Mimosoidea and is closely related to the *Acacia* species. The genus *Albizia* has over 100 species, and is named after an Italian nobleman called Filipo degli Albizia who introduced them to cultivation 200 years ago (Noad and Birnie 1989). The genus is widespread throughout the tropics. In many ways they resemble acacias, with their flattened pods, and bipinnate compound leaves. However, they do not have spines, or thorns, and the flowers are born in an inflorescence that is a half-spherical head, never a spike. The **purple-leaved albizia** (*A. antunesiana*) is another species commonly found in woodland (miombo) savannahs. Both these albizias are small to medium trees, 5 to 11 m in height. Other members of the Mimosoidea, occurring in 'tree and shrub' savannah, are the **sickle bush** (*Dichrostachys cinerea*) a small, spiny acacia-like tree to 6 m, and **mesquites** (*Prosopis* spp.), thorny shrubs/trees that grow to about 3 m, but can reach 15 m. *Prosopis africana* occurs from Senegal in the west, to Sudan and Kenya in the east.

The sub-family Caesalpinioideae includes some spectacular flowering trees as well a climbers. Most are forest species, but several genera (*Baikiaea, Bauhinia, Brachystegia, Burkeae, Colophospermum, Guibourtia, Julberlandia*) are important species in many 'wooded' savannahs (chapter 1). The **camel's foot trees** (genus *Bauhinia*), contains over 300 species throughout the tropics. However, two species, the **yellow tree bauhinia** (*Bauhinia tomentosa*) and the **white bauhinia** (*Bauhinia petersiana*) are low shrubs, or small trees up to 5 m, often found in East African 'tree and shrub' savannahs. Another species found in this type of savannah, further south, is the **Rhodesian teak** (*Baikiaea plurijuga*), dominating the so-called baikiaea open woodlands (chapter 1). This is a medium to large tree (8 to 16 m), with a large, dense, spreading crown. It has very attractive, large, golden-brown flowers and flattened, woody, pods. The **wild green-hair tree** (*Parkinsonia africana*) is widespread in arid savannahs.

The **African mahogany** (*Afzelia africana*) is a large tree with a spreading crown; its height varies from 10–20 m and in west Burkina Faso mean average height was found to be 10 m for a tree with a 36 cm diameter. Scaly grey-brown bark, with a pink or red slash. Twigs and leaves are hairless (glabrous), the latter up to 30 cm long, with 7–17 pairs of leaflets (5–15 × 3–9 cm) with an obtuse tip. Flowers with one single white petal, with red stripes, 1.5 cm long set in terminal panicles up to 20 cm long. Pods flattened 12–17 × 5–8 × 3.5 cm, glabrous, black, woody, persistent bursts open at maturity spreading the seeds. Seeds poisonous, with a sweet edible covering. The tree is associated with ectomycorrhizal fungi. Found from Senegal to Uganda.

The sturdy, spreading, **miombo tree** or **Prince-of-Wales' feathers** (*Brachystegia boehimii*) (5 to 16 m in height) gives its name to the miombo 'woodland' savannahs of central African (Chapter 1). *Brachystegia* have pinnate leaves, small inconspicuous flowers, and flat woody pods that split explosively. Apart from *B. boehimii*, miombo woodlands are dominated by two other species of *Brachystegia. Brachystegia spiciformis*, also known as **miombo** or **msasa**, is a medium to large tree (8 to 15 m in height) and in Zimbabwe and Mozambique it is the most widespread species of *Brachystegia.* It hybridizes readily with *B. glaucescens*. The **mountain acacia** (*B. glaucascens*) is a spreading, beautiful tree (up to 15 m in height) that often develops a flat top, hence its common name. Coexisting in miombo woodlands are the closely related **munondo** (*Julberlandia globiflora*) and **large-leaved munondo** (*J. paniculata*). *Julberlandia* also have pinnate compound leaves, inconspicuous, small, white/cream flowers and dark brown, velvety, pods. The **burke** (*Burkeae africana*) is a widely distributed, medium sized tree (8 to 10 m). It is found in both the wooded grasslands of the 'forest savannah mosaic' and the 'tree and shrub' savannahs. Once again, the leaves are pinnate and the fruit a thin, flat, pod. Emphasizing the similarity of many of these leguminous savannah trees, *Burke* is often confused with *Albizia*; a tree in another sub-family. The **mopane** (*Colophospermum mopane*) is a medium to large tree (4 to 18 m) dominant over many areas south of the miombo woodland savannahs. The **large false mopane** (*Guibourtia coleosperma*) is a medium to large, evergreen, tree (6 to 20 m) with a rounded crown. The **small false mopane** (*Guibourtia conjugata*) is a small to medium tree (7 to 9 m).

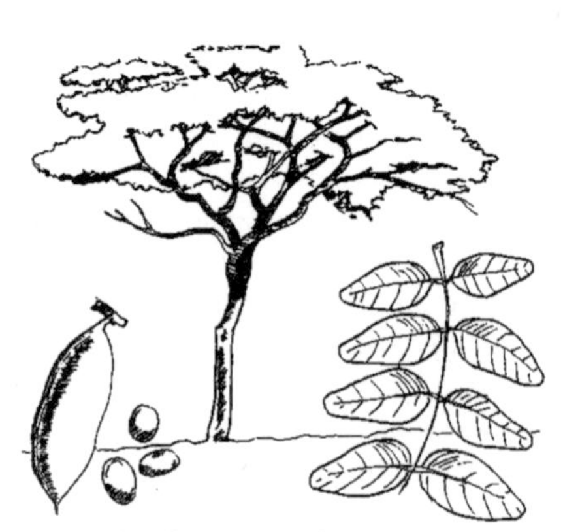

Brachystegia spiciformis

Desert date family (Balanitaceae)

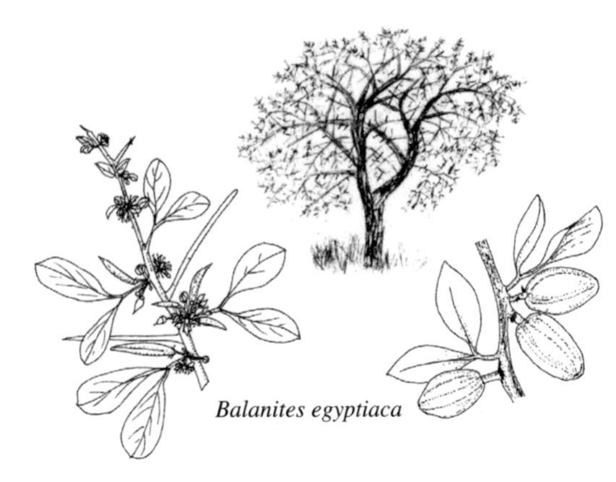
Balanites egyptiaca

A small tropical family with only one genus, *Balanites*, of which there are several African species. They are shrubs or trees, with spines that are simple or forked, and a fleshy fruit. A species frequently found in dry savannah grassland is the **desert date** (*Balanites aegyptiaca*), a slow growing, evergreen, tree usually 3 to 6 m in height, occasionally growing to 10 m. It is armed by stout green or yellow spines, up to 8 cm long. The date-like fruit is long (5 by 2.5 cm) and yellowish red when mature. The fruits are even produced in very dry years. Both fruit and foliage are browsed by goats and camels and also by game, especially giraffe. It is widely distributed in both 'grass and shrub' and 'tree and shrub' savannahs, from northern and eastern Africa, to southern areas.

Myrrh family (Burseraceae)

Commiphora africana

A tropical family found in Africa, Malaysia, and Central America, notable for its aromatic resins. Even in biblical times these where used for incense and perfumes. Frankincense comes from the resin of *Boswellia carteri* and other species found in Somalia and Ethiopia, and myrrh is extracted from *Commiphora abyssinica*. The largest African genus in this family, with about 30 species, is *Commiphora*. These are shrubs or trees, often with spines, and are leafless for most of the year. The **angular-stemmed commiphora** (*Commiphora karibensis*) is a small to medium tree (up to 7 m in height) often associated with the **soft-leaved commiphora** (*Commiphora mollis*) (8 m in height) in 'tree and shrub' savannah. The **poison-grub commiphora** (*Commiphora africana*) is a deciduous thorny shrub or small tree, usually about 5 m high, with a grey-green bark that peels in papery scrolls. Although it is widely distributed in dry savannahs, its leaves, although they look a good source of browse, are rarely eaten because of the bitter tannins they contain. The larva of the beetle *Diamphidia*, from which the bushmen make their arrow poison, feeds exclusively on this tree.

Euphorbis family (Euphorbiaceae)

The euphorbias comprise a very large family of herbs, shrubs, and trees, with their distribution centred in the tropics. Economically important species include the **cassava plant** (*Manishot esculenta*). There are approximately some 5,000 species, in 300 genera. A widespread savannah species is the **snowberry tree** (*Securinega virosa*), a many-stemmed, bushy shrub 2 to 3 m in height. In Tanzania the roots and fruits are believed to be an effective snakebite remedy (Coates Palgrave 1983). The **heart-fruit** (*Hymenocardia acida*) is a shrub or small tree (up to 6 m) often found with *Brachystegia* species and *Uapaca* species, in 'woodland' miombo savannah. *Uapaca kirkiana*, known as **mahobohobo** in southern Africa, and the **narrow-leaved mahobohobo** (*U. nitida*), are just two of a number of species. They are medium sized trees of about 10 m, with leathery leaves, creamy-green inconspicuous flowers, and an edible fruit. One of the best known savannah species is the **candelabra tree** (*Euphorbia candelabrum*). It grows up to 15 m, with a characteristic crown of massive ascending branches that make it look like a complex candelabra. The spined, green, stems are succulent and rather cactus-like. It is widely distributed in savannah areas from East Africa to southern Africa. Although a characteristic savannah species, it is not used by animals because its stems produce an extremely toxic latex. Some euphorbias, such as the non-succulent **western woody euphorbia** (*Euphorbia guerichiana*) are woody shrubs that have fleshy leaves, rather than succulent stems.

Mango family (Anacardiaceae)

A world-wide, but mainly tropical family, of about 600 trees, shrubs, and vines. It includes a number of commercially important species, such as the cashew, pistachio, and mango. Savannah species including the marula (*Sclerocarya birrea*), a medium sized tree up to 10 m in height found in 'tree and shrub' savannah. The fruits are fleshy, up to 3.5 cm in diameter, and yellow when mature. There are very old stories that elephants, who eat them eagerly, become intoxicated after eating the over-ripe, fermenting, fruits lying on the ground. However, elephants show a clear preference for marula fruit still on the tree and it is unlikely that a three-ton elephant, gorging itself quickly on nothing but marula fruit, would be able to ingest enough ethanol to reach a blood alcohol content indicative of inebriation (Morris, Humphreys and Reynolds 2006). Observations on baboons, which also like these fruits, suggest that they also prefer fresh Marula fruit, and because the pulp is digested and the seeds passed within a 24-hour period, internal fermentation is impossible.

Baobab family (Bombacaceae)

A small family of tropical flowering trees, whose best-known species is the **baobab** (*Adansonia digitata*) with its massive bottle-shaped trunk, up to 15 m in height. Like many trees in this family, the swollen trunk of the baobab stores water, allowing it to survive in very dry savannahs. It can be bare of leaves for as much as nine months of the year, giving it an odd 'upside down' appearance. Elephants are quite fond of its soft pithy wood, yet despite this apparent fragility it is one of the longest lived trees in the world. Radio carbon dating has estimated that trees 5 m in diameter might be 1,000 years old and the very largest trees as much as 3,000 years old. These extraordinary trees are surrounded by a wealth of African myth and legend, including the story that God planted the tree upside down and that a lion will devour anyone that picks a flower from the baobab, because the flowers are inhabited by spirits.

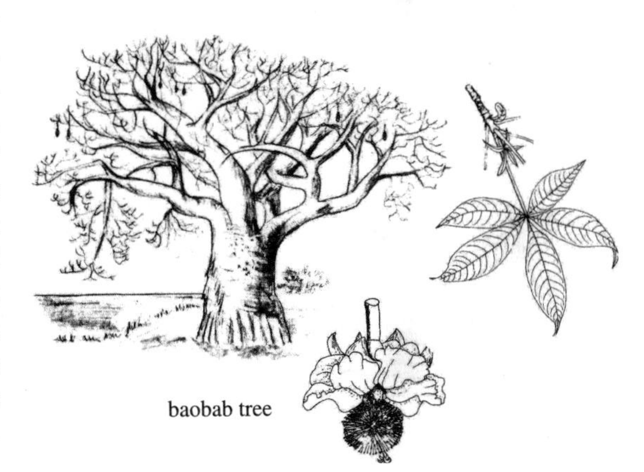
baobab tree

Terminalia family (Combretaceae)

A mainly tropical family of trees, shrubs, and climbers, all with simple leaves and flowers in clusters, often producing abundant nectar. The main African genus is *Combretum*, with over 30 species. They are shrubs, climbers, or trees. They have woody, four, or five winged, fruit. In contrast the related genus, *Terminalia*, with twelve indigenous species, has woody fruit with only two wings, often reduced to ridges. They are small to medium sized trees. In both genera the flowers have no petals and it is the stamens which are conspicuous. They are mainly found in the open 'tree and shrub' savannahs, north and south of the miombo 'woodland' savannah.

The **red bushwillow** (*Combretum apiculatum*) is a small to medium tree (3 to 10 m in height), occasionally shrub-like. The flowers are yellow to creamy-green and heavily scented. The wood is very resistant to termites. The **lead-wood** (*C. imberbe*) is sometimes a shrub, but more frequently a small to large

Combretum molle

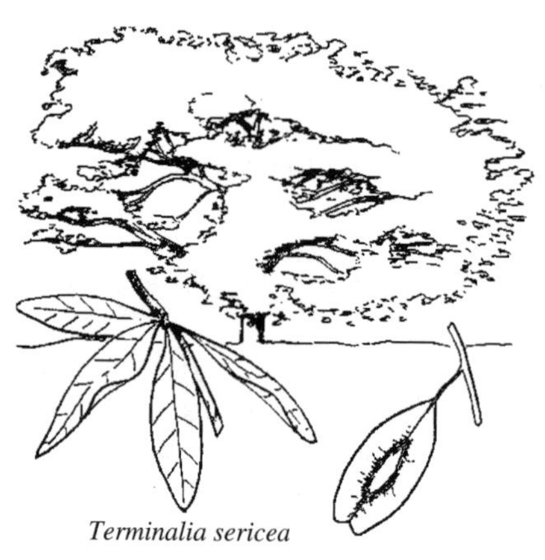
Terminalia sericea

tree (7 to 15 m in height). The bark is dark grey and rough, with characteristic deep longitudinal fissures. Again the flowers are heavily scented. The **velvet-leaved combretum** (*C. molle*) is a small to medium tree (up to 10 m in height). The flowers are heavily scented, attracting insects. The **variable combretum** (*C. collinum*) is a spreading deciduous, small to medium, tree (4 to 12 m in height). The bark is light grey and rough, and the flowers are cream to yellow and sweetly scented. The genus *Terminalia* gets it name from the Latin *terminus*, referring to the fact that the leaves appear at the very tips of the shoots. The **silver terminalia** (*Terminalia sericea*) is a small to medium tree (4 to 6 m in height), with silvery silky leaves. The small cream flowers are unpleasantly scented. In the drier wooded savannahs of East Africa, the **muhutu** (*T. brownii*), is a widely distributed deciduous tree of 4 to 5 m (in Kenya it is the most common *Terminalia*). It is both drought and termite resistant, with a heavy leaf fall. The **purple-pod terminalia** (*T. prunoides*) is a shrub or medium tree (3 to 7 m in height). They have dark green leaves and the cream flowers have a strong unpleasant smell. The **Kalahari sand terminalia** (*T. brachystemma*) is a small bushy tree (3 to 5 m in height) with the branches often horizontal. It has leathery green leaves and pale yellow flowers. It hybridizes with *T. sericea* and the hybrids are widespread. The **large-leaved terminalia** (*T. mollis*) (10 m in height) has large leaves (up to 32 by 13 cm) and larger than usual greenish-white flowers.

Ebony family (Ebenaceae)

This is a largely tropical and subtropical family named after the ebony tree (*Diospyros ebenum*) which is native to the rainforests of southeast Asia. However, several species of *Euclea* and *Diospyros* are found in African savannahs. Both the **large-leaved euclea** (*E. natalensis*) and *E. pseudebenus*, also confusingly called the **ebony tree**, are widespread southern African species.

The giant diospyros (*Diospyros abyssinica*) (30 to 40 m in height) is a widespread African species, but really a forest, rather than a savannah, tree. It is however the co-dominant species, with the **mutanga** (*Elaeodendron buchanii*), in zone VI of the Serengeti (see below).

Jacaranda family (Bignoniaceae)

The jacaranda family is a tropical group of trees and shrubs with showy flowers, such as the striking mauve-blue of the introduced jacaranda tree (*Jacaranda mimosifolia*). The **sausage tree** (*Kigelia africana*) is a medium to large tree (up to 18 m in height), noticeable for its unusual sausage-shaped, greyish-brown (up to 1 m by 18 cm), hanging fruit, much loved by elephants. Each fruit can weigh up to 10 kg. Under this tree, David Livingstone pitched camp just before he saw the Victoria Falls for the first time (Coates Palgrave 1983). The genus *Rhigozum* are small, spiny, shrubs or small trees. The **western rhigozum** (*R. brevispinosum*) (2 to 4 m in height) is widespread in southern Africa in the mopane 'tree and shrub' savannah. The flowers are golden-yellow and sweetly scented.

Local vegetation patterns: the Serengeti–Mara ecosystem

In Chapter 1 we looked at the distribution of savannah vegetation on a World and, for Africa, on a continental scale. Rainfall, its yearly amount and annual distribution, was important in determining this continental pattern. However, the large areas of savannah types shown in Figure 1.9 are not uniform stands of vegetation. There is widespread local variation in species and the different types of savannah often form a mosaic at a local scale. It would be impossible to talk about this local scale for the whole of Figure 1.9 and so to illustrate this I will now examine the vegetation of one local geographical area of savannah that has been very well studied. This is the Serengeti–Mara ecosystem, in northern Tanzania and southern Kenya. Here we will see the local interplay of rain, topography, soil, and fire producing local changes in vegetation.

Rainfall and green biomass

The Serengeti–Mara ecosystem is an area of approximately 25,000 km^2 on the border of Tanzania and Kenya, in East Africa (Fig. 2.9), and essentially defined by the movement of the migratory wildebeest (see Chapter 5). The system has a long history as a protected area. In 1929 an area of 2,286 km^2 was established as a game reserve, in what is now southern and eastern Serengeti. In 1959 the northern extension to the Kenyan border was added and the eastern area was removed as the Ngorongoro Conservation Area. In 1965 the northern 'Lamai Wedge' was added, and the Mara National Reserve was formed. The mean maximum monthly temperature is relatively constant between 27°C and 28°C at Seronera. The minimum temperature

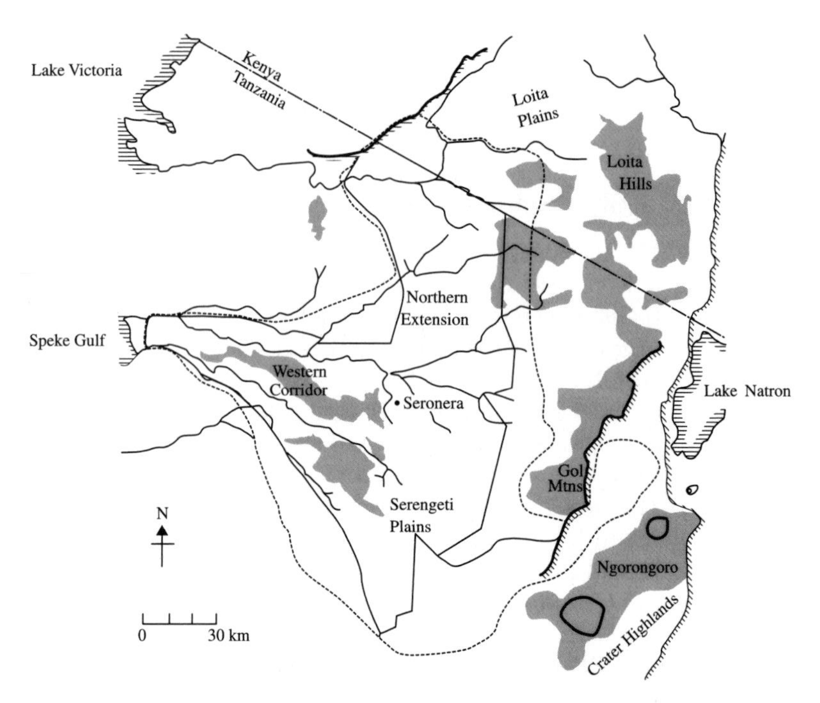

Fig. 2.9 The Serengeti–Mara ecosystem defined as the area used by the migratory wildebeest (broken line). The Serengeti National Park is outlined by a solid line. Hills are shaded and lakes are horizontal hatched (simplified from Sinclair 1979, University of Chicago Press.).

varies between 13°C and 16°C. However, the most important climatic variable is rainfall. Rainfall is seasonal (Fig. 2.10a: four histograms) because of the movement of the Intertropical Convergence Zone (Chapter 1); modified by local influences. The rainfall pattern is influenced by the crater highlands, which rise to over 3,000 m and form a rain shadow immediately to the northwest. This produces a gradient of increasing rainfall from the southeast grass plains, to the northwest woodlands on the Mara river. This gradient is present in both the wet and dry seasons, and Figure 2.10 shows the rainfall contours (isohyets) for both seasons.

Lake Victoria also modifies the rainfall pattern. The lake affects northern and western Serengeti by producing its own convergence zone and increasing rainfall between June and November. One consequence of this gradient in rainfall is a corresponding gradient in green grass biomass (there is also brown, or dry, grass biomass which of course is related to the biomass of new green grass). The widespread relationship between rainfall and green grass biomass, shown earlier in Figure 2.1, is also found more locally within the area of the Serengeti. Figure 2.11 shows the results of Braun (1973) for 69 different sites, over short grass (II), intermediate grass (III), long grass (IV) and savannah woodland (V and VI) (see vegetation zones in Figure 2.10b and below). Aerial surveys of green biomass during 1974 and

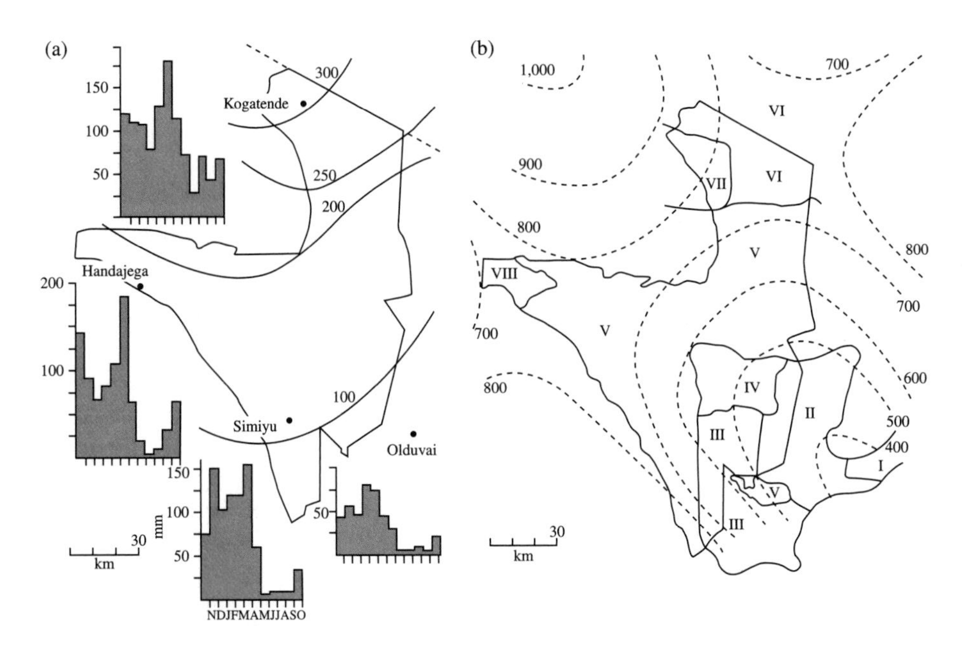

Fig. 2.10 Mean rainfall isohyets (mm) for the Serengeti. (a) dry season, plus four annual profiles, (b) wet season, with vegetation zones (see text) (from Sinclair 1979, University of Chicago Press.).

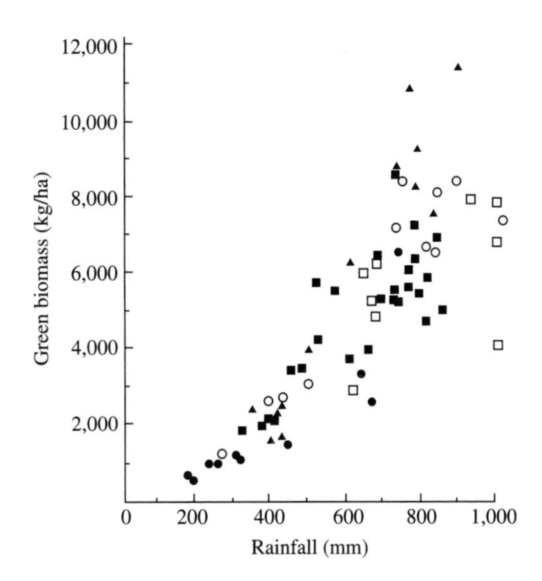

Fig. 2.11 Serengeti grass yield with increasing rainfall. Data are for short grassland (●), intermediate grassland (▲ Andropogon greenwayi stands, ○ mixed stands), long grassland (■), and savannah woodland (□) (from Braun 1973).

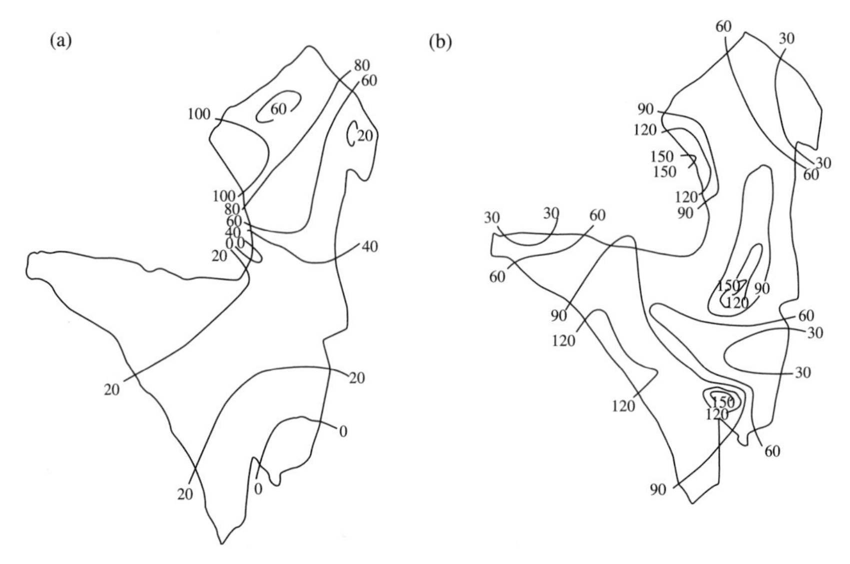

Fig. 2.12 Isoclines of green biomass for the Serengeti. (a) 6 September 1974 in the middle of the dry season, (b) March 1975 in the middle of the wet season (from McNaughton 1979, University of Chicago Press.).

1975 (McNaughton 1979) give a remarkable insight into how Figures 2.10 and 2.11 combine to produce the local spatial distribution of available grazing vegetation, that must be typical of many 'uniform' savannah areas (Figure 2.12). During the dry season (Figure 2.12a) the southern grass savannahs have little or no available green forage. This is a major factor in the northern migration of wildebeest, plains zebra, and Thomson's gazelle. With the arrival of the rains these ungulates can migrate back to the southern grasslands which not only now have a good supply of green grass, but are also rich in minerals (Murray 1995) and lower in predators (Fryxell 1995). These three attributes of the southern grasslands, in the wet season, combine to produce ideal conditions for the production of young, which takes place at the beginning of the wet season. However the important message here is that the grass in savannahs is not a uniform 'lawn'. Green biomass can vary quite markedly over relatively short distances, and grazing animals must respond to this spatial variation. In fact this 'local' variation is not simply confined to green grass, most of the vegetation responds to rainfall, soil and fire (Sinclair 1979).

Soils, fire and vegetation

In the south eastern plains, the soils of the Serengeti are highly saline and alkaline as a result of their recent volcanic origin in the Crater Highlands (Figure 2.9). These highlands are volcanoes of Pleistocene age that produced aerially discharged material which was blown westward and formed the Serengeti Plains. The youngest soils, in zone I (Fig. 2.10b), are highly porous, course-grained sands, which combined with the low rainfall of this area results in virtual desert conditions that are on the margin of true savannah. Stable dunes support deep rooted grasses like *Sporobolus consimilis*. West of these dunes are the 'short grasslands' (zone II) characterized by dwarf growth forms of grasses such as *Digitaria macroblephora* and *Sporobolus marginatus*, and the grass-like sedges of the genus *Kyllinga*. The 'short' growth forms are produced by the peculiarities of the soil. Rain leaches salts (calcium) out of the highly porous, sand, top layers and redeposits them as a calcium carbonate hardpan about 100 cm below the surface. This impermeable hardpan restricts root growth and the soil above it is highly saline and alkaline. Only plants that can withstand these conditions and have short roots are able to survive here. Grass cover is sparse, ranging from 10 per cent to 30 per cent, but typically nearer 10 per cent. The top layers of soil are susceptible to erosion, which can be made worse by grazing.

West and south of the short grasslands the rainfall is higher and the soils become deeper. Because they are further from their volcanic origins, the soils are finer, hold water longer (therefore less leaching, and a less continuous hardpan), and are less alkaline and saline. They therefore present a milder environment for the vegetation. These are the 'intermediate grasslands' (zone III). Taller growth forms of grasses found in zone II are found here and some patches of *Andropogon greenwayi* can have 100 per cent grass cover, although 30 per cent is more typical. In both zone II and zone III there are extensive areas of the herbs *Indigofera basiflora* and *Solanum incanum*.

The 'long grass' plains (zone IV) have deep soils (about 2 m) of fine volcanic ash mixed with local material to form silty clay with low alkalinity and salinity, with no hardpan. Here there are grasses such as *Themeda triandra* and *Pennisetum mezianum*. Basal cover is now about 50 per cent. Herbs are infrequent and there are no shrubs or trees. One of the features of both the 'intermediate grasslands' and the 'tall grasslands' is the mosaic of *Cynodon dactylon*, and *D. macroblephora* patches in a surrounding matrix of *A. greenwayi* and *T. triandra*. This mosaic is caused by termites (Sinclair 1979). When the termites build their mounds, they tend to use weatherable subsoil material of high salinity and alkalinity. These termite mounds produce patches about 3 m across on which salt- and alkali-tolerant species such as *C. dactylon*, and *D. macroblephora* can grow.

In the 'tree and shrub' savannahs of the north and west (zones V, VI, and VII), the soils tend to be formed from the parent granite or quartzite rocks, although towards the east there is increasing volcanic ash content. Salinity and alkalinity are low and there is no hardpan, allowing trees to grow. Grass cover varies from 20 per cent to 60 per cent, with an average of 45 per cent. The typical topography of these areas is a series of undulations which, together with the associated soil and vegetation, are know as 'catenas' or 'catenary sequences'. Rain, falling on an undulation, gravitates from the top, via the slopes, to the bottom, or sumps. Here it usually drains away along drainage lines or occasionally, if drainage is impeded, it accumulates. In its passage, the water carries with it soluble material and small soil particles.

The top of the catena therefore becomes progressively leached of organic matter, and the finer soil particles, and comes to consist of shallow, course, sandy soil resembling the parent underlying rock. In contrast the lower levels come to consist of deep clay, with high organic content and a high capacity for water retention. The slopes show a gradient between these two extremes. Grass species are adapted to the different condition, so that small perennials and annuals like *D. macroblephora* and *Chloris pycnothrix* predominate on the sandy soils. On the slopes, the deeper soils and good drainage provide optimal conditions for *Themeda triandra*, which is one of the dominant grasses of these 'shrub and woodland' savannahs. As Figure 2.13 shows, these western and northern areas are particularly prone to bush fires, and *T. triandra* and associated grasses are fire-tolerant. These savannahs are frequently regarded as fire-induced (see Fig. 2.3). In some of the drainage areas, coarser grasses like *Pennesetum mezianum* predominate, and in those areas that remain flooded for long periods very tall grasses such as *Panicum maximum* grow. At the end of the western corridor there are soils derived from old beds of Lake Victoria (zone VIII) that support flood plain grasslands with few trees.

Herlocker (1975) described the woody vegetation of these 'tree and shrub' savannahs of the north and west Serengeti (zones V, VI, and VII). Two broad zones can be recognized. A thorn-tree zone (V) dominated by 38 species of *Acacia*, but with 10 species accounting for 45 per cent of the woody vegetation. The most extensive is the whistling thorn (*A. drepanolobium*), with *A. clavigera* and *A. tortilis* also common. *Commiphora trothae* is the only other common tree that occurs in this *Acacia* dominated zone.

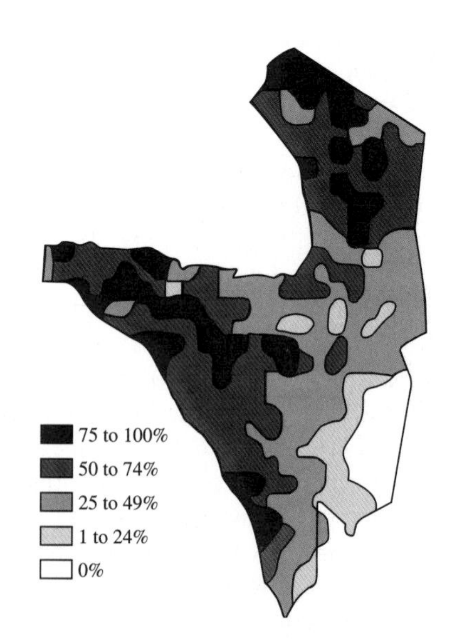

Fig. 2.13 Distribution of the incidence of fire in the Serengeti between 1963 and 1972 (modified from Sinclair 1975).

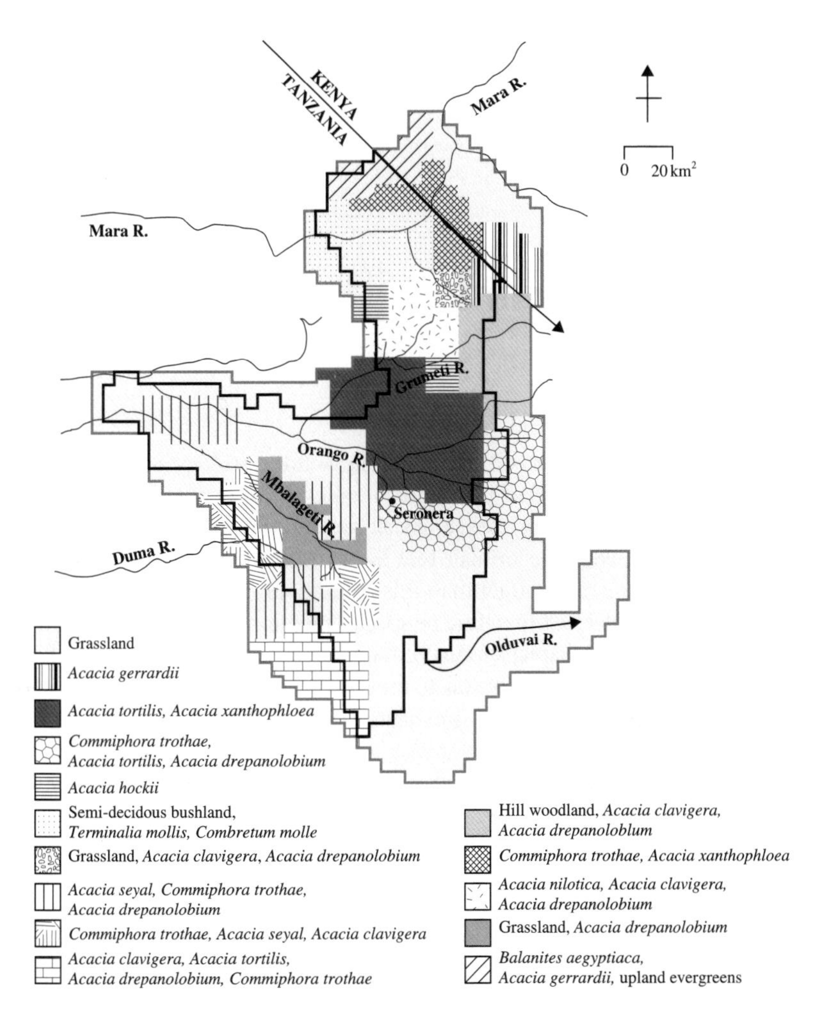

Fig. 2.14 Vegetation types within the Serengeti–Mara ecosystem (data from Herlocker 1975).

North of the Grumeti River there is open 'woodland' savannah (zones VI and VII). The dryer zone VI is essentially an evergreen thicket of *Croton*, *Techlea* and *Euclea*, surrounded by grassland with *A. clavigera* and *A. gerrardii*. Zone VII is open woodland savannah with the broad-leaved trees *Combretum molle* and *Terminalia mollis*. A more detailed view of the distribution of the woody vegetation, which is a simplified version of Herlocker's map, is shown in Figure 2.14.

The map of African savannahs in Figure 1.9 therefore shows broad areas where a particular savannah type predominates. Locally there is a patchwork of plant species assemblages and areas of other types of savannah. The animals, particularly the large mammals so characteristic of African savannahs, are frequently more wide-ranging. I consider these next.

3 The animals

This chapter will take a brief look at the ecology and behaviour of a selection of animals that play a major role in the dynamics of the African savannah ecosystem. Of course not all species of animal can be mentioned. Even just listing all the animals that live in African savannahs would require a volume many times larger than this entire book. Most of the species discussed in this chapter will therefore be species mentioned in the studies and topics detailed in later chapters. Many of these are very prominent, often common, and contribute in a major way to the dynamics of the savannah system. Several species that are prominent and have important interactions with other species can be treated as a group, rather than individual species (e.g. grasshoppers and rodents). Many species are visually prominent, but as individual species do not greatly influence the savannah ecosystem (e.g. most birds). Below I describe very briefly the diversity of the savannah avifauna, and mention two quintessential savannah species. Vultures are mentioned briefly in Chapter 5. Reptiles, although important savannah predators of small birds, are not included. They do not figure in the later chapters even though species such as the leopard tortoise (*Geochelone pardalis*), the Nile crocodile (*Crocodylus niloticus*), and snakes such as the boomslang (*Dispholidus typhus*), the black mamba (*Dendroaspis polylepis*), the puff adder (*Bitis arietans*), and the Egyptian cobra (*Naja haje*) are familiar, common, and widespread.

Of course what makes African savannahs so unique and exciting is the wealth of 'big game': its mammalian megafauna. Many of these charismatic animals have been extensively studied and feature prominently in the examples in subsequent chapters. These 'large' mammals are given extensive coverage.

The insects

There are thousands of insect species living in African savannahs, and *en masse* they can play an important role in the ecology of this biome. For example, they remove the tonnes of dung that would otherwise completely cover large areas of savannah ecosystems. Anderson and Coe (1974) counted 16,000 dung beetles arriving at a 1.5 kg heap of elephant dung in East Africa, which was eaten, buried, and rolled away in two hours. In West African savannahs, Cambefort (1984) estimated that beetles bury one metric tonne of dung per hectare per year. Another important role for insects, in savannahs, is that of pollinators. Of course this is an insect role that is not confined to these ecosystems. Globally, some 40,000 plant species have been recorded as being important food resources for honeybees of the genus *Apis*, and races of the African honeybee are undoubtedly important pollinators for a wide variety of plants in savannahs systems. In the Mkomazi Game Reserve of northern Tanzania, Stone *et al.* (1998) record over 94 insect species from just six species of *Acacia*, including the widespread *Apis mellifera*. Termites are instrumental in removing the dry, old, grass and therefore are important detritivores. Because they remove the dry combustible grass they influence the intensity of fires and therefore indirectly influence tree recruitment (Chapters 2 and 5). Ants are frequently involved in mutualistic associations (Chapter 5) with *Acacia* trees, helping in their defence against herbivory.

Grasshoppers (Orthoptera) are by far the most abundant insect herbivores in many savannah systems. As an example we can look at the work of Sinclair (1975). He estimated the total annual grass production in three separate parts of the Serengeti ecosystem. These were long grass (characterized by *Themeda triandra*, *Pennisetum mezianum*, and *Sporobolus pyramidalis*), short grass (characterized by *Digitaria macroblephora*, *Andropogon greenwayi*, *Sporobolus pellucidus*, and *S. spictus*) and kopjes, or inselbergs, with grasses from the surrounding long grass area. He also estimated the total annual grass production consumed by each of four animal groups, and also destroyed by fire. His results are shown in Table 3.1. A total of 38 species of grasshopper was recorded in the Serengeti grasslands. However, *Mesopsis abbreviatus*, *Coryphosima stenoptera*, *Afrohippus taylori*, *Rhaphotitha* species,

Table 3.1 Total annual production and off-take in different parts of the Serengeti ecosystem.

Off take by	Long grassland		Short grassland		Kopjes	
	kg/ha/yr	% total	kg/ha/yr	% total	kg/ha/yr	% total
Annual grass production	5,978		4,703		5,978	
Ungulates	1,122	18.8	1,597	34.0	122	2.0
Small mammals	69	1.2	4	0.1	259	4.3
Grasshoppers	456	7.6	194	4.1	484	8.1
Detritivores (Termites)	1,146	19.2	2,322	49.5	1,683	28.2
Burning	3,185	53.3	586	12.5	3,430	57.4

Data from Sinclair 1975.

and *Acrida* species comprised 80 per cent of the long grass population and *A. taylori*, *Acrotylus elgonensis*, and *A. patrrualis* were the dominant species in the short grass areas. In total these grasshoppers removed about 1,134 kg of grass, per hectare, per year, amounting to about 7 per cent of the total grass production, and approximately 26 per cent of the green grass consumption.

The birds

The avifauna of Africa is large and diverse, and although most birds have little impact on savannah systems as individual species, *en masse* they can be important, and certainly intriguing. Many species consume vast quantities of plant seeds and insects, many frugivores are important dispersers of seeds, and many scavengers, such as vultures, disperse essential nutrients and important diseases. There is a rich ensemble of avian predators that consume an array of reptiles, birds and small mammals, and there are idiosyncratic individuals, like the ostrich and secretary bird, that are quintessential savannah sights.

There are about 1,850 species of birds in Africa, although only about 1,450 are resident south of the Sahara. Two orders are endemic to sub-Saharan Africa, the Coliiformes (containing six species of mousebirds) and the Struthioniformes containing the ostrich (*Struthio camelus*). Thirteen other taxonomic groups are also endemic to sub-Saharan Africa. These are the hammerkop (*Scopus umbretta*), shoebill (*Balaeniceps rex*), secretary bird (*Sagittarius serpentarius*), guinea fowls (7 species), turacos (18), wood-hoopoes (6), bush-shrikes (39), helmet-shrikes (9), rockfowls (2), sugar-birds (2), buffalo-weavers (2), parasitic weavers (9), and oxpeckers (2). Although not specific to Africa, several other groups are especially well represented. These include the bustards (16 species in Africa out of 22), honeyguides (11/13), larks (47/69), shrikes (55/74), *Cisticola* grass warblers (36/37), and ploceine weavers (101/112).

Table 3.2, from Brown, Urban, and Newman (1982), show their estimates of the distribution of bird populations over various African habitats. Rainforests are clearly the most productive for birds at an estimated 8,000 individuals per km^2, followed by freshwater habitats at 7,000 per km^2. However, savannahs are clearly not devoid of birds. Densities range from 125 km^2 for the very arid scrub savannah, through 1,500 km^2 for the dry *Acacia* savannah, to 6,000 km^2 for the rich forest–savannah mosaic. Adding together all their habitats that in this book I call savannah (2 to 5), results in a figure of 2,565 km^2. Below I describe two characteristic species.

The **ostrich** (*Struthio camelus*) (height up to 2.4 m) is a fairly common bird in dry open grass savannah, often at densities of 0.2–20 km^2. Four races/species have been recognized (Brown, Urban, and Newman 1982; Zimmerman *et al.* 1996). It is mainly diurnal and roosts squatting on the

Table 3.2 Distribution of African bird populations.

Habitat	Total Area (km²)	Density (km²)	Numbers in millions
1. Desert	8,547,000	25	214
2. Sub desert scrub savannah	2,530,000	125	316
3. *Acacia* grass savannah	4,671,000	1,500	7,007
4a. Dry woodland savannah	3,869,000	2,500	9,673
4b. *Isoberlinia-Brachystegia* woodland savannah	3,979,000	3,500	13,927
4c. Moist woodland savannah	1,304,000	4,500	5,868
5. Forest–savannah mosaic	1,502,000	6,000	9,012
6. Lowland rainforest	2,515,000	8,000	20,120
7. Other habitats, montane etc.	1,082,000	2,000	2,164
8. Lakes, rivers, swamps	301,000	7,000	2,107

Modified after Brown, Urban and Newman 1982.

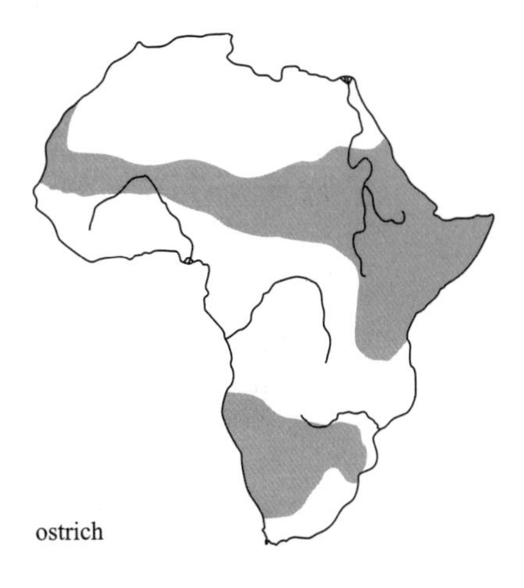
ostrich

ground at night. It eats both herbs and grasses, but prefers the former. Outside the breeding season it lives in rather open groups of 2–5 individuals. In the breeding season both males and females are usually solitary, and groups are made up of immature birds. Although in moister areas it may be almost sedentary, in more arid areas it is more or less nomadic. There are two kinds of female. Major hens that mate with a territorial male and incubate the eggs in a nest, and minor hens that wander through many male territories and lay eggs in several nests. A clutch of 5–11 eggs is laid into a ground nest, by a major hen, the nest site having been chosen by the male. The female lays on alternate days, and incubation begins about 16 days after the first egg is laid. The major hen's clutch may comprise less than 50 per cent of the nest contents. The incubation period is 45–46 days. Broods from several nests may join together to form one large crêche, guarded by several adults.

The **secretary bird** (*Sagittarius serpentarius*) (height up to 1.3 m) is an unmistakable resident of almost all African savannahs. It roosts and nests in trees. When hunting during the day it walks steadily through grassland and catches its prey with its beak, but sometimes it kills with its feet. It eats insects, amphibians, lizards, snakes, young birds, and small mammals.

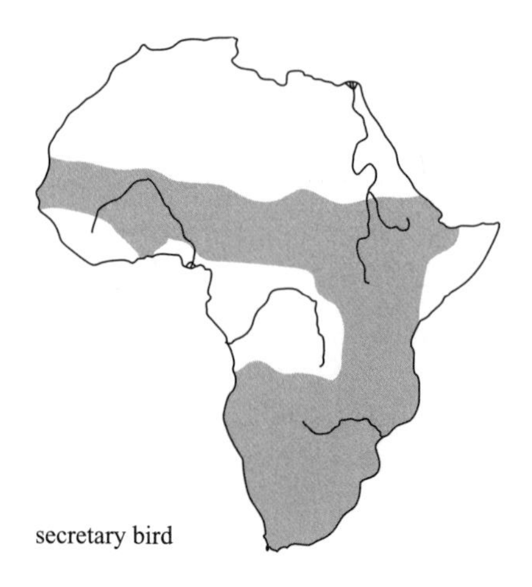

Although it is often viewed as a snake killer, it probably feeds mainly on grasshoppers and small rodents. In the breeding season, pairs defend large home ranges (5,000–6,000 ha). The nest is a large flat structure of twigs and 1–3 pale bluish or white eggs are laid at daily intervals. The incubation period is 43–44 days. The young leave the nest at 75–85 days, without parental stimulus. Although decreasing in many areas, due to habitat loss, it is still locally common.

The mammals

In this section I will look at the ecology and behaviour of several individual large mammals: ungulates, carnivores, and primates, and consider rodents as a 'group' of species. Modern opinion (MacDonald 2001) now places placental mammals into four groups: the Euarchordata (+ Glires), Laurasiatheria, Xenartha, and Afrotheria. Each of the last three almost certainly share a common ancestor (they are monophyletic), but it remains uncertain if the Euarchordata and Glires have a common, or separate, ancestral root. Figure 3.1 shows members of these groups, with some Africa species, referred to in the section below, highlighted. The Xenartha are an entirely New World group.

The term 'ungulate' is used for those groups of mammals that have hooves. The primitive mammalian limb ends in five digits, all of which are placed on the ground during locomotion. Ungulates have reduced this basic number of digits and also lengthened and compressed the metapodial bones (the long bones in our hands and feet). Ungulates therefore walk on tiptoe, rather like a ballerina. In addition there is restricted movement in the joint surfaces so that the main movement of the limbs is forward and backward, rather than sideways. This 'suite' of ungulate limb characteristics, appears to be associated with a terrestrial, herbivorous lifestyle, often in open savannah habitats, in which both fast and sustained galloping locomotion are useful. Ungulates are classified into two different groups, the perissodactyls and artiodactyls, depending on whether they have an odd (Greek: *perisso-*) or even (Greek: *artio-*) number of toes (Greek: *dactyla*). In artiodactyls the axis of the foot passes between digits 3 and 4, while in perissodactyls (zebras and rhinoceroses) it passes through the third middle digit (Fig. 3.2). The early artiodactyls had four hooves (digits 2 to 5),

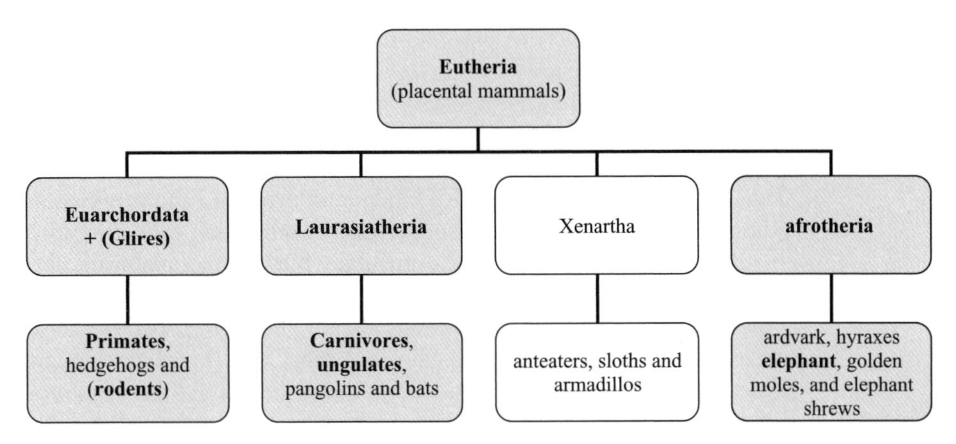

Fig. 3.1 A simplified classification of placental mammals with some Africa savannah species highlighted.

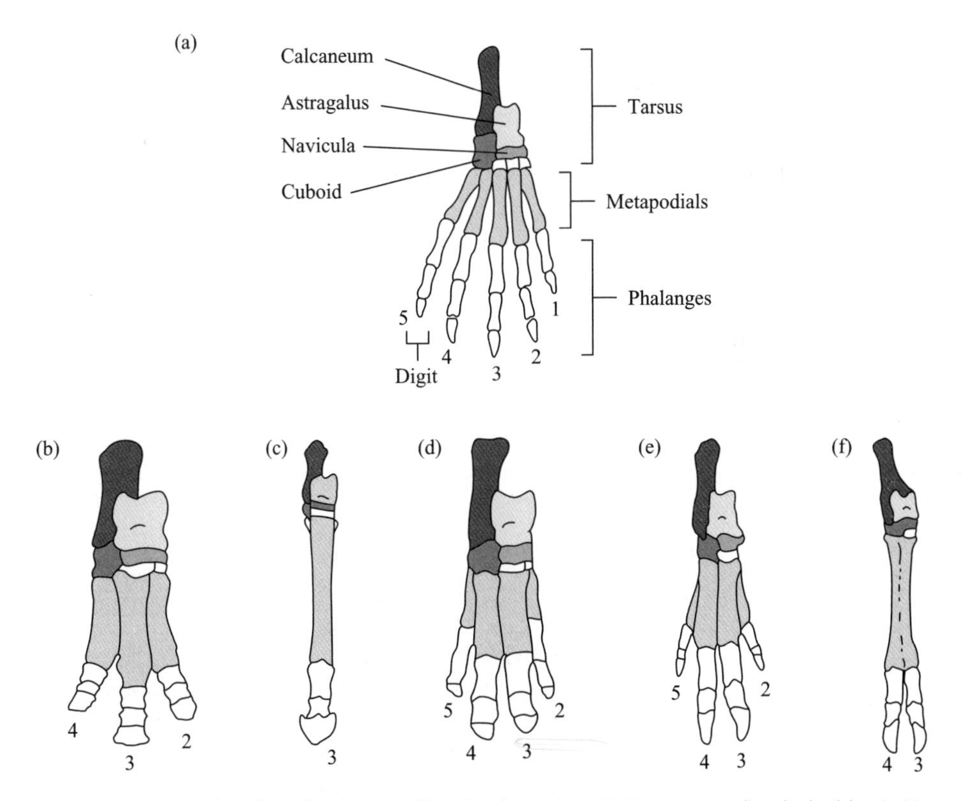

Fig. 3.2 Ungulate feet showing modification from the primitive mammalian limb. (a) primitive mammalian foot, (b) rhinoceros, (c) horse, (d) hippopotamus, (e) pig, (f) buffalo (modified from MacDonald 2001).

a condition that persists in pigs and hippos. The number of hooves has been reduced to one pair in the horses although the vestiges of digits 2 and 5 are still present as the false hooves.

Ungulates have also evolved two rather different systems for converting cellulose, a major constituent of all plant tissues, into digestible carbohydrates. These two systems are called hindgut fermentation and rumination. In both systems, the conversion from cellulose to digestible carbohydrate is not directly brought about by the ungulate, but by microrganisms that achieve the breakdown by a process of fermentation. The two systems are closely associated with the two groups already mentioned, perissodactyls and artiodactyles. The perissodactyles (e.g. zebra and rhino) are hindgut fermenters while most of the artiodactyles (e.g. giraffe, buffalo, antelopes, and gazelles) are ruminants. Only pigs, hippos and camels in the artiodactyles are not ruminants. Figure 3.3 shows a simplified comparison of the

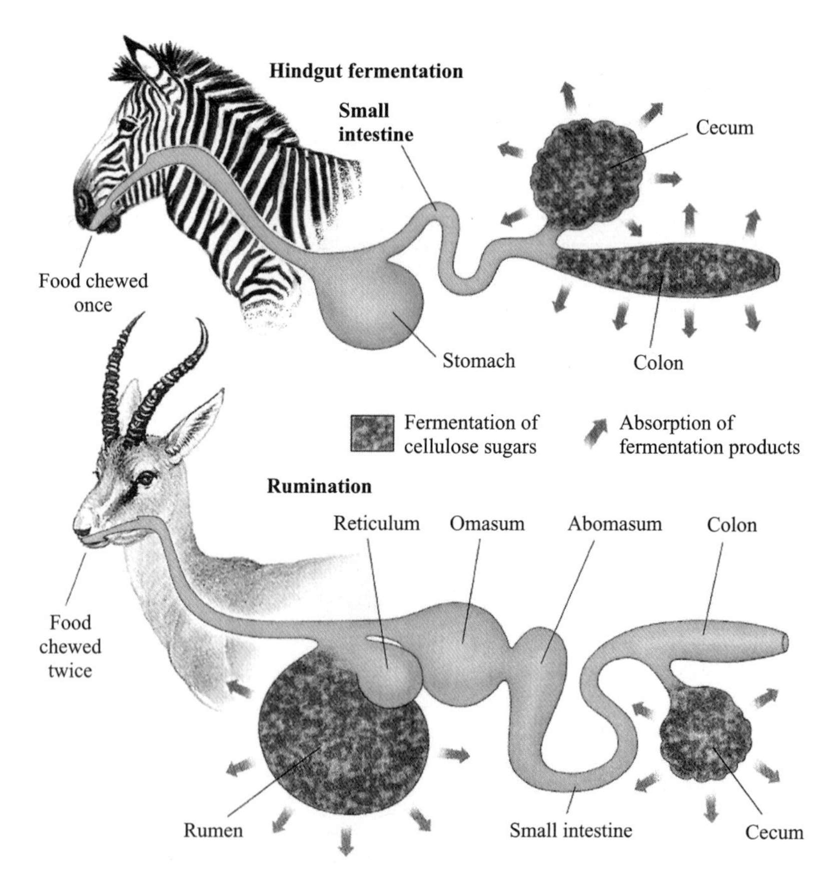

Fig. 3.3 Stylized comparison of 'hindgut fermentation' (zebra) and 'rumination' (gazelle) digestion (modified after MacDonald 2001).

two systems. In the hindgut fermenters, who have a simple stomach like ourselves, food is chewed once and plant cell contents are digested in the stomach. The food is then passed to the caecum (our appendix is a reduced vestige), and colon, where the cellulose of the plant cell wall is fermented by microrganisms. Ruminants, apart from the four species of primitive ruminants called chevrotains, have a more complex four-chambered stomach. Vegetation is swallowed and initially enters the first chamber or rumen. The coarsest plant particles (the cud) float to the top of the rumen, are regurgitated by its contractions, and rechewed in the mouth (rumination). The rechewed food is then returned to the rhythmically contracting second stomach, or reticulum, for 'sorting'. Food particles are eventually pumped into the third stomach, or omasum (also known as the 'book organ' because of the leaf-like plates that line it), and filtered. The digestive process is completed in the fourth or 'true stomach' (abomasum) and small intestine, and additional fermentation and absorption takes place in the caecum. In ruminants therefore cellulose digestion takes place, before normal digestion, in the rumen. In hindgut fermenters, cellulose digestion takes place after normal digestion, in the caecum. Ruminants are therefore able to extract more nutrients from their plant food (cellulose utilization in a cow is about 80 per cent), but pass it through their digestive system very slowly (rate of passage in a cow is about 80 hours). They require high quality food. Hindgut fermenters are less efficient (cellulose utilization in a horse is about 50 per cent), but pass food through their digestive system faster (rate of passage in a horse is about 48 hours). They can therefore use low quality food. Much more fibre is left undigested in the hindgut fermenter, as can be observed in the coarse dung of zebra, rhino, elephant, and of course horse. In comparison, the dung of ruminants is fine grained, as seen in the buffalo, wildebeest, and domestic cow.

Ruminants can also recycle urea (normally a waste product excreted in the urine) and can therefore conserve water. In the ruminant stomach, the protein in their plant food is converted to ammonia, and transported in the blood to the liver. Here it is converted to urea and transported back to the stomach where the microrganisms use it as food. In addition to their role in cellulose fermentation, these microrganisms excrete fatty acids which are used by their host. They also provide them with protein, as they pass down the ruminants digestive system, mixed with the vegetable food (Demment and Van Soest 1985). However, although ruminants have a more efficient system for dealing with the cellulose component of plant food, and gain protein and fatty acids from the microrganisms themselves, their one major disadvantage is that they cannot process food quickly. The more fibrous the food, the longer the process can take. When the protein content of their plant food falls below about 6 per cent, ruminants cannot process their food fast enough to maintain their weight and condition (McNaughton and Georgiadis 1986). This will be a particular problem in the savannah dry season, when per cent crude protein in many grasses drops below this value (Table 3.3). It is therefore important for many ruminants to be selective in their feeding (see Chapter 5).

Another feeding distinction that has arisen in ungulates is that between grazers (roughage eaters), browsers (selective eaters), and mixed feeders (Hofmann 1968; Estes 1991). Grazing ungulates feed on monocotyledonous plants (essentially grasses, but also reeds and rushes), while browsers feed on dicotyledonous plants (mainly tree foliage, but also fruits and herbs). Mixed feeders (e.g. elephant and impala) are both grazers and browsers, often switching their foraging activity between the wet season (grass) and the dry season (foliage). Some small browsers, such as klipspringer and bushbuck, are extremely selective feeders, choosing food with the highest protein content. They tend to feed in short bouts and rarely fill the rumen more than half full before stopping to ruminate. Larger browsers such as giraffe, gerenuk, and lesser kudu need to take in more food and cannot afford to be as choosy. They often have larger and more subdivided stomachs designed to slow down the passage of food and allow more time for fermentation. In contrast to these selective browsers, bulk grazers (Estes 1991) such as waterbuck, wildebeest, and buffalo, tend to keep eating until the rumen is full, often regardless of grass quality. The different stomach chambers are more muscular and subdivided in order to churn, pump, sieve and filter the ingested food. The omasum, which in the selective browsers is simply a straining chamber, is highly developed. So called roughage grazers (Estes 1991), such as hartebeest and topi, are frequently more selective in the parts of the grass they eat and can consequently often survive on poorer grass swards. These dietary differences may aid the coexistence of herbivores (Chapter 5).

Other parts of the ungulate body that show adaptation to their herbivorous way of life are the teeth and jaw muscles. The generalized placental

Table 3.3 Percentage crude protein in two grass species, in the Serengeti, at different growth stages.

Growth stage and grass species	Green leaf	Dry leaf	Stem and sheath
Young green growth			
Pennisetum mezianum	20.0	—	15.0
Digitaria macroblephora	16.6	—	10.5
Mature			
Pennisetum mezianum	9.8	4.3	4.4
Digitaria macroblephora	10.0	4.0	3.8
Dry season			
Pennisetum mezianum	6.8	3.9	3.1
Digitaria macroblephora	11.1	4.9	2.5

Data from Duncan 1975.

mammal has four types of teeth. The position and number of these teeth can be represented by a dental formula, which in humans is $\frac{2.1.2.3}{2.1.2.3}$. The upper numbers represent the teeth on one side of the upper jaw, and the lower numbers the teeth on one side of the lower jaw. The four successive numbers indicate the number of incisors, canines, premolars and molars. Typical formulae for a ruminant would be $\frac{0.0.2.3}{3.1.2.3}$ or $\frac{0.0.3.3}{3.1.3.3}$. The upper incisors and canines have been lost and the lower canines look like incisors forming a row of eight chisel-shaped teeth. The cheek teeth (premolars and molars) are adapted for grinding vegetable matter and their grinding surface has complex folds and sharp ridges. There is typically a gap, called the diastema, between the cutting incisors/canines and the grinding cheek teeth (Fig. 3.4, giraffe and bushbuck).

In hindgut fermenters the presence and absence of incisors and canines is more variable, but all have the grinding cheek teeth. In male zebra the dental formula, is $\frac{3.1.3.3}{3.1.3.3}$ (Fig. 3.4), but in females the canines are rudimentary or absent. In both black and white rhinos, all incisors and canines are absent and the formula is $\frac{0.0.3.3}{0.0.3.3}$ (Fig. 3.4). In the hippopotamus the canines and incisors are continually growing and the lower canines are enlarged as tusks (Fig. 3.4). The adult dental formula is $\frac{2.1.3.3}{2.1.3.3}$ but while the cheek teeth have the usual grinding function, the protruding position of the incisors and canines renders them useless for feeding. They are used as weapons only, and the cropping of grass is taken over by the hard edges of the lips. In the pigs, as represented by the warthog (Fig. 3.4) the cheek teeth have rather flat uncomplicated surfaces. The number of teeth is more variable in this species than in any other ungulate. A full complement for adults is $\frac{1.1.3.3}{2\text{–}3.1.2.3}$, but in old adults it may reduce to $\frac{0.1.0.1}{0.1.0.1}$. The canines are enlarged as persistently growing tusks.

It was stated earlier that the cheek teeth of most ungulates have complex grinding surfaces, with many folds and sharp ridges (apart from the pig family). Vegetation in general, and grass in particular, is very tough material to grind and over a lifetime of continual abrasion these cheek teeth are very likely to be worn down to the gum. To compensate for this, some ungulates (mainly grazers) have evolved high crowned (hypsodont) cheek teeth. A simple way to produce a high crowned tooth would be by simple elongation of the tooth—a higher tooth would give longer grinding life. Some early fossil mammals actually took this path but it is relatively unsuccessful at prolonging the grinding life of the tooth. Once the hard enamel surface of the tooth is worn away, most of the wear comes on the soft and easily abraded internal dentine, with the resistant enamel remaining only as a thin rim. In the modern hypsodont tooth the height is attained by a growth of each cusp or ridge on the tooth. These hard peaks are fused together by a growth of hard cement over the entire tooth surface, producing a tall tooth very resistant to grinding. This problem of extreme tooth wear in grazing animals is solved in an entirely different way by the

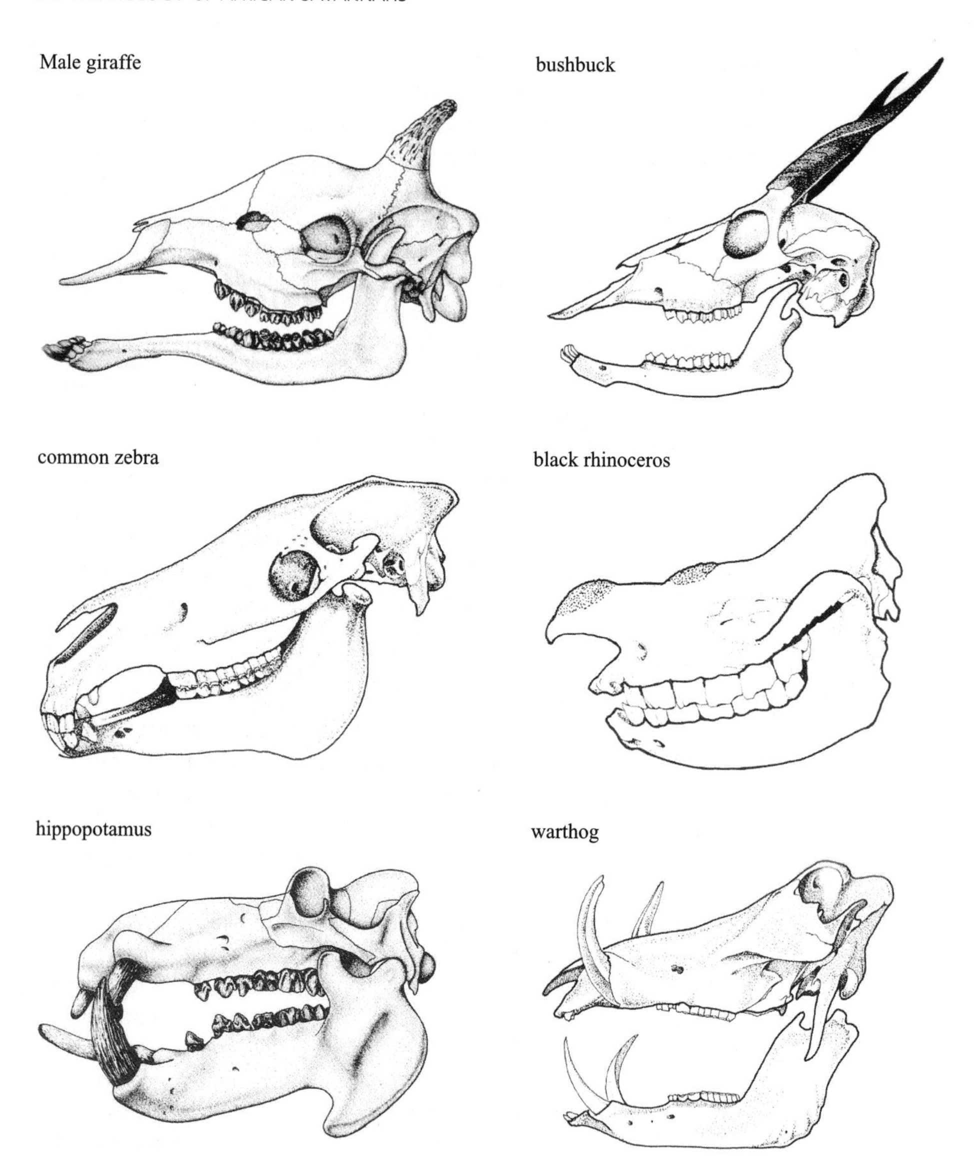

Fig. 3.4 Skulls of six ungulates. Giraffe and bushbuck are ruminants, the other four are non-ruminants (from Skinner and Smithers 1990).

elephant. The infamous elephant tusks, frequently responsible for their premature death at the hand of hunters are upper incisors, that grow continuously throughout life. At 60 years they can average 61 kg in a bull and 9.2 kg in a cow. Like many other grazers, they have six cheek teeth (all molars) in each quarter of the mouth, but they appear in succession throughout the life of the elephant. Only two molars normally occur together on each side of the upper and lower jaw, and are in use at the same time. As each successive tooth comes into use, it moves forward in the jaw, becoming worn and breaking down towards its forward edge, the roots being reabsorbed. The next one then moves forward from the back of the jaw to replace it. Laws (1966) shows that the six molars develop and erupt at the following approximate ages: molar 1 (1 year), molar 2 (2 years), molar 3 (6 years), molar 4 (15 years), molar 5 (28 years) and molar 6 (47 years). By the age of about 60 only the last molar remains. Once this last grinding tooth breaks down the elephant dies of starvation.

Ungulates dominate terrestrial communities, making-up about 80 per cent of mammal species over 50 kg. The great majority of these ungulate species are artiodactyls (about 92 per cent). This percentage is reflected in the African fauna, with only three zebra and two rhinoceros species that are perissodactyls. In the next two sections we look at some of these successful ungulates, that inhabit African savannah ecosystems. However, a word should be said about distribution maps. In all field guides (e.g. Estes 1991; Kingdon 1997, 2004; Stuart and Stuart 1997) the distribution of an African species is shown as a shaded area on a map of Africa. Intriguingly this shaded area often differs in detail, for the same species, in different guides. The probable reason for this is that a cross-section through a species distribution is not likely to be rectangular in shape. Species are not usually equally abundant across their distribution range, with a sudden drop to zero at the edge. Cross-sections across the shaded area of these maps are more likely to look like a bell-shaped curve (often skewed), with high abundance towards the centre of the range and a gradual decrease towards zero at the edge (unless the edge is the sea). Some authors give a distribution map that includes all the range, some authors I suspect give only that area in which you are quite likely to see the animal. In other words they miss out the edge of the species distribution. The distribution maps in the species accounts below are a compilation from Estes (1991); Stuart and Stuart (1997); Kingdon (1997), and for the 'antelopes', East (1998). In the species accounts, the Conservation Status is also stated. These are IUCN categories indicating the threat to a species. The categories are: *extinct*—no doubt that the last individual has died, *critically endangered*—a species faces an extremely high risk of extinction in the immediate future, *endangered*—a species faces a high risk of extinction in the near future, *vulnerable*—a species faces a high risk of extinction in the medium-term future, and *lower risk* or *not threatened*—none of the above.

Ungulates: nonruminants

Zebras (Family Equidae)

There are three species of zebra in Africa but only two are detailed here. The **common, or plains, zebra** (*Equus burchellii*) (Fig. 3.5a) (shoulder height: 1.3 m, weight: 175–340 kg) is Africa's most abundant and widespread wild

horse. The species includes several subspecies, which in older books are frequently called species. These include Grant's, Crawshay's, Chapmans, and the Damara zebra. They can be found in a wide range of savannah habitats, from short or tall grassland, to open woodland. They are adaptable grazers, taking the grasses that are most frequent. They also occasionally browse. In the great migrations of the Serengeti they are the first species to move, often grazing the tall flowering grasses ahead of the wildebeest and Thomson's gazelle. They live in small family groups (harems) of four to six animals. These comprise a male, with several females and their young. Males that have not got a family group join together, in bachelor herds. They are not territorial, and rather nomadic. Females produce a single foal (30–35 kg), after a gestation period of about 375 days. Birth usually occurs at the start of the rainy season when green grass is available. It is not threatened, although numbers are declining.

Africa's largest equid is **Grevy's zebra** (*Equus grevyi*) (Fig. 3.5b) (shoulder height: 1.5 m, weight: 350–430 kg). It was the 'hippotigris' paraded in Roman circuses, in 211–217 AD, pulling carts. Less water dependent than the common zebra, it occupies dry savannahs in northern Kenya. It is predominantly a grazer, often making use of a tough grass *Pennisetum schimperi*, not fully used by other grazers. Up to 30 per cent of the diet can be browse. This zebra is territorial and migratory. Mature males (over six years) establish core territories as large as 12 km^2 in the wet season, which they actively defend against other males. Ten or fewer mares, with their young, make up nursery herds which are allowed to pass through the male territories unchallenged. During the dry season Grevy's zebra are often forced to roam over vast home ranges, up to 10,000 km^2. Birth occurs at any time of the year, but peaks at the start of the wet season. After an exceptionally long gestation period of about 400 days, a single foal is produced. Its conservation status is *endangered*.

Fig. 3.5 Savannah mammals. (a) common or plains zebra, (b) Grevy's zebra, (c) black rhinoceros, (d) warthog, (e) hippopotamus, (f) reticulated giraffe, (g) African buffalo, (h) common eland (photgraphs a, b, d, e, f, g, h, by Bryan and Jo Shorrocks, photograph c by Tory Bennett).

Rhinoceroses (Family Rhinocerotidae)

The hook-lipped or **black rhinoceros** (*Diceros bicornis*) (Fig. 3.5c) (shoulder height: 1.6 m, weight: 800–1,100 kg) is not black but does have a hook-lip which is used to grasp leaves and twigs. The two horns are made of matted hair-like filaments and are attached to the skin rather than the bone. Once very common in savannah habitats it is now found only in protected areas. It requires savannahs with shrubs and trees to a height of 4 m, to provide both food and shade. It is a browser, taking a wide range of leguminous herbs and shrubs. Some 200 species from 50 families have been recorded in their diet. A female, and her young, is the basic social unit. The males are usually solitary but it is not clear if they have territories or simply home ranges. In Natal, radio-tracked bulls had 'territories' of 3.9–4.7 km^2, and female ranges were 5.8–7.7 km^2 (Hitchins 1968, 1969). Bulls and cows only come together briefly for mating. After a gestation period of some 450 days one calf (40 kg) is dropped. The calf may stay with the mother for between two and four years. Black rhinos are *critically endangered* having undergone catastrophic declines in numbers over the past few decades. Only about 3,000 remain in protected areas.

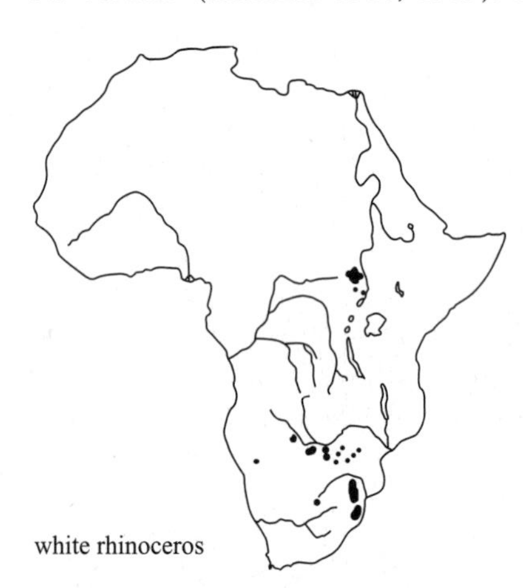

The square-lipped or **white rhinoceros** (*Ceratotherium simum*) (shoulder height: 1.8 m, weight: ♂ 2,000–2,300 kg, ♀ 1,400–1,600 kg)

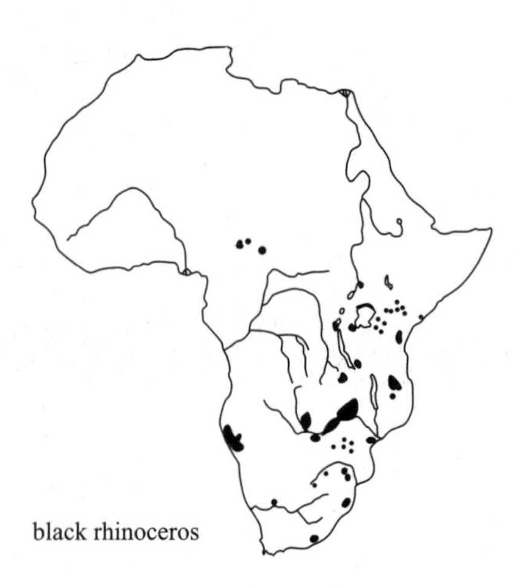

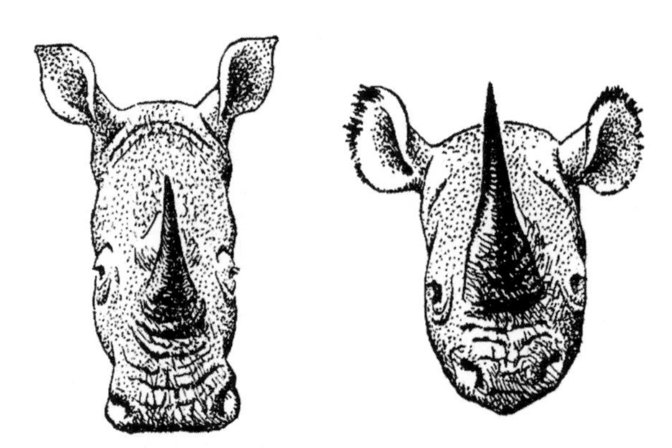

Fig. 3.6 Head of square-lipped or white rhinoceros (grazer) (left) and the hook-lipped or black rhinoceros (browser) (right) showing the different shape of the lip (from Dorst and Dandelot 1972).

is not white. It is much the same colour as the black rhinoceros. The name is derived from the Dutch '*weit*' (wide) referring to the mouth (Fig. 3.6). It is twice the weight of the black rhinoceros and the second largest land mammal after the elephant. It prefers areas of short-grass savannah with access to trees for shade. It's a selective grazer with a preference for short grasses such as *Cynodon*, *Digitaria*, *Heteropogon* and *Chloris* species. Females and their offspring occupy large (4–60 km^2), overlapping home ranges and sometimes associate in larger groups. About a ⅓ of males defend territories that can be quite small (3 km^2) but their size depends on the productivity of the habitat and population density. The remaining males live as satellites on the territories. After a gestation period of some 480 days a single calf (40 kg) is dropped. A calf stays with the mother for between two and three years. The conservation status of the southern African race is *lower risk*. The northern race is *critically endangered.*

Pigs (Family Suidae)

Africa has six species of pig, only two of which will be mentioned here. The **common warthog** (*Phacochoerus africanus*) (Fig. 3.5d) (shoulder height: 60–70 cm, weight: ♂ 60–105 kg, ♀ 45–70 kg) is the most abundant member of the family, and is distributed widely throughout the savannahs of Africa. In the wet season it grazes on grass, particularly species such as *Sporobolus*, *Cynodon*, *Panicum*, and *Brachiaria*. In the dry season it eats underground rhizomes of perennial grasses and sedges, and bulbs and tubers. Characteristically it frequently 'kneels' when grazing. They are heavily preyed upon by carnivores such as lion, leopard, and cheetah. When

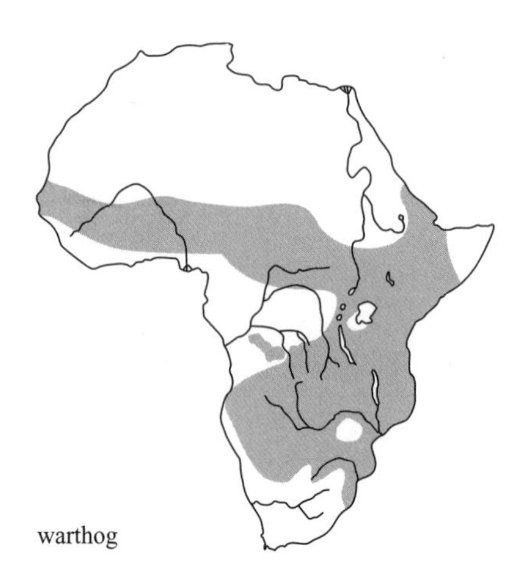
warthog

threatened by predators they retreat to burrows, previously excavated by them, or another species such as the aardvark. These burrows are used at night. The basic social unit, or sounder, consists of a female with her young, sometimes with a second female. Boars only accompany sounders containing oestrus females, otherwise adult males are usually solitary. Both sexes tend to remain in their natal area, so that clans of related sounders develop, often using the same network of burrows. These clan areas average about 4 km^2. The young are born in a burrow, after a gestation period of 160–170 days. Litters average two or three, but can go up to eight. Weaning takes between nine weeks, and five months. The **desert warthog** (*P. aethiopicus*) is slightly smaller than the common warthog, but otherwise indistinguishable in the field. It is confined to the arid savannahs of Somalia and NE Kenya. It also eats grass, and roots. Neither species is on the IUCN threatened list.

Hippopotamuses (Family Hippopotamidae)

Africa has two members of this family. The **pygmy hippopotamus** (*Hexaprotodon liberiensis*) is not a savannah species and has a restricted range in West Africa. It will not be detailed here. The more familiar, and much larger, **common hippopotamus** (*Hippopotamus amphibius*) (Fig. 3.5e) (shoulder height: 1.5 m, weight: ♂ 1,000–3,000 kg ♀ 1,000–2,500 kg) is commonly observed in, or near, rivers that run through savannah areas.

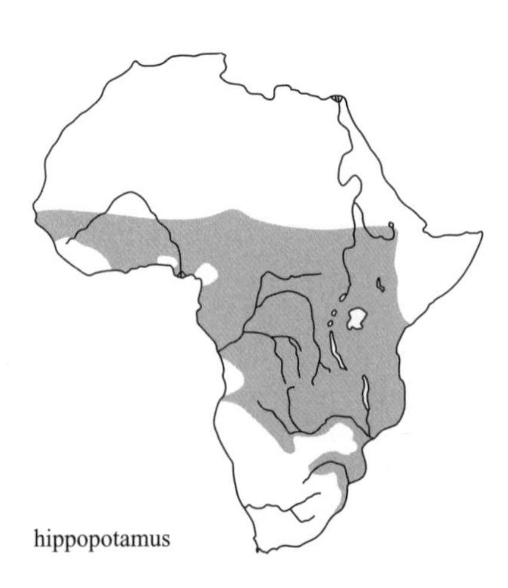
hippopotamus

Their skin is quite unique. They have a thin epidermis, with no sweat glands, and lose water at several times the rate of other mammals. Because of this they spend most of the day in water. At night they emerge and commute to their feeding grounds, which may be a few hundred metres, to several kilometres away. Here they graze on grasses (about 40 to 60 kg per night), frequently reducing the grass in such areas to 'grazed lawns'. At high density, hippos can have a devastating effect upon the vegetation and soils bordering savannah rivers. Both creeping and tussock grasses are eaten, including *Cynodon*, *Panicum*, *Brachiara*, *Themeda*, *Chlois* and *Setaria* species. They usually live in groups of between five and fifteen individuals. These herds are usually dominated by an adult male who holds tenure over a number of females and their young. Males vigorously defend their group against intruding males and serious fights can take place. Females conceive at about nine years, and calve at two-year intervals. Mating takes place in the water. A single calf (30 kg) is born after an 8-month gestation period, usually in the rainy season. Although they have declined over the last fifty years, estimates of numbers in excess of 150,000 would suggest that their conservation status is *not threatened*. Zambia is thought to have the largest 'national herd' with an estimated 20,000 along the Luangwa River.

Ungulates: ruminants

Giraffes (Family Giraffidae)

The **giraffe** (*Giraffa camelopardalis*) (Fig. 3.5f) (shoulder height: 2.6–3.5 m, height to top of head: 3.9–5.2 m, weight: 970–1,400 kg) is the tallest animal. The name probable comes from the Arabic *zarāfa*, which in turn comes from the Ethiopic *zarat* meaning 'slender'. Considerable uncertainty surrounds the validity and geographical limits of the described subspecies. Between six and nine depending on the author. On the map I have indicated six groups: the western giraffe (W) (*G. c. peralta, antiquorum*), the Nubian/Rothschild giraffe (N) (*G. c. camelopardalis, rothschildi*), the reticulated giraffe (R) (*G. c. reticulatus*), the Masai giraffe (M) (*G. c. tippleskirchi*), Thornicroft's giraffe (T) (*G. c. thornicrofti*), and the southern giraffe (S) (*G. c. giraffa, angolensis*). They like dry savannahs where *Acacia*, *Commiphora* and *Terminalia* are abundant trees. They are browsers, and have been recorded feeding from over 100 species of plant. However, *Acacia* and *Combretum* form the bulk of the diet. Most of the diet is made up of leaves, flowers, shoots, and some pods. These are stripped from the tree by the powerful 45 cm long tongue and by the lips. They form loose, open, herds of usually between four and thirty individuals. They do not defend territories, but occupy large home ranges of 5 to 650 km^2, depending on habitat quality and giraffe density. However, like many savannah herbivores they disperse widely during the rains and concentrate along

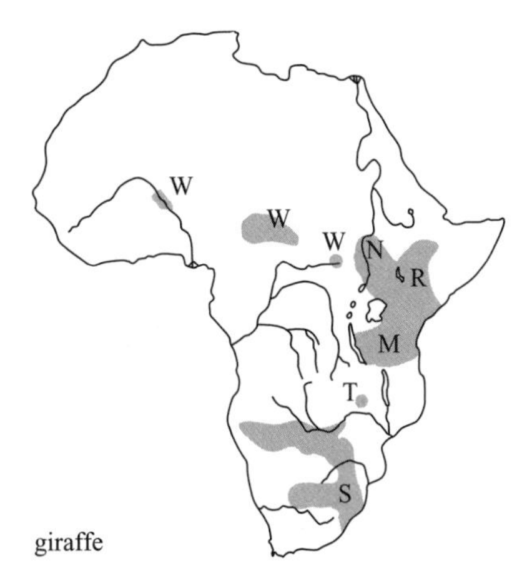

giraffe

watercourses in the dry season, where high quality food continues to be available. A single calf (100 kg) is born after a long gestation period of 15 months. The conservation status of the various subspecies is variable. The Masai (46,210) and southern (31,700) giraffe are not endangered. Thornicroft's giraffe (1,160) has low, but stable, numbers. The western (3,000), Nubian/Rothschild's (445), reticulated (27,680) are all declining and regarded as endangered. Population estimates are taken from East (1998).

Buffalo (Family Bovidae, Tribe Bovini)

There are two subspecies of African buffalo, the larger **savannah buffalo** (*Syncerus caffer caffer*) (Figure 3.5g) (shoulder height: 1.4 m, weight: ♂ 700 kg, ♀ 550 kg) and the smaller **forest**, or **red, buffalo** (*S. c. nanus*), found in the rain forests of the Congo basin. Only the former subspecies is detailed here. The savannah buffalo prefers open wooded savannah, with abundant drinking water, but also inhabits montane forest in areas such as the Aberdares and Mount Kenya in East Africa. They are predominantly (95 per cent) bulk grazers, taking a wide selection of grass species. They are found in herds of 20 to 2,000, depending on habitat quality, although the larger herds frequently split into smaller 'family clans' for part of the time. These clans comprise several related cows, and their offspring, to which are attached a number of adult and subadult bulls. Within these family groups, both adult males and females establish dominance hierarchies. Old bulls are often solitary, and rather sedentary (3–4 km^2). Herds occupy clearly defined home ranges which rarely overlap. Most calves (40 kg) are born in the wet season, after a gestation period of about 11 months. Birth intervals of two years are normal. The calf can stand within a few hours. The African buffalo was very badly hit by the rinderpest epidemic that started in 1890. Some estimates say that 10,000 died for every one that survived. Although greatly reduced in recent years, by hunting for meat, neither subspecies is threatened.

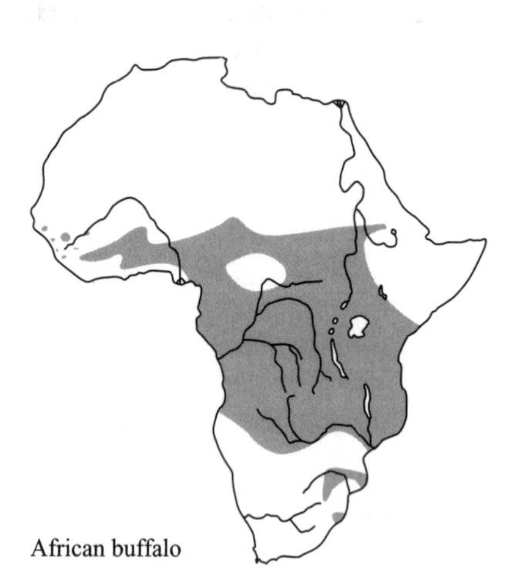
African buffalo

Spiral-horned antelopes (Family Bovidae, Tribe Strepsicerotini)

The **common eland** (*Taurotragus oryx*) (Fig. 3.5h) (shoulder height: ♂ 1.7 m, ♀ 1.5 m, weight: ♂ 700–900 kg, ♀ 450 kg) is the largest living antelope (along with the giant or Derby's eland). It is very adaptable, being found in all types of savannah from subdesert to miombo woodland. They are predominantly browsers of foliage and herbs, but also graze green grass

(50 per cent–80 per cent of wet season diet), fruits, pods, seeds, and tubers. Food is gathered by the lips, not the tongue. Although the eland is less desert adapted than the oryx, or some of the gazelles, it can go indefinitely without drinking. They can allow their body temperature to rise as much as 7°C during the day and let the night temperature cool them down. This would be equivalent to using about 5 litres of water per day in evaporative cooling. Elands form open groups of between one and 500 individuals, the latter usually in the rainy season. The only regular associations are between mother and calf. It is one of the most mobile antelopes, with home ranges varying between 200 and 1,500 km^2. A single calf (22–36 kg) is born after a gestation period of 9 months. The calf remains hidden in bush cover for the first two weeks. Its conservation status is *not endangered.*

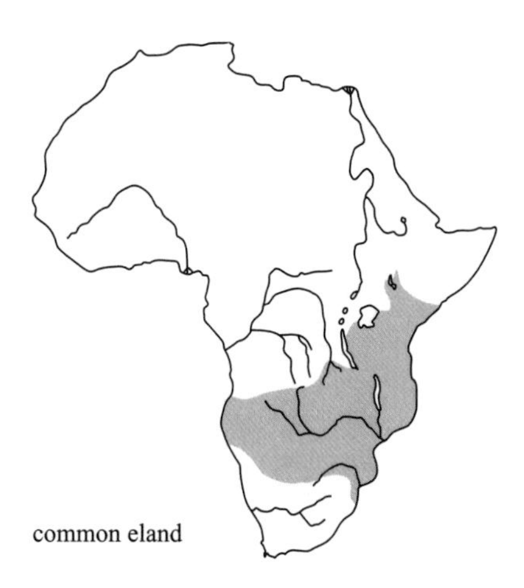
common eland

Africa has two species of kudu. The large and elegant **greater kudu** (*Tragelaphus strepsiceros*) (shoulder height: 1.4–1.6 m, weight: ♂ 250 kg, ♀ 180 kg) is one of few large antelopes that has extended its range in recent years, mainly the southern African subspecies. This is largely because of its secretive nature and its ability to survive in settled areas with sufficient cover. It prefers wooded savannahs and is predominantly a browser. However, a wide range of foliage, herbs, vines, flowers, fruits, succulents, and a little new grass, are taken. They can survive without water if browse is sufficiently moist. It is nonterritorial, but lives in small, nursery, herds of 2 to 15 individuals: cows, their young and sometimes an associated male. Males may be solitary, or form temporary bachelor herds of 2–10 individuals. The home range of nursery herds varies from 1 to 25 km^2 depending on the quality of the habitat, but males may have home ranges of 50 km^2. In the wet season a single calf (16 kg) is born after a gestation period of nine months. The young lie up for about three weeks, but are weaned and independent by about six months. Overall it is not endangered but

greater kudu

lesser kudu

has declined in the northern part of its range (Somalia, Chad, Uganda, and Kenya) where it may be *endangered* or *vulnerable.* The smaller **lesser kudu** (*Tragelaphus imberbis*) (shoulder height: 1 m, weight: ♂ 100 kg, ♀ 62 kg) is restricted to northeastern savannahs where it inhabits arid, *Acacia-Commiphora*, woodland and scrub. It is a browser, with a wide and varied diet of leaves, flowers, fruits, and seed pods. In the rainy season a little green grass is eaten. They are water independent, and eat succulents such as wild sisal in the dry season. The basic social unit is 1–3 females with their offspring. These female associations tend to be exclusive, and relatively long lasting (three known females stayed together for 4–5 years). Adult males spend much of their time alone, only associating with females for reproduction. They are not territorial and quite sedentary. Most births occur in the rainy season, when a single calf (7 kg) is born after a gestation period of seven months. Although it suffered a substantial decrease as a result of the 1990 rinderpest outbreak, it appears to be recovering and estimated numbers are high (about 22,000, East 1999). It is not thought to be *endangered* at present, although they are hunted for meat, and this may be a continuing problem.

The **bushbuck** (*Tragelaphus scriptus*) (Fig. 3.7b) (shoulder height: ♂ 80 cm, ♀ 70 cm, weight: ♂ 45 kg, ♀ 30 kg) is a very variable species, both in colour and markings, with over 29 subspecies described. They like woodland and scrub savannah near water (riverine habitats). It only leaves cover, for more open areas, to find choice food items, water and other bushbuck. It eats tender new grass, but is predominantly a browser on herbs and shrubby leguminous plants. In some areas they cause some damage to agricultural crops (fruits and flowers) and are considered a pest. They are solitary, nonterritorial and rather sedentary. The only regularly associated individuals are a female and her latest offspring. A single calf (4 kg) is produced after a gestation period of 6 months. The young lie

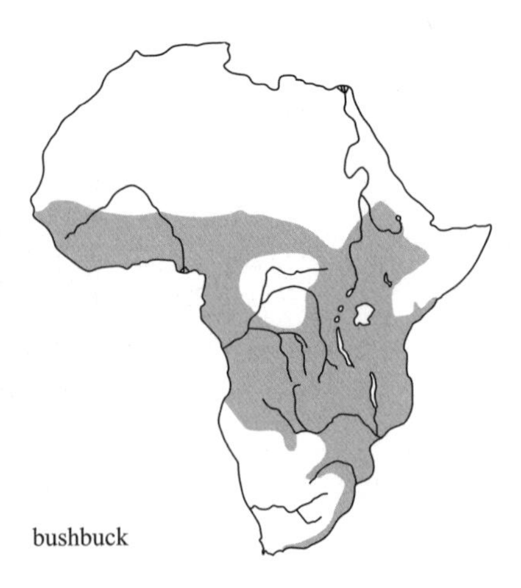
bushbuck

Fig. 3.7 Savannah mammals. (a) sable antelope, (b) bushbuck, (c) beisa oryx, (d) defassa waterbuck, (e) topi, (f) common wildebeest, (g) impala, (h) Grant's gazelle (all photgraphs by Bryan and Jo Shorrocks).

up in cover for up to four months, before moving around with the mother. Some montane subspecies (Mt Elgon and northern Uganda) are *vulnerable*, but overall it is widespread and abundant, and not *endangered*. It occurs in a larger number of African countries (40) than any other antelope species.

Horse antelopes (Family Bovidae, Tribe Hippotragini)

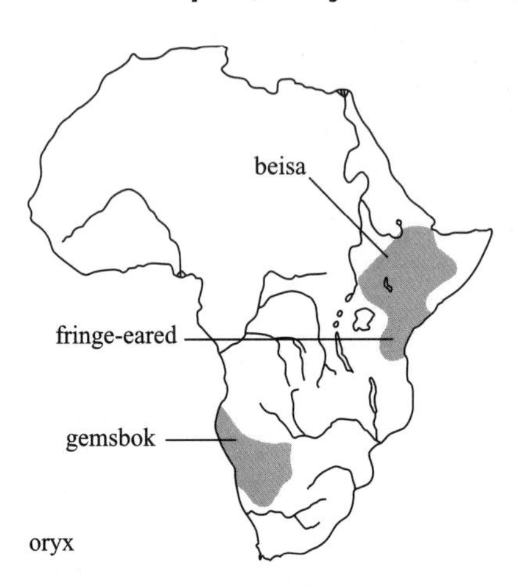

The oryx (*Oryx gazella*) (Fig. 3.7c) (shoulder height: 1.2 m, weight: ♂ 240 kg, ♀ 210 kg) has five subspecies. These include the **beisa oryx** (*O. g. beisa*), the **fringe-eared oryx** (*O. g. callotis*) and the **gemsbok** (*O. g. gazella*). The fringe-eared oryx, of southern Kenya and Tanzania, inhabits *Acacia-Commiphora* savannah that is less open, and less arid than the scrub savannahs occupied by the other two subspecies. One of the best desert adapted large mammals. It prefers green grass, but also grazes dry grass, and sometimes browses on wild fruits and tree pods (*Acacia*). It will dig up roots, bulbs and tubers for moisture. Like other desert mammals, when deprived of drinking water it employs various methods to minimize its water needs. These include allowing the body temperature to rise during the day (from a normal 35.7°C to 45°C) before beginning evaporative cooling, and concentrating the urine. It is nomadic, gregarious (mixed herds of up to thirty) and the males are probably territorial. A single calf is born after a gestation period of about nine months. Like other species that hide their newborn calves, oryx cows isolate themselves from their herd just before calving. The young remain hidden for three to six weeks, after which they join the herd. The northern subspecies have declined in recent years but none of the subspecies are endangered.

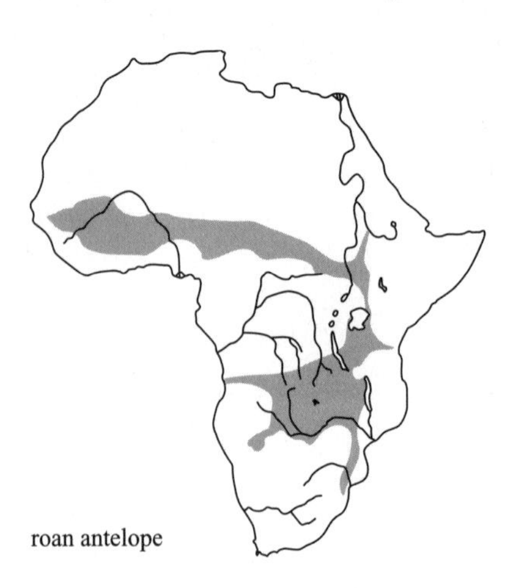

The horse-like **roan antelope** (*Hippotragus equinus*) (shoulder height: 1.1–1.5 m, weight: 220–300 kg) was historically one of the most wide ranging antelopes in sub-Saharan Africa. Its present distribution is now largely divided between the northern and southern savannahs. It inhabits open or slightly wooded savannah with long grass. It is a grazer. Herds of between 6 and 20 individuals (adult cows and their young), led by an adult bull, are formed. These groups share a traditional and exclusive home range, which in Kruger N.P. is 60–120 km^2). In Tanzania, a herd of 14 animals was observed to use an area of 12 km^2 over a period of seventeen years. Ranges are often smaller, and herds larger, in the dry season. Herd males appear to simply defend an area around the moving herd. Between years three and six young males form bachelor herds. A single calf is born (16–18 kg) after a gestation period of about nine months. The female leaves her herd before giving birth and the young roan remains hidden for about two weeks, before joining the herd. Numbers have been severely reduced in recent years, but 1998 estimates suggest about 39,000 roan in sub-Saharan Africa. It is classified as *lower risk*, but conservation dependent.

Like the roan antelope, the **sable antelope** (*Hippotragus niger*) (Fig. 3.7a) (shoulder height: 1.2–1.5 m, weight: ♂ 200–270 kg, ♀ 190–230 kg) is a large, strongly built animal with magnificent curved horns. There are four subspecies. The giant sable, *H. p. varianii*, noted for its large horns and confined to Angola, *H. n. roosevelti*, now confined to Shimba Hills N.P. on the East African coast, *H. n. niger*, the black sable found south of the Zambezi river and *H. n. kirkii* centred in the Miombo woodlands. It inhabits open woodland with medium to tall grass. They are mainly grazers of medium height grasses, but forbs and foliage make up 20 per cent of their diet. Herds of 15–25 females and young are common, but larger groups of 30–75 are not uncommon, especially in the dry season. Females tend to remain in their birth area so that local 'clans' tend to develop, and knowledge of the area is passed on from generation to generation. A single calf (13–18 kg) is produced after a gestation period of 9 months. Females leave the herd to give birth, and the calf remains hidden for several weeks. The giant sable is classed as *critically endangered*, the rest are regarded as *lower risk*, but conservation dependent.

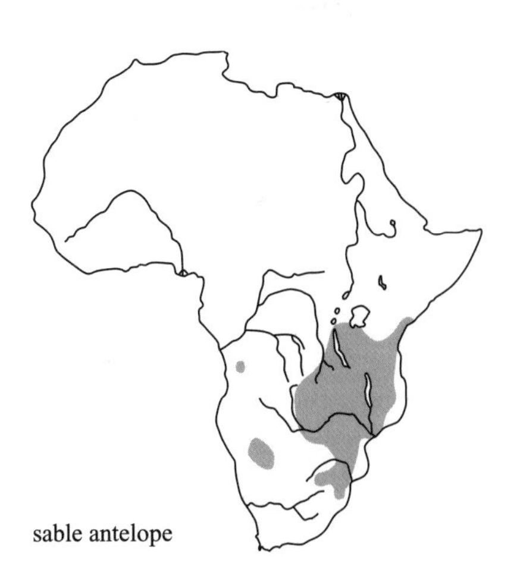
sable antelope

Reedbucks, kob and waterbuck (Family Bovidae, Tribe Reduncini)

The large, shaggy, and robust waterbuck (*Kobus ellipsiprymus*) (shoulder height: 1.3 m, weight: ♂ 200–260 kg, ♀ 160–215 kg) is one of the heaviest antelopes. It has two subspecies, the **common waterbuck** *K. e. ellipsiprymus*), with a white ring encircling its rump, and the **defassa waterbuck**

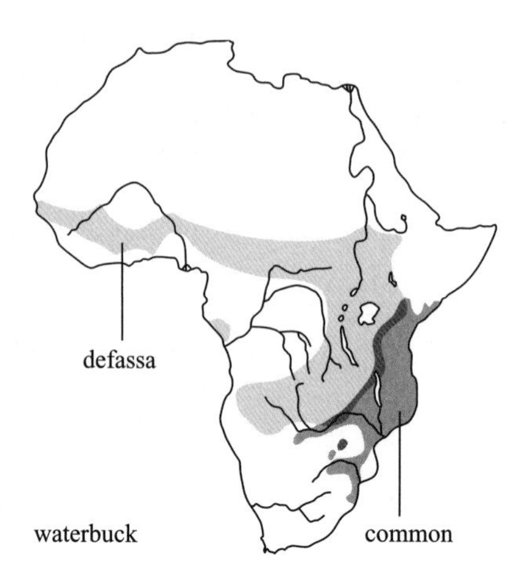

(*K. e. defassa*) (Fig. 3.7d), with a white patch on its rump. It is possibly the most water dependent of all antelopes and consequently occupies savannah habitats near to water. It is a grazer of short to medium length grass but acquires additional protein by browsing on herbage such as acacias and *Caparris* for up to 21 per cent of its feeding time. At about six years, males establish territories (0.5–2.8 km^2) which they hold for an average of 1½–2 years (the range is a few months to several years). Females associate in small herds of 5–10 individuals. The home ranges of these female 'nursery herds' may encompass the territories of several males. Young males live in small bachelor herds after being chased out of the nursery herd by territorial males. A single calf is produced (13 kg) after a gestation period of about nine months. It is not endangered.

The **kob** (*Kobus kob*) (shoulder height: ♂ 92 cm, ♀ 78 cm, weight: ♂ up to 120 kg, ♀ 60 kg) is rather similar to the impala, but more heavily built. There are 3 main subspecies, the **white-ear kob** (*K. k. leucotis*) in Sudan, the **Ugandan kob** (*K. k. thomasi*) in Uganda and Zaire and **Buffon's kob** (*K. k. kob*), in the northern savannah. It is a grazer that inhabits grassy floodplains. At low density, kobs have conventional territorial behaviour with males spaced at least 100–200 m apart. Females, with their calves, live in small, loosely structured, herds (5–15 individuals). Young males join bachelor herds. However, studies on the Ugandan kob (Leuthold 1966) have shown that at high densities ⅓ of males cluster on traditional breeding arenas, or leks, where territories are extremely small (15–30 m). The continued close presence of males and transitory females removes most of the grass cover from such arenas, leaving nothing substantial to eat. However, since 80–90 per cent of females visit these leks, it is worth the strongest males defending them, for an obvious reproductive advantage. A single young (4–5 kg) is produced after a gestation period of seven months. The kob's status is *lower risk*, but conservation

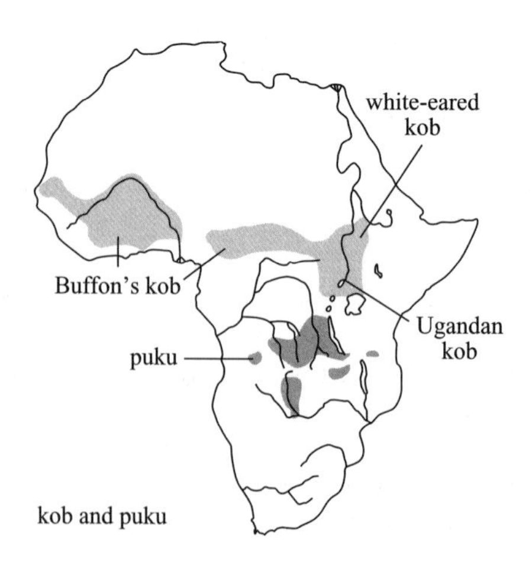

dependent. The **puku** (*Kobus vardonii*) is a slightly smaller, but very similar, species that replaces the kob ecologically in the central and southern savannahs. They are classified as separate species because there are no hybrids between them, a reflection of their geographical isolation. It is also *lower risk*, but conservation dependent.

The three reedbuck species are all rather similar. The **bohor reedbuck** (*Redunca redunca*) (shoulder height: ♂ 70–90 cm, ♀ 65–80 cm, weight: 45–55 kg) and **common, or southern, reedbuck** (*R. arundinum*) (shoulder height: ♂ 96 cm, ♀ 80 cm, weight: 43–51 kg) frequent grasslands, with tall grass. The **mountain reedbuck** (*R. fulvorufula*) (shoulder height: 72 cm, weight: 30 kg) inhabits hill country with scattered trees. All three are grazers, although the common reedbuck will also browse. Studies of their digestive system suggest that reedbucks can take grasses that are unpalatable to most other antelopes. Reedbucks bridge the gap between solitary and gregarious territorial systems. The common reedbuck lives in monogamous pairs while the mountain, and bohor, reedbuck live in small herds of three to eight females and young. Males defend territories in all three species. Reedbucks produce a single young (3–5 kg) after a gestation period of seven to eight months. The status of all three species is *lower risk*, but conservation dependent.

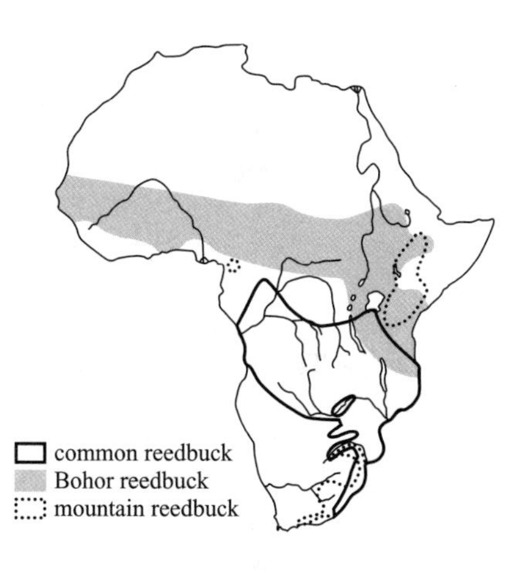

Hartebeests, topi, wildebeests and impala (Family Bovidae, Tribe Alcelaphini)

The **common hartebeest** (*Alcelaphus buselaphus*) (shoulder height: 1.25 m, weight: ♂ 150 kg, ♀ 120 kg) is a typical open savannah antelope. There are several subspecies, the more easily recognized being **western hartebeest** (*A. b. major*) (Senegal to western Chad), **lelwel** (*A. b. lelwel*) (south Sudan, Ethiopia, Uganda and Kenya), **tora** (*A. b. tora*) (east Sudan, Ethiopia), **Swayne's hartebeest** (*A. b. swaynei*) (Ethiopia and Somalia), **Jackson's hartebeest** (*A. b. jacksoni*) (central Kenya), **Coke's hartebeest** (*A. b. cokii*) (Kenya, Tanzania), and Cape or **red hartebeest** (*A. b. caama*) (southern Africa). **Lichenstein's hartebeest** (*A.* (*Sigmoceros*) *lichensteinii*) (miombo woodland savannah of central East Africa) is sometimes regarded as another subspecies, but more often as a separate species. Hartebeest prefer open savannah but will enter open woodland and bush savannah more readily than wildebeest or topi. They are almost entirely grazers, often associated with medium grasslands dominated by red-oat grass (*Themeda*

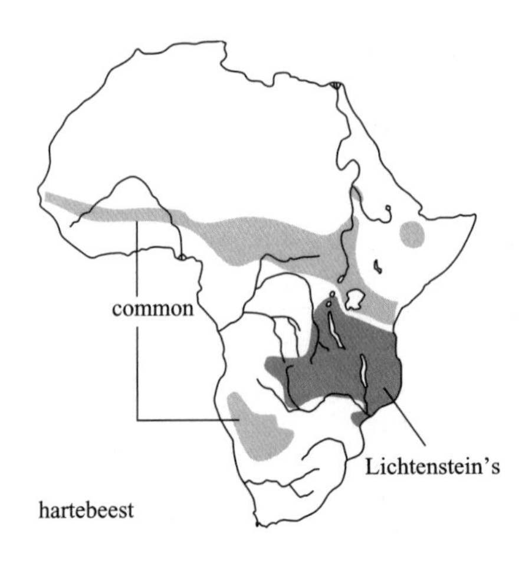

triandra). In Nairobi N.P., in the 1960s and 1970s, female herds (6–15 ♀♀ + young) had home ranges of 3.7 to 5.5 km² and wandered through male territories. Prime males (4–7½ years) held small territories in the best habitats. A single calf (15 kg) is born after a gestation period of eight months. The calf usually remains hidden until it is strong enough to follow the herd. Swayne's hartebeest (only about 200 individuals) and tora hartebeest (numbers unknown) are regarded *as endangered* and are in danger of extinction. The other subspecies are *lower risk*, but conservation dependent.

Damaliscus lunatus (shoulder height: 1.2 m, weight: ♂ 140 kg, ♀ 126 kg) has several subspecies, all known by a unique common name. In northwest Africa it is the **korrigum** (*D. l. korrigum*), in north east Africa it is the **tiang** (*D. l. tiang*), in East Africa it is the **topi** (*D. l. jumela*) (Figure 3.7e) and in southern Africa it is the **tsessebe** (*D. l. lunatus*). It resembles a hartebeest, but is smaller and darker. It inhabits open woodland and scrub savannah and eats grass. The long narrow muzzle and mobile lips enable it to select the more tender green blades. Like many other grazers it likes the new green flush after rain, but cannot use short grass as efficiently as bulk feeders such as wildebeest, waterbuck and zebra. Social and reproductive organization is more variable than any other antelope. In grassland patches in more wooded savannahs where population density may be low, males hold small territories (¼–4 km). Each territory holds the resources sufficient for one herd of 2–6 females (rarely over 10) and their young. The resident male has exclusive rights to the females. Where *Damaliscus lunatus* live at higher densities, on more open extensive grasslands (for example, Queen Elizabeth N.P. in Uganda), temporary territorial networks can be established wherever the circulating mass of animals settle for a few hours. In Akagara N.P., in Rwanda, leks are established by the males, on traditional arenas. Females and young circulate in large herds

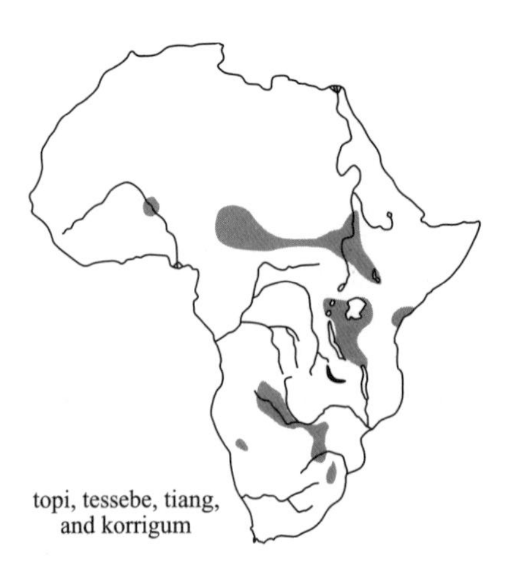

(200–300), and females enter the lek arena singly or in small groups. A single young (10 to 12 kg) is born after a gestation period of eight months. Calves are either *hiders* (remain hidden after birth for several days) or *followers* (keep up with the herd from birth) depending on the habitat and social organization. The korrigum subspecies (about 2,500 individuals) is *vulnerable*, the other subspecies are *lower risk*, but conservation dependent.

common wildebeest

The **common wildebeest** (white-bearded wildebeest, blue wildebeest, brindled gnu) (*Connochaetes taurinus*) (Figure 3.7f) (shoulder height: ♂ 1.5 m, ♀ 1.3 m, weight: ♂ 250 kg, ♀ 180 kg) has a number of subspecies. The most widespread and abundant (East 1999) are the **white-bearded wildebeest** in East Africa and the **blue wildebeest** in southern Africa. This species dominates some open short-grass and shrub savannahs. In the Serengeti ecosystem, of northern Tanzania, the annual migration of 1.5 million white-bearded wildebeest, from the southern grass savannahs to the northern Masai Mara, of Kenya, is renown. The animals move off the grasslands as the dry season begins and return with the advent of the rainy season. However, large migratory populations of blue wildebeest also occur, or occurred, on the extensive short-grass plains and *Acacia* savannahs of Botswana, Namibia, southern Angola and southwestern Zambia. Some populations, for example those resident in the northern part of the Serengeti ecosystem, and those in Ngorongoro Crater, do not migrate. They graze a wide variety of grasses that form short swards. They are bulk feeders and their broad muzzle, with its wide incisor row, enables them to rapidly close-crop the short grass. Resident populations form small, segregated, herds of females that are relatively sedentary and overlap the smaller territories of several males. When on the move, migratory males simply defend a zone around their cows. A single calf (14 kg) is born after a gestation period of about eight months. The calving season varies from area to area. They are not endangered, and classified as *lower risk* but conservation dependent. In southern Africa there is also a **black wildebeest** (*C. gnou*) which is slightly smaller, but also a grazer.

Regarded as the quintessential antelope, the **impala** (*Aepyceros melampus*) (Fig. 3.7g) (shoulder height: 90 cm, weight: ♂ 50 kg, ♀ 40 kg) is difficult to classify taxonomically. Sometimes it is placed near gazelles, and sometimes near hartebeests. There is one easily recognized subspecies, the **black-faced impala** of southwest Africa. It inhabits open and lightly wooded

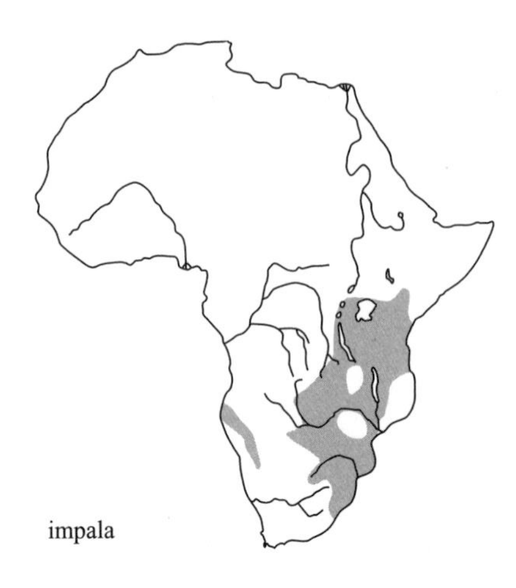

savannah with access to water. It is a mixed feeder. During the wet season it eats green grass and during the dry season it browses on foliage, forbs, shoots, and seedpods. For example, in the northwest of Zimbabwe their diet was observed to change from 94 per cent grass, in the wet season, to 68 per cent browse, in the dry season. Mature males alternate between bachelor and territorial status and rarely hold a territory for more than a few months. In the Serengeti only about one third of adult males were territorial at any one time, however, in East Africa, territorial behaviour is seen most of the year. In southern Africa, vigorous territorial behaviour is limited to only a few months, during the annual rut. During the rest of the year males and females can associate in mixed herds. Male territory size varies with population density, and habitat quality. Groups of females, and their offspring, associate in herds of anything from 6 to 100 individuals. These herds, or clans, occupy traditional home ranges of 80–120 ha. A single calf (5 kg) is born at the beginning of the rainy season after a gestation period of 6½ months. The common impala is one of the most abundant antelopes in Africa. It is not endangered. The black-faced impala is *vulnerable*, mainly due to the risk of hybridization with the common impala.

Gazelles and dwarf antelopes (Family Bovidae, Tribes Antilopini and Neotragini)

The 'giraffe-necked antelope' or **gerenuk** (*Litocranius walleri*) (shoulder height: 95–100 cm, weight: 30–50 kg) is one of the most easily identifiable antelopes. It is an inhabitant of the arid grass and shrub savannah of the Horn of Africa. It is a strict browser of trees, not known to eat grass or herbs. Although 87 different types of shrubs and trees have been recorded in its diet, within a tree it feeds quite selectively using its very narrow muzzle to take choice foliage from even thorny branches. When browsing, it frequently stands on its hind legs and uses its long neck to reach foliage inaccessible to other browsers, apart from the giraffe. Although they can be found in small groups of two to twelve individuals it is frequently seen alone (in Tsavo N.P., 42 per cent sighted were single individuals). Males become permanently territorial when mature and rarely move off their territory, which may be up to 4 km^2. Most gerenuks are local and very sedentary. A single young (3 kg) is produced after a gestation period of seven months. It is not endangered, its conservation status being *lower risk*.

The pale **Grant's gazelle** (*Gazella granti*) (Fig. 3.7h) (shoulder height ♂ 85–95 cm, ♀ 80–85 cm, weight: ♂ 55–80 kg, ♀ 35–50 kg) is found over a wide range of savannahs from semi-desert, through arid scrub, to open woodland savannah. It is a mixed feeder, but takes grass only when it is green. Herbs and foliage are preferred during the later wet and dry seasons. The fruits of both *Balanites* and *Solanum* have been recorded in the diet. Grant's gazelle is gregarious, territorial and migratory. A study in the Serengeti suggests that size and composition of social groups varies with habitat. On the Serengeti grasslands, nearly half the herds were mixed, with the most common group size being about fifty animals. In woodland savannah, only 12 per cent were mixed groups, most being harems of one male with several females and their young. Here the most common herd size was only about ten individuals. When males reach three years they tend to leave their natal herd and become territorial. After a gestation period of 6½ months, a single young (2 kg) is dropped. Although most populations are in decline, this gazelle remains widespread in East Africa and is not presently endangered. Its conservation status is *lower risk*.

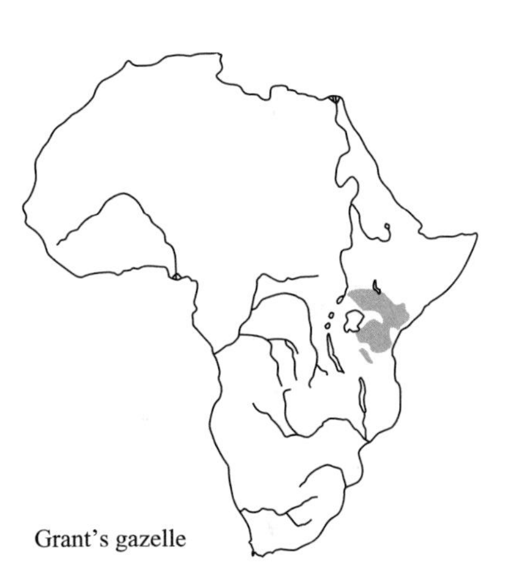
Grant's gazelle

The 'Tommy' or **Thomson's gazelle** (*Gazella thomsonii*) (shoulder height: 55–65 cm, weight: 15–25 kg) is an abundant species (about 550,000) with a restricted East African range. The subspecies in Sudan (*G. t. albonotata*), is known as the **Mongalla gazelle**. Thomson's gazelle inhabits open grassland shrub savannah, although it prefers areas with short grass. Unlike Grant's gazelle, this species is predominantly a grazer of green grass, requiring regular access to drinking water. Some herbs and seeds are taken in the dry season. Like the previous species this gazelle is gregarious, territorial and migratory. Social groupings are rather loose and open, especially during long migrations, with females leaving and entering female herds. During the mating season,

males defend small territories (100 to 300 m) and attempt to mate with any females that enters the area. Towards the end of the rainy season, a single young (2 to 3 kg) is born, after a gestation period of about six months. It is not endangered, its conservation status being *lower risk*.

The **springbok** (*Antidorcus marsupialis*) (shoulder height: 75 cm, weight: 26–41 kg) is the only gazelle in southern Africa, where it is abundant (estimated 670,000). When alarmed the springbok, like many antelopes, employs a distinctive bouncing gait called stotting. However, with this antelope it is particularly dramatic, reaching a height of 2 m, and gives the species its common name. It inhabits open arid grass and shrub savannah, but has considerable habitat tolerance. It can be found in savannahs with rainfall of up to 750 mm, to the Namib Desert with 0–100 mm. A mixed grazer/browser that takes young and tender grasses, and a variety of shrubs and succulents. In the Transvaal reserve, springboks fed on 68 different species, of which nine grasses and eleven shrubs formed the major part of the diet. Its social structure, and territorial behaviour, is very similar to that of Thomson's gazelle. In the past, mass migrations of springbok have been recorded. Four such migrations, involving upwards of 15,000 individuals were recorded between 1946 and 1959, usually at the end of the dry season in October/November. Springbok in the Kalahari moved southwest, into Cape province, and were shot in their thousands. These journeys (trekbokkens) still occur from time to time, in Botswana, but on a much smaller scale. A single young (about 4 kg) is born after a gestation period of about 5½ months. Like the previous two gazelles, this species is not endangered, its conservation status being *lower risk*.

The dwarf antelopes are small, delicate, antelopes with the female slightly larger than the male. They are largely crepuscular or nocturnal and live singly or in small family groups. There are thirteen species but only the four dik diks, steenbok, oribi and two grysboks

are savannah species. The klipspringer and beira inhabit rocky areas, often within savannah ecosystems. Only three species of dwarf antelope are outlined here. The **steenbok** (steinbok) (*Raphicerus campestris*) (shoulder height: 50 cm, weight: 11 kg) occurs widely in dry savannahs, in two separate regions either side of the miombo woodland savannah. It inhabits open savannah grasslands, with bush and light woodland which can be used for cover. The steenbok is a selective mixed feeder on high quality browse, and tender new grass. It browses on the leaves and shoots of low shrubs and trees, forbs, seeds and seed pods, berries and fruit. In the Kalahari it also digs for roots and tubers. It is usually seen singly or in loosely-maintained pairs. A single young (900 g) is produced after a gestation period of 5½ months. The young remain hidden for the first few weeks after birth. It is abundant, and not endangered, with an estimated total population of 663,00 (East 1999). **Kirk's dik dik** (*Madoqua kirki*) (Fig. 3.8a) (shoulder height: 38 cm, weight: 5 kg) and **Guenther's dik dik** (*M. guentheri*) (shoulder height: 34–8, weight: 3.7–5.5 kg) are virtually impossible to distinguish in the field, and are simply treated here as a single entity. The two species overlap in the Laikipia region of Kenya and may well interbreed. The other two species are also very similar, and confined to the dry savannahs of the Horn of Africa. Like the steenbok, Kirk's dik dik has a disjunct distribution. They inhabit relatively dry shrub savannah and are browsers. Because of their relatively large surface area, these small antelopes have water-loss problems in arid environments—they cannot afford to sweat. Evaporative cooling of the blood is achieved through nasal panting. To develop this mechanism, the nose has become swollen and elongated, a characteristic feature of all dik diks. They live in closely associated pairs, in territories defended by the male. There is some evidence that they mate for life. This territory varies in size depending upon the quality of the habitat. A single young (600–750 g) is born after a

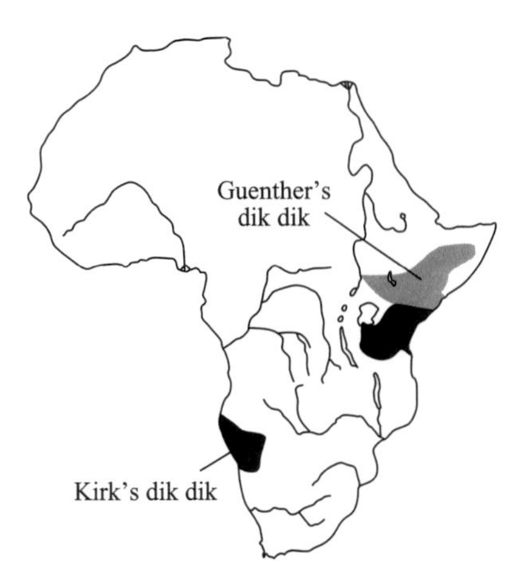

gestation period of about 5½ months. Both species are common and not endangered.

Subungulates: mammals out of Africa

Molecular biology and anatomy have shown that elephants (three species), hyraxes (eleven species), and the aardvark, although looking very different, may well be closely related. They appear to have evolved in Africa during the continents isolation in the Eocene period. They have short, nail-like hooves, and are collectively called the subungulates. Only the two African elephants are described here.

African elephant (Family Elephantidae)

There are two African elephants, the savannah elephant and the forest elephant. They were originally regarded as subspecies, but recent molecular work (using biopsy samples collected by dart from 195 free-ranging elephants, in eleven African countries) (Roca, *et al.* 2001) suggests that they should be treated as separate species. The savannah populations in southern, eastern and north central Africa, although widely separated, were genetically indistinguishable, while the forest animals showed more genetic diversity.

The paucity of gene introgression (exchange) between forest and savannah populations, even near regions of potential physical contact, suggests that hybridization in nature is rare. Most ecological and behavioural studies have probably been carried out on the savannah species. The **savannah elephant** (*Loxodonta africana*) (Fig. 3.8b) (shoulder height: ♂ 3.2–4 m, ♀ 2.5–3.4 m, weight: ♂ 5,000–6,300 kg, ♀ 2,800–3,500 kg) is the largest land mammal. Their range of habitats is extremely wide, from semi-desert, through all types of savannah, to forests and swamps. It is a mixed feeder, taking a wide variety of plants. In the wet season they eat green grass, and in the dry season they browse on woody and herbaceous species. Post reproductive survival is rare in mammals, but is found in both elephants and man, presumably because leadership and experience play an important role in social organization. Female elephants form matriarchal herds (2 to 24 individuals, 9 to 11 being typical) of related individuals. Direction and rate of movement are set by the matriarch. Mature males form separate herds, or are single. Many populations are nomadic/migratory, with very large home ranges. Females come

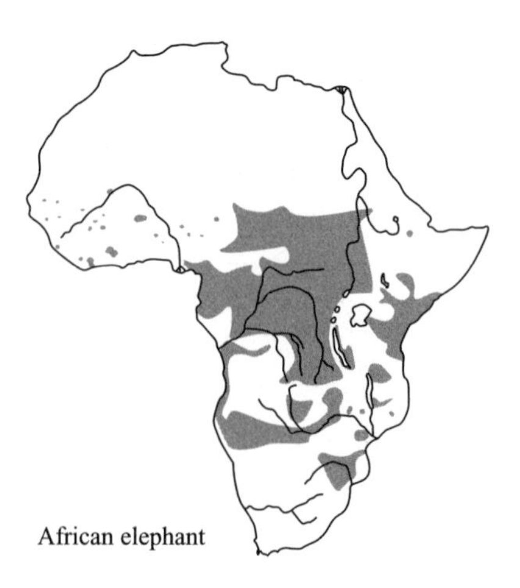
African elephant

Fig. 3.8 Savannah mammals. (a) Kirk's dik dik, (b) elephant, (c) lion, (d) cheetah, (e) serval, (f) spotted hyaena, (g) wild dog, (h) black-backed jackal (photgraphs by Bryan and Jo Shorrocks).

into oestrus for 2–6 days, after a period of between three and nine years. Males that are in 'musth' (in a sexually active or rutting phase) search out these rare females by responding to their calls. A single calf (120 kg) is born after a gestation period of 22 months. Its conservation status is *endangered.* The **forest elephant** (*L. cyclotis*) (shoulder height: ♂ 2.35 m, ♀ 2.1 m, weight: ♂ 2,800–3,200 kg, ♀ 1,800–2,500 kg) is smaller and darker than its savannah relative and has characteristically rounded ears. Forest elephants appear to live in much smaller groups, of between five and eight individuals. Its conservation status is also *endangered.*

Carnivores

Although very variable in outward appearance, most African carnivores share certain features. The body is typically long, supple, and agile, and the legs are well muscled, with movement in all directions. They are usually equipped with strong claws, which in the case of the cats (except the cheetah), and some genets and civets, are retractable. Probably associated with the advantage of a long stride when running, is the presence of a reduced, and unattached collar bone. Like the herbivores, they have special teeth that help them with their diet. The last upper premolar, and first lower molar, have sharp, shearing edges for cutting through meat. These two carnassial teeth (Fig. 3.9), on either side of the jaw, act against each other like the blades of a pair of scissors. However, few carnivores (except cats and weasels) are strictly carnivorous, including in their diet some plant material, particularly fruit. Some carnivores have even specialized in a diet of insects and, as a consequence, the teeth are less specialized. In the aardwolf, which almost exclusively eats harvester termites, the teeth are simple and peg-like. Most carnivores will scavenge dead meat if it is available, although the cheetah appears to be an exception. Table 3.4 shows some values for percentage of the diet scavenged, for large carnivores in the Serengeti ecosystem.

Most carnivores are small. Of the 66 African carnivores, 50 per cent are less than 50 cm in length (head + body) and 90 per cent are under 1 m. Weight ranges from the 300 g dwarf mongoose to the 185 kg male lion. They are however intelligent, possessing the mental alertness to outwit, capture and kill other animals. Their senses are all very well developed. In particular, they have an acute sense of smell and use scent not only to find prey but to communicate with each other. The secretions of several skin glands, urine, and faeces are all used to leave behind signals for other members of their species. We know that in at least one mongoose these secretions can even convey information about the identity of other individuals, thereby yielding information about their sex, age, and social status.

Although animal food is easier to digest than plant food, it is harder to catch. Although hunting behaviour can be quite varied, there are two principal types, stalking and/or ambush, and cursorial. The stealthy style of hunting seen in many cats (including domestic cats) falls into the first

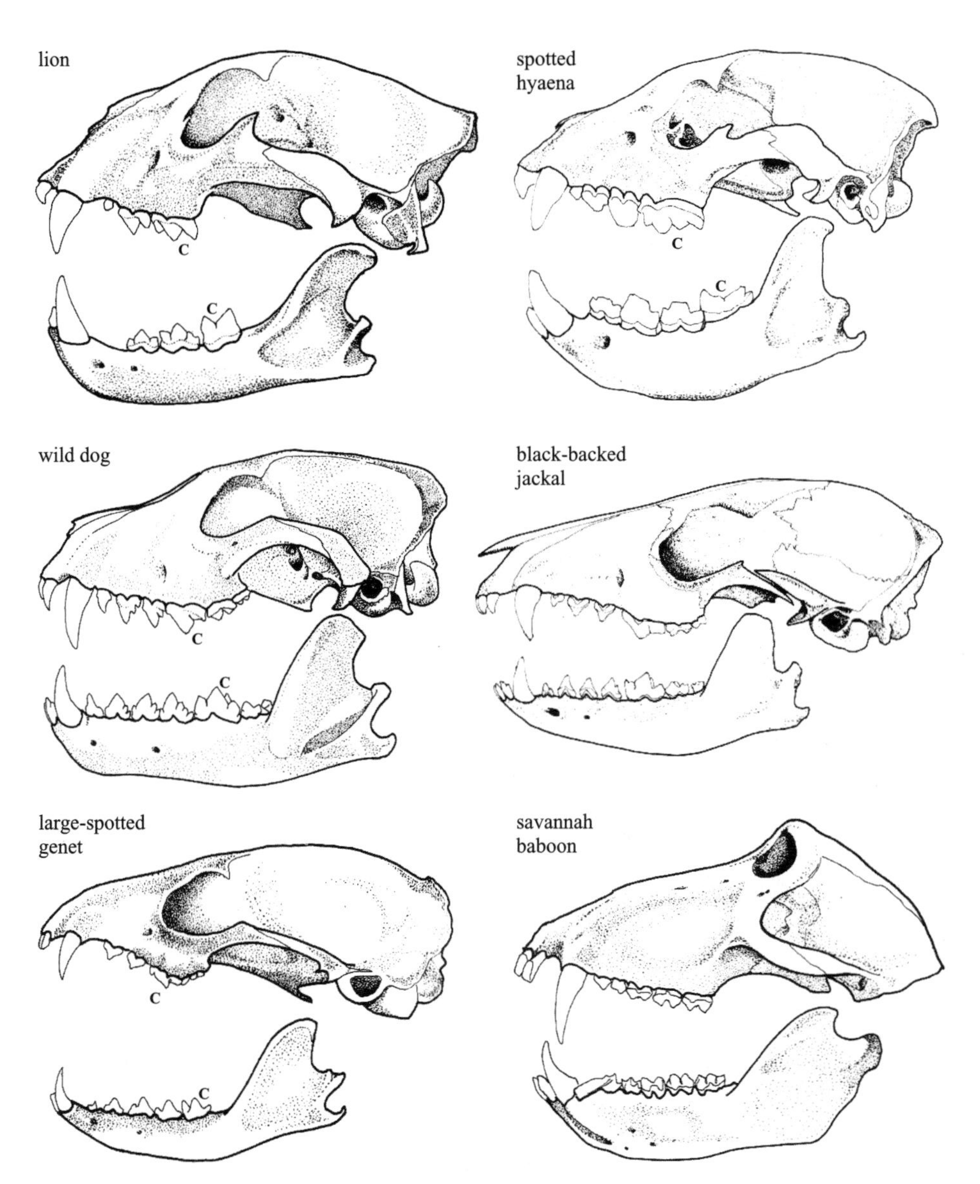

Fig. 3.9 Skulls of five carnivores, and one primate. The carnassial shearing teeth are indicated by a letter **c**, on the carnivores. (from Skinner and Smithers 1990).

category and the long chase of the wild dog, or spotted hyaena, falls into the second. Perhaps surprisingly many attempts at catching prey appear to fail. Table 3.4 shows some percentage failures for large carnivores in the Serengeti ecosystem. Plant food is also more numerous and concentrated than animal prey, and as a consequence carnivore home ranges tend to be larger than

Table 3.4 Percentage of the diet obtained by scavenging, and measured % success rate once hunting has been initiated.

	cheetah	leopard	lion	spotted hyaena	wild dog
% Scavenged	0	5–10	10–15	33	3
% Success	37–70	5	15–30	35	50–70

Data from Bertram 1979.

those of herbivores. A pair of jackals, each weighing 10 kg require a territory of at least 2 to 4 km^2 in order to survive. In contrast a whole herd of topi, each weighing 100 kg, could subsist in an area less than 1 km^2.

Most African carnivores are solitary, hunting and living as single individuals (e.g. leopard, cheetah, serval, slender mongoose and palm civet). Females of course will associate, for a time, with their cubs. However, some species have adopted group living. These include the lion, wild dog, spotted hyaena, and some mongooses, and their social groupings are often among the most complex, and cooperative, of any mammals. Two explanations have usually been put forward to explain why some carnivores go around in groups. First, some species may hunt together in order to be more efficient at capture (e.g. lions) or in order to bring down larger prey (e.g. wild dogs). Second, some species may live together in order to improve their vigilance against other predators, and also to collectively repel such larger predators (e.g. some mongooses). This latter 'antipredator' explanation can also be extended to defending a carcass. Several spotted hyaenas can protect their kill against a lion, as can several wild dogs against a spotted hyaena. In fact the necessary ratio of defenders to 'robbers' (in both examples) appears to be about 4 to 1. However, some species of carnivore live in 'groups' but hunt and travel as single individuals. For example, brown hyaenas which hunt at night and are therefore not frequently observed, were thought to be 'asocial'. However, radio-tracking of individuals has found that they often share roughly the same home range. They occur together in the same area more often than you would expect by chance. Of course, they may simply be using the same best habitats, but they may also be said to live in 'spatial groups' (MacDonald 2001). In some carnivores it is not just the parents that look after the young and a third explanation, or benefit, of group living may be shared parenthood. In the black-backed jackal, for example, an older female offspring may delay reproduction and stay with her parents to help rear another, younger, set of pups. In lions the females of a pride (who are sisters) help to rear all the cubs in the pride. In a wild dog pack, one pair are reproductively active and subordinate dogs help to maintain the offspring of this dominant pair. This behaviour may appear disadvantageous to the helpers who are delaying their own reproduction. They may benefit however, by learning parenthood and therefore being more successful when they have young of their own. Also, if the breeding pair are monogamous, the helper may in fact be just as closely related to the groups young, as they would be to their own young. From an evolutionary point of view they are helping to ensure the survival of their own genes.

The same comments apply to the carnivore distribution maps below, as applied to the herbivore maps above. However, with carnivores our estimates of numbers are often more unreliable than for the larger, more visible herbivores (see Chapter 4). The distribution maps in the species accounts are based on Estes (1991); Stuart and Stuart (1997); Kingdon (1997); Nowell and Jackson (1996)(cats); Mills and Hofer (1998)(hyaenas); and Woodroffe *et al.* (1997)(wild dog).

Cats (Family Felidae)

The **lion** (*Panthera leo*) (Fig. 3.8c and Fig. 3.9) (shoulder height: ♂ 1.2 m, ♀ 1 m, weight: ♂ 150–240 kg, ♀ 122–182 kg) is not just an African species although its Asian distribution is now sadly restricted to the Gir Forest Sanctuary in India. It is the largest of Africa's three 'big cats'. They use a very wide range of habitats, from semi-desert, through grassland, to wooded savannahs. Their prey are medium to large mammals, such as buffalo, giraffe, zebra, warthog and antelopes. Some prides learn to specialize in certain prey. For example, a pride that foraged along Namibia's Skeleton Coast learned to prey upon Cape fur seals. They usually (but not always) hunt at night. Prides consist of a stable core of related females (sisters and daughters) and coalitions of males that are unrelated to the females. The males in a coalition are related, having been born in the same pride. Despite maternal defence, infanticide is common when males take over a pride. Pride home ranges usually vary from 26 to 220 km^2 but can go up to 2,000 km^2. One to four cubs (1.5 kg) are born after a gestation period of 3¼ to 4 months. Birth can take place at any time of the year, although adult females in a pride often conceive at the same time so that cubs can suckle from any female that is lactating. Males tend to leave their natal pride at 2–4 years. Most young females join their natal pride. There are no sound estimates of the total number of lions in Africa. Guesstimates range from 30,000 to 100,000 (Nowell and Jackson 1996). Their scavenging behaviour makes them vulnerable to poisoned carcasses put out to eliminated predators. Its IUCN status is *vulnerable*.

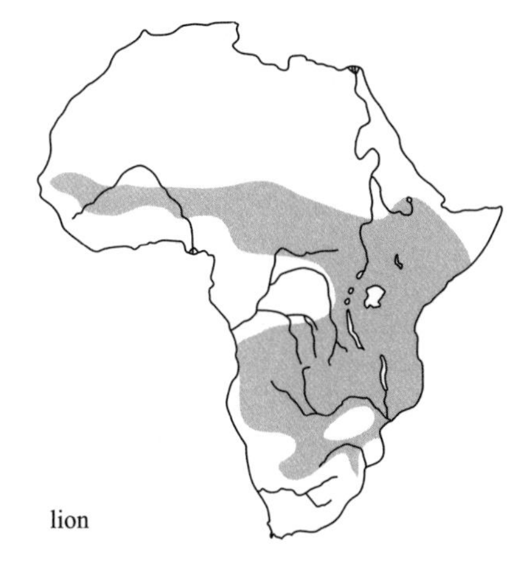

lion

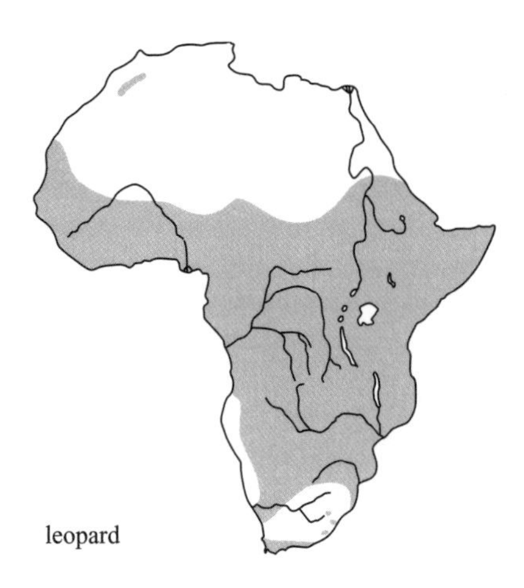

The **leopard** (*Panthera pardus*) (shoulder height: 70–80 cm, weight: ♂ 20–91 kg, ♀ 17–60 kg) is the quintessential cat. The most catlike of all cats. It has the widest distribution of all felines, and has been found in virtually all African habitats, from desert to savannah, and from rain forest to high mountains (a carcass was found on the rim of Mt Kilimanjaro's Kibo Crater in 1926). It also takes a huge variety of prey, from dung beetles to adult male eland. Bailey (1993) found that at least 92 prey species have been documented in the leopard's diet from sub-Saharan Africa. It escapes kleptoparasitism (stealing of its prey) by stashing it up a tree. It is solitary and territorial. Adults associate only long enough to mate, and males defend territories by calling and demarcating with urine, dung and tree-scratching. One to four cubs (500 g) are born, at any time of year, after a gestation period of 3⅓ months. The leopard is not currently at risk in sub-Saharan Africa, although the Barbary leopard (Morocco) is *critically endangered*.

The **cheetah** (*Acinonyx jubatus*) (Figure 3.8d) (shoulder height: 80 cm, weight: 30–72 kg) is the fastest land animal, attaining speeds of 95 km/h (60 mph). Because a final, fast, sprint is an integral part of its hunting technique, it prefers to live in relatively open savannah, with some cover. It hunts by day and takes mainly medium sized antelopes such as Thomson's gazelle, puku, springbok and impala, although it will also eat smaller prey such as hares and birds. It escapes some kleptoparasitism by avoiding areas with lions. Females are usually solitary, while males can form coalitions, or also be solitary. Where their main prey is migratory, they can have very large home ranges. For example, in the Serengeti, (main prey Thomson's gazelle) these averaged about 1,000 km^2. In other areas they may be only 12–36 km^2. One to five cubs (250 to 300 g) are born, at any time of year, after a gestation period of about three months. For the first six weeks they are usually hidden in dense cover. For a large

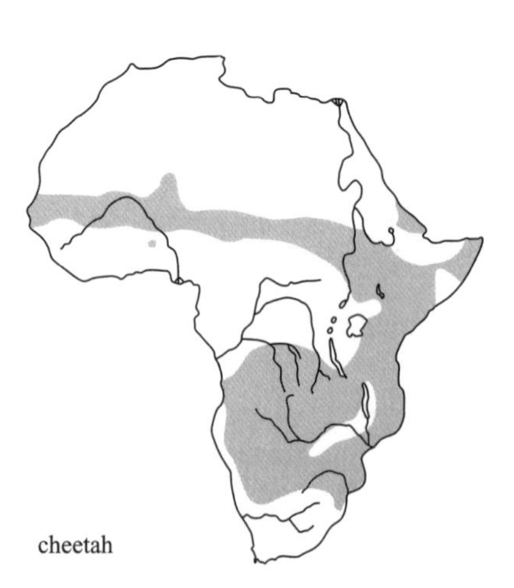

carnivore, cheetahs have a high mortality rate. In the Serengeti, 95 per cent of young cheetahs never reach independence, largely due to lions killing the cubs. It is *endangered* throughout its range. Between 5,000 and 15,000 may remain in sub-Saharan Africa.

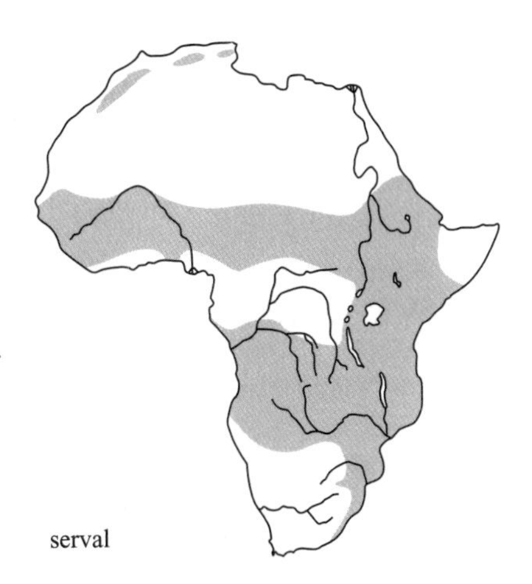
serval

There are seven species of smaller cat found in Africa although only five are sub-Saharan, and only four are found in savannahs. The **black-footed cat** (*Felis nigripes*) is found in South Africa, Botswana and Namibia where it inhabits semi-arid scrub and grassland savannah. The **African wild cat** (*Felis silvestris*) looking like our domestic cat, is found everywhere in sub-Saharan Africa, except the rain forests of the Congo Basin. The **serval** (*Felis* (*Leptailurus*) *serval*) (Fig. 3.8e) (shoulder height: 60 cm, weight: 8–13 kg) is found in all types of savannah, especially with long grass. The serval locates prey in tall grass mainly by hearing and takes a variety of small rodents, hares, birds, reptiles, invertebrates, and probably the young of small antelopes. The final pounce is a high leap, which may span 1–4 metres and be over one metre high. Servals are solitary and territorial, although they may well fall into the 'spatial groups' category mentioned in the carnivore introduction. In other words, they share an area but avoid contact. In a study carried out in the Ngorongoro Crater of northern Tanzania, up to seven servals occupied an area of 7.5 by 4 km over a 2½ year period (Geertsema 1981). One to three, rarely five, young (200 g) are produced, mainly during the wet season, after a gestation period of 2½ months. It is common over most of its range and not listed as endangered by IUCN. They appear very tolerant of agricultural development which encourages increased rodent densities. The **caracal** (*Felis* (*Caracal*) *caracal*) (shoulder height: 40–5 cm, weight: 7–19 kg) is the largest of the small cats. It inhabits all types of savannah, although to some extent the serval and caracal replace each other. Servals tend to be found in wetter habitats, caracals in more arid. They take a wide range of prey including rock hyraxes, reedbuck,

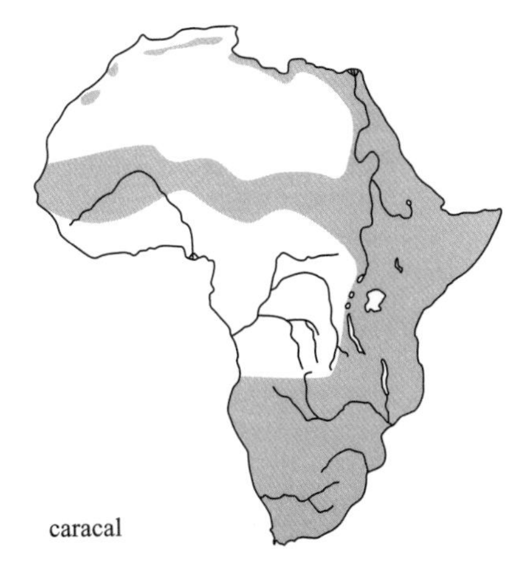
caracal

springbok, steenbok, hares, rabbits, small rodents and birds. It uses the same killing technique as the big cats—a suffocating bite to the neck. Like the serval they are solitary and territorial. One to four young (250 g) are produced after a gestation period of 2½ months. The conservation status of the caracal is satisfactory in sub-Saharan Africa. It is most abundant in South Africa and Namibia.

Hyaenas (Family Hyaenidae)

Hyaenas are dog-like carnivores which are however, more closely related to cats. There are four species, in four genera. Three species are 'typical' hyaenas, while the fourth, the aardwolf, is rather different. The **aardwolf** (*Proteles cristatus*) (shoulder height: 50 cm, weight: 6–11 kg) eats mainly harvester termites and has reduced, peg-like, teeth, rather than the massive cheek teeth of the three hyaenas that are used for crushing bone. It prefers relatively open savannahs. It is a nocturnal, solitary forager and probably monogamous. Evidence from observations of scent marking would suggest that male and female aardwolves share a 1–2 km^2 territory together with their most recent offspring. Between one and four pups (500 g) are produced after a gestation period of 2 months. Its conservation status is *lower risk* although data is poor for the northeastern populations. The **brown hyaena** (*Parahyaena brunnea*) (shoulder height: 80 cm, weight: 45 kg) is the dominant large carnivore in the south west arid savannahs and even penetrates the Namib Desert. It is a nocturnal scavenger, killing only 6 per cent of its food. It will eat almost anything and over 58 different kinds of food have been identified in its droppings. Since most of the large mammals that inhabit the arid savannahs of the southwest are migratory, the composition of the diet of brown hyaenas changes with the season. In the dry season the amount of large mammal material drops to 17 per cent and fruit and vegetables increase. They cache surplus food. Brown hyaenas live in clans of 1–4 adults, 0–5 subadults and 0–4 cubs, but forage alone. Clan members mark and defend territories (Mills 1990). Two to three cubs are born after a gestation period of three months, in a birthing den, separate from the communal den of the clan. After about three months they are brought to the clan den where they are reared communally. Its status is *lower risk*, although poisoning, trapping and hunting have had a detrimental effect on populations. Estimated total population is between 5,000 and 8,000 (Mills and Hofer 1998). The **striped hyaena** (*Hyaena hyaena*)

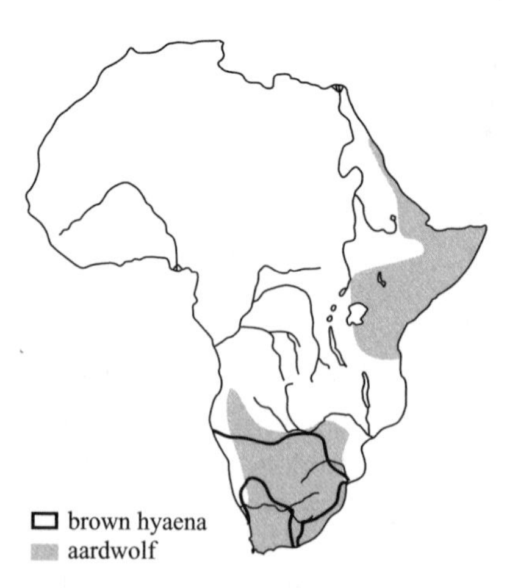

(shoulder height: 72 cm, weight: 40–55 kg) has a northern African distribution, spreading across the Middle East into India. There are at least five subspecies. It inhabits arid scrub savannah. Its diet is mainly mammalian carrion, with some invertebrates, eggs, wild fruit, and organic human waste. They probably have a similar clan structure to that of brown hyaenas, although they are relatively unstudied. They hunt alone, at night. Two to four cubs are born, at any time of year, after a gestation period of three months. Its conservation status is *lower risk*. The estimated African population is between 3,500 and 6,500, with the greater part being in Egypt and Kenya (about 60 per cent) (Mills and Hofer 1998).

The large, dog-like, **spotted hyaena** (*Crocuta crocuta*) (Figure 3.8f and Figure 3.9) (shoulder height: 85 cm, weight: 60–80 kg), is a predator, rather than a scavenger. In fact, it is one of the few predators that may influence the numbers of its prey. It is found in semi-desert, arid scrub savannah, woodland savannah, and even mountainous forest up to 4,000 m. It usually hunts in small groups, running down its prey in a prolonged chase. These prey include small, medium, and large antelopes, buffalo, zebra, warthog and the young of giraffe. However, the composition of the diet varies depending upon the area. In Kruger N.P. the most important prey are wildebeest, buffalo, zebra, greater kudu and impala. In the southern Kalahari they take gemsbok, wildebeest and springbok and in the Masai Mara, 80 per cent of the prey are topi and Thomson's gazelle. Unlike most carnivores that waste up to 40 per cent of their kills, the spotted hyaena consumes almost everything, including the bones. It is highly social, living in clans of up to 80 individuals, with a strict dominance hierarchy. Females have high levels of testosterone, and are dominant over males. Territory size is variable from less than 40 km^2 in the Ngorongoro Crater to over 1,000 km^2 in the Kalahari. One to two cubs (1.5 kg) are born after a gestation period of about 3½ months. Its conservation status is *lower risk*, and estimated numbers are between 27,800 and 48,200 (Mills and Hofer 1998).

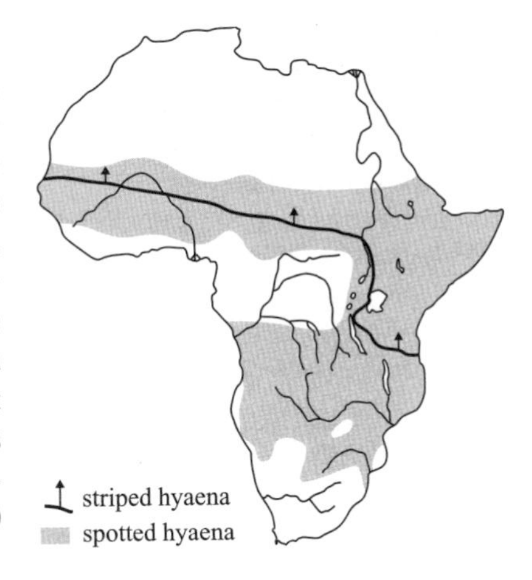

Genets, civets, and mongooses (Family Viverridae and Herpestidae)

These are small to medium-sized carnivores, with an elongated body and face, a pointed muzzle, short legs (apart from the African civet), and generally a long and furry tail. They are thought to be related to the cats and hyaenas. Between 10 and 12 species of African **genets** (shoulder height: 18–25 cm, weight: 1.2–3.5 kg) are recognized, depending on the author (Duff and Lawson 2004; Macdonald 2001; Stuart and Stuart 1997). They all look rather similar and most people would have some difficulty separating them in the field. Only three species could be said to use savannah type habitats, the rest are forest species. These are the **common**, or **small-spotted, genet** (*Genetta genetta*) (including the feline genet, *G. felina*), the **large-spotted genet** (*G. tigrina*) (including the forest genet, *G. maculata*) and the **Hausa**, or **Villier's, genet** (*G. thierryi*). The last species is restricted to the drier savannahs of West Africa, while the small-spotted and large-spotted occur over most of sub-Saharan Africa. They eat insects, small rodents, reptiles, birds, and fruit. They are nocturnal foragers. Between two and five young (50–80 g) are born after a gestation period of about two months. None are listed as endangered. Two species of civets can be found in savannahs. The **African civet** (*Civettictis civetta*) (shoulder height: 40 cm, weight: 9–15 kg) and the **African palm civet** (*Nandinia binotata*) (shoulder height: 22 cm, weight: 1.5–3 kg). Both prefer forested savannah and forage at night. Their diet is similar to that of genets. They produce two to four young after a gestation period of about two months. Both are common and not endangered.

At least 22 species of mongoose live in Africa, of which fourteen frequent savannahs. Of these, only nine are common (Table 3.5), although the Egyptian mongoose is uncommon but very widely distributed. All nine are diurnal foragers except the white-tailed mongoose, which is nocturnal, and the Egyptian mongoose which can be both nocturnal and diurnal. All mongooses depend on scent for communication and to mark territories. None of the mongooses that inhabit savannahs is listed as endangered.

Table 3.5 Common African savannah mongooses. Social structure: s = solitary, f = family groups and c = colonial. Distribution: sub-Saharan implies everywhere except rainforest of Congo Basin and horn = the Horn of Africa.

Name	Scientific name	Weight (kg)	Main diet	Behaviour	Distribution
Egyptian	*Herpestes ichneumon*	2.4–4.0	Small rodents, invertebrates and fruit	s and f	North, sub-Saharan, and Nile valley
slender	*H. sanguineus*	0.5–0.7	Invertebrates, small vertebrates and fruit	s	Sub-Saharan
banded	*M. mungo*	1.0–2.0	Insects, birds, reptiles, rodents	c	Sub-Saharan
white-tailed	*Ichneumia albicauda*	3.0–4.0	Invertebrates, rodents and fruit	s and f	Sub-Saharan
dwarf	*Helogale parvula*	0.2–0.4	Insects, small reptiles and birds	c	East and central
desert dwarf	*H. hirtula*	0.2–0.4	Insects, small reptiles and birds	c	Horn
yellow	*Cyntis penicillata*	0.3–0.8	Invertebrates, vertebrates and carrion	c	South
cape grey	*H. pulverulenta*	0.5–1.0	Small rodents and invertebrates	s and f	South
meerkat	*Suricata suricata*	0.6–0.9	Insects, and other invertebrates	c	South

Dogs and jackals (Family Canidae)

Dogs are more closely related to bears, seals, and mustelids than to the cats and their allies mentioned above. There are eleven species ranging in size from the fennec fox (1.0–1.5 kg) to the African wild dog (17–36 kg). Four of these are not savannah species. The red fox (*Vulpes vulpes*) is confined to the northern coast and Nile basin, the fennec fox (*V. zerda*) and Ruppell's fox (*V. ruppelli*) are found in the Sahara desert and the Ethiopian wolf (*Canis simensis*) is confined to the alpine moorlands of the Ethiopian Highlands. The other seven species are found in savannah habitats.

The **wild dog** (*Lycaon pictus*) (Fig. 3.8g and Fig. 3.9) (shoulder height: 65–80 cm, weight: 17–36 kg) was once found throughout the savannahs of sub-Saharan Africa but its numbers are now greatly reduced and its distribution very fragmented. It will inhabit a range of habitats from semi-desert to miombo woodland savannah. It hunts a wide range of medium-sized ungulates, which vary with the region. In Kruger N.P. it takes mainly impala, in the Kalahari it takes springbok and in the Serengeti ecosystem its main prey are Thomson's gazelle and wildebeest. Wild dogs have the same evil folk-lore associated with them as do wolves in Europe, and they have been hunted systematically by humans. They are diurnal, cooperative hunters. Packs consist of a breeding pair and several nonbreeding adults (often twenty or more dogs) that assist in provisioning the lactating mother and pups. Young males, when they mature, stay with the natal pack while females emigrate. Hunting ranges are often huge, with estimates in the Serengeti of 1500–2000 km^2. Two to 19 pups are born after a gestation period of 2 to 2½ months. The young are born in dens, often the abandoned burrows of other animals, and remain close to this den for about three months. Woodroffe *et al.* (1997) estimate that there are between 3,000 and 5,500 wild dogs, in perhaps 600–1,000 packs, remaining in Africa. The major causes of their historic decline are persecution by humans, habitat fragmentation, loss of kills (kleptoparasitism) and road deaths. Disease from domestic dogs is another major threat. They are listed as *endangered.*

The three savannah jackals are all rather similar in size (shoulder height: 30–48 cm, weight: 6–15 kg). The **black-backed jackal** (*Canis mesomeles*) (Fig. 3.8h and Fig. 3.9) is the most commonly seen jackal of East and southern Africa. The **side-striped** (*C. adustus*) is rather uncommon while the **golden** (*C. aureus*) is the most desert adapted and restricted to Africa north of the side-striped.

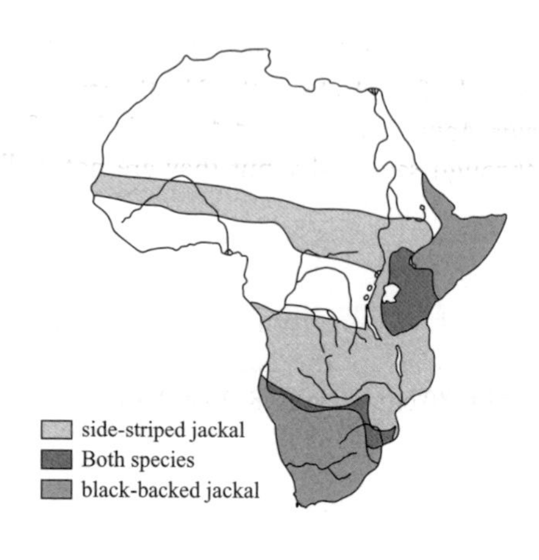

They have a wide tolerance of habitats from quite arid semi-desert to more open wooded savannah. They are opportunistic feeders and take a wide variety of food items, from small mammals, to carrion and fruit. The black-backed is perhaps more carnivorous taking young antelopes, rodents, birds and reptiles. They are both nocturnal and diurnal. All three species are thought to have similar social structure, although the side-striped is largely unstudied. They are monogamous and territorial, with yearling offspring helping to rear the next set of young. Three to six pups (occasionally one to nine) are born after a gestation period of 2 to 2½ months. None are listed as endangered.

Of the three savannah foxes, the **Cape fox** (*Vulpes chama*) is confined to the grassland and arid scrub savannahs of southern Africa while the **pale fox** (*V. pallida*) is found only in the dry Sahel, from the Atlantic to the Red Sea, bordering the Sahara Desert. The **bat-eared fox** (*Otocyon megalotis*) (shoulder height: 30–40 cm, weight: 3–5 kg) is very much a savannah species being found in both southern and eastern areas. It likes open grassland and light acacia woodland. Its food is mainly insects, particularly the harvester termite (*Hodotermes mossambicus*). It will occasionally eat small vertebrates and fruit. In a sample of 72 bat-eared foxes from Botswana, 88 per cent of stomachs contained mainly insects. The most common food items were termites, followed by beetles, grasshoppers, scorpions, rodents, and reptiles. They live as monogamous, nonterritorial pairs, sometimes with female helpers that are daughters. One to six pups are produced after a gestation period of 2 to 2½ months. They are not endangered.

Badgers, weasels, polecats, and otters (Family Mustelidae)

These are small to medium-sized carnivores, with plantigrade feet. That is they walk flat-footed like elephants, hyraxes, bears and man. They include the weasels, badgers and otters. There are ten African species, three of which are the Eurasian otter (*Lutra lutra*), weasel (*Mustela nivalis*) and polecat (*M. putorius*) that are only found along the North African coast. Three otters (*Lutra maculicollis, Aonyx capensis*, and *A. congica*) are found in rivers that sometimes run through savannahs, but they are not really savannah animals and the North African banded weasel (*Poecilictis libyca*) is largely confined to the Sahara Desert and North African coast. Of the three truly sub-Saharan savannah species the African striped weasel (*P. libyca*) is restricted to parts of central and southern Africa, and is uncommon. The two common savannah species are the **honey badger** or **ratel** (*Mellivora capensis*) (shoulder height: 30 cm, weight: 8–14 kg) and the **striped polecat** or **zorilla** (*Ictonyx striatus*) (shoulder height: 10–15 cm, weight: 0.6–1.4 kg). Both have an extensive range from the Sahel, through dry scrub savannah to Miombo wooded savannah. The honey badger eats a wide range of items, including invertebrates, rodents, reptiles, birds, carrion, and wild fruit. It will break into bee colonies and eat the honey and larvae, a behaviour that gives it one of its common names and its scientific name of *Mellivora*. They are mainly nocturnal and crepuscular. It appears to have a mutualistic relationship with a bird, the greater honey guide (*Indicator indicator*), which it follows to bee hives (Chapter 5). Its social behaviour is largely unstudied, but it is probably monogamous. One to four young are produced after a gestation period of about six months. It is not endangered. The zorilla eats mainly insects, but takes some rodents and other small mammals. It is nocturnal and solitary. Two to three young (15 g) are produced after a gestation period of 36 days. It is not endangered.

Primates

This order of mammals, to which we belong, has approximately 59 African species, although whether some of these are species or subspecies is debated. However, in general they are not inhabitants of savannah ecosystems, except perhaps for the forest savannah mosaics that surround the Congolean basin of central Africa. They range in size from the gorilla (*Gorilla gorilla*) (♂ 140–180 kg, ♀ 70–100 kg) to the Demidoff's galago (*Galagoides demidoff*) (60 g). Primate characteristics include, a rather generalized, highly mobile, skeleton, five fingers and toes equipped with flat nails instead of claws, a large collar bone, eyes placed frontally giving binocular vision, and in many species a large and complex brain. The nails protect sensitive finger tips, and in most species one digit (e.g. the thumb or big toe) is opposable, enabling the hand and foot to grasp objects. Clearly many of these primate features are associated with life in trees, grasping branches and judging distances. Two species, commonly seen in savannahs, are detailed below. A third species, the **patas monkey** (*Erythrocebus patas*), inhabits the dry savannahs of West Africa, from Senegal eastward to Ethiopia, but is less frequently seen. It is thought to be closely related to the vervet monkey and is probably the

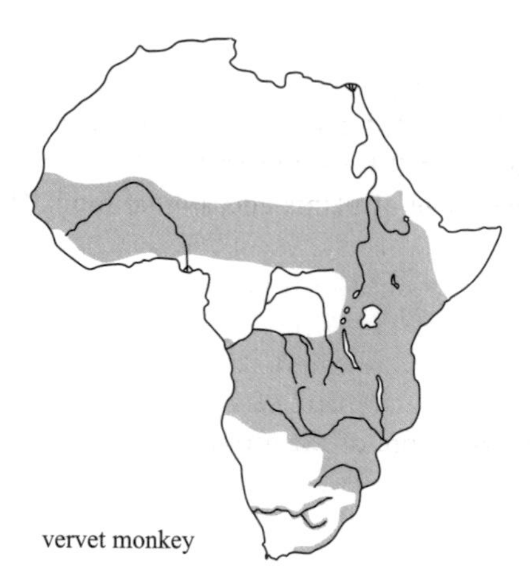

most terrestrial of primates, apart from the baboons. It feeds on *Acacia* fruit, galls, leaves, insects, and tree gum.

The **vervet**, **savannah**, or **green monkey** (*Chlorocebus aethiops*) (weight: ♂ 4.3 kg, ♀ 3 kg) is common and widespread in sub-Saharan Africa. Several geographical races have been described, which some people give specific status. It is by far the most numerous and widespread monkey in Africa, and in many urban and agricultural areas is regarded as a pest. It is found in all types of savannah, but usually never far from water and tall trees, in which it spends the night. Vervets are opportunistic omnivores and take a wide range of fruits, flowers, leaves, gum, seeds, and even insects. It will forage on the ground, turning over small logs to search for beetles and other invertebrates. They live in troops with a social structure rather similar to that of baboons. Long-term, successive, studies in Amboseli N.P. suggest these troops can vary in size from eight to fifty individuals, with anything from one to eight males. The troops are in fact a hierarchy of families whose members sleep, forage, and rest together. In Amboseli, troops lived in a mosaic of stable territories that varied in size from about 18 to 76 ha. A single young (300–400 g) is born after a gestation period of about seven months. Vervet monkeys are not endangered.

The **savannah baboon** (*Papio hamadryas*) (Fig. 3.9) (shoulder height: 40–60 cm, weight: ♂ 16.9–25.1 kg, ♀ 9.9–13.3 kg) occurs in a number of distinct races that have always been given their own name. In the lowlands of East and Central Africa is the yellow baboon (*P. h. cyanocephalus*) with, as its name suggests, yellowish fur. In the highlands of East Africa, with olive-greenish fur, is the olive baboon (*P. h. anubis*). The southern African form has dark grey fur and is known as the chacma baboon (*P. h. ursinus*), and finally the form found on the extreme West Coast is known as the Guinea baboon (*P. h. papio*) and has brown fur. The hamadryas baboon (*P. h. hamadryas*), which used to be classed as a separate

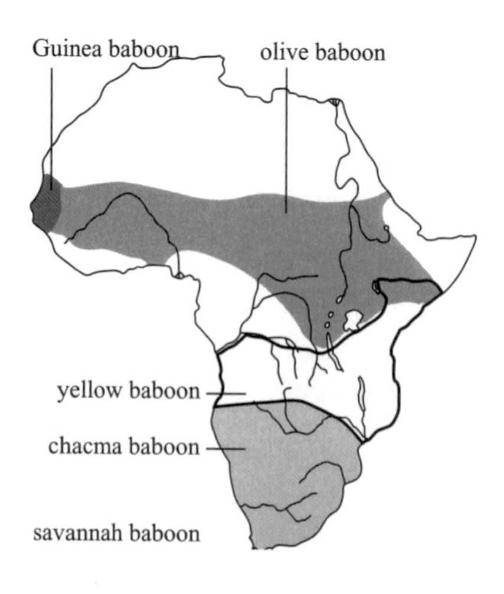

species and is confined to the rocky hill country of Eritrea, Ethiopia and Sudan, is now classed as a subspecies of the savannah baboon. They are found in most savannahs, limited only by the availability of water and secure sleeping sites, such as tall trees and cliffs. Like the vervet they are omnivores taking a wide range of plants and animal food. These include flowers, seeds, fruits, resin, leaves, roots, bulbs, invertebrates, and even young antelopes, rodents, birds and reptiles. In agricultural areas they will raid crops. Baboons live in groups (troops), usually of between thirty and fourty individuals, although the observed range is between eight and 200 individuals. Adult females tend to remain their whole lives in one troop and therefore make up the stable core of the group (comprising about 30 per cent of individuals) with immature classes making up about 50 per cent. Males emigrate during adolescence (after about age four) and often transfer repeatedly, between troops, even when mature. There is a strict hierarchy between adult females, with the young taking on the rank of their mother. A lower ranking adult female will therefore defer to the infant of a higher ranking female. Home ranges vary between 400 ha and 4000 ha, dependent on the quality of the habitat, and not the size of the troop. A single young is produced after a gestation period of about six months. Its IUCN classification is *lower risk*.

Rodents

Of all the mammal orders, rodents contain the largest number of species world-wide. In Africa, they include the ground squirrels (five species), the rope squirrels (nine species), the bush squirrels (ten species), the giant and sun squirrels (nine species), the spring hare, the gundis (four species), the dormice (two species), the blesmols (five species), the porcupines (three species), the cane-rats (three species), jerboas (three species), and the rat-like murids (1,330 species). One of the main characteristics of rodents is their continuously growing, gnawing incisors, and chewing molars, separated by a gap (the diastema) left by the absence of the canine and premolar teeth. They consume a great variety of plant material and have a relatively large caecum containing bacteria that facilitate the digestion of cellulose. Many, of course, have a prodigious capacity for reproduction. Most species are pregnant for just 19–21 days, mate again within two days of giving birth, and the young can begin breeding when six weeks old. Theoretically, a single breeding pair of mice could produce 500 mice in 21 weeks. Although many of these rodents, such as the ground and bush squirrels, porcupines, cane-rats, and the murid rats and mice are abundant in savannahs, there are simply too many species to pick out any for individual description. However, their combined effect upon savannah ecosystems can be noticable. They are both consumed and consumers. Many of the smaller carnivores, described above, take rodents as part of their diet. In the Serengeti study mentioned earlier, Sinclair (1975) (Table 3.1) estimated that rodents consumed about 69 kg/ha/yr in the long grasslands, about 4 kg/ha/yr in the short grasslands and about 259 kg/ha/yr in the kopjes. While this is much less than that consumed by the total grasshopper and termite populations, and considerably less than that consumed by the total mammalian herbivore population, it is none-the-less significant.

4 Single species populations

The next three chapters will look at the numbers of savannah organisms, mainly large mammals, at the level of 'one species', 'two species', and 'many species'. This division is obviously artificial because if you are trying to determine what influences the numbers of, say wildebeest, one of the factors is grass and you immediately have a 'two species' situation. Another major influence would be the disease organism rinderpest, and buffalo may compete for food, and so on. However this artificial division is not just convenient, and traditional in ecology, but necessary. The reason is simply one of detail. If you are looking at a specific animal population, in order to try to understand its numerical changes, and possibly predict future numbers, you need to know many details. You need good estimates of population size, for several consecutive years. This, in itself, will be very time consuming. You need details of birth rates and death rates, and how they change over time. You need to know what agents are responsible for these rate changes, such as food shortage, disease, and predation. The final description may well involve a detailed mathematical model, or computer simulation, of the population. Only a few single species populations, of savannah animals, have received this amount of detailed attention. For single species populations we can therefore talk about population estimation, birth rates, death rates, and such things as k-factor analysis, and I do this in this chapter. Once you have two interacting species the amount of detail, and work, doubles and there are even fewer detailed studies available. At this level of detail, ecologists talk about types of interactions between species, such as competition, predation, and mutualism. Many of the descriptions, or models, of these interactions are more general, with less specific detail, and less predictive power. Ecologists look more for evidence of these interactions, rather than attempting detailed description, and speculate on their past or future effects. This approach to savannah organisms is dealt with in Chapter 5. With more than two species, the approach is frequently even less analytical and more descriptive. Ecologists now talk about species richness (numbers), species diversity (numbers plus frequency), body size-abundance relationships, and species–area relationships, and try to formulate 'assembly rules' (Diamond 1975; Prins and Olff 1998). All are an attempt to reduce the complexity of analysing many species, and to search for repeated natural patterns that will make sense of nature. This I do, for savannahs ecosystems, in Chapter 6.

This chapter, on savannah populations, is divided into four sections: how population size is estimated; what regulates populations, using two detailed studies; what regulates populations, using two population models; and a survey of the possible regulatory factors for other species.

Estimating numbers

The aim of this section is to give readers an appreciation of some of the common methods that have been used to estimate animal and plant numbers in savannah ecosystems. Many of these numbers will inform our discussions in this and later chapters, and it is useful to have an appreciation of the effort required and the accuracy of the estimates obtained. An appreciation of the techniques is the aim of this section, not a detailed introduction that would allow the reader to estimate population size. For a more detailed coverage of these techniques, the reader should look at Norton-Griffiths (1987); Wilson *et al.* (1996); and Buckland *et al.* (2001).

Population estimates are conventionally divided into absolute population estimates (e.g. 10 lion km^2) or relative population estimates (e.g. more buffalo in that area than in this area). Absolute estimates may be shown as either population density (numbers per area), or population size (total numbers in an area). Since population size, divided by area, equals population density, it may seem that the two measures are essentially the same. In most cases they are. However, with some conservation issues it is important to appreciate that they are not always equal. For example in Figure 4.1, if each circle represents a single individual (animal or plant) and its spatial

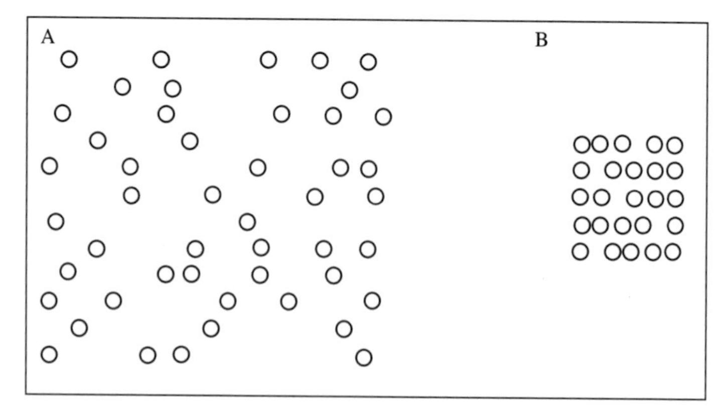

Fig. 4.1 Population size versus population density. A dilemma in measurement?

position, then a curious dilemma may occur. In A, the population size is 48 and in B the population size is 25, but A occupies a much larger area than B. The population size is greater in A than B, but density is clearly greater in B than A. If A and B were two species, which is the one more in need of conservation measures? Which is rarer?

In most cases the population sizes, or densities, obtained are estimates. It is very rarely that all animals or plants can be counted. Of course if individuals can be recognized repeatedly, when observed on subsequent occasions, then very good population data can be obtained. With static organisms, like trees for example, this usually involves simply tagging the individual tree with a marker and recording its position with a Global Positioning System (GPS). Birkett (2002) carried out this type of census on individual *Acacia drepanolobium* trees, in Sweetwater Game Reserve, in the Laikipia area of Kenya. In this particular study it was the growth rate of the trees, and how this was affected by browsing, that was of interest, not the population size or density of whistling thorn. Trees were selected to give a height-stratified sample. Random trees were chosen, within all the available height categories. Some trees were in an area protected from browsers (elephant, giraffe, and black rhinoceros) and some were not. Birkett found that in the protected area, the 78 selected trees had a mean annual growth rate of 19.1 ± 2.1 cm, while the 879 selected trees in the unprotected areas had a mean annual growth rate of only 7.5 ± 0.5 cm.

With animals, that are mobile, identifying individuals can be more difficult although not impossible. Schaller (1972), in his study of Serengeti lions, found that he could recognize about 60 lions by their individual natural markings such as a missing ear, tail or other deformity. However, many more lions were marked by placing coloured metal tags in their ears. In a study of cheetahs, Caro (1994) used the pattern of spots to identify individual cheetahs. He took black and white photographs of each side of the face and tail, which he kept in his vehicle, so that animals could be matched in the field. In Mole N.P., in northern Ghana, bushbuck were uniquely identified using a combination of their spot and line body markings (Dankwa-Wiredu and Euler 2002). Some animals, such as zebra, have the equivalent of bar-codes on their body and these have been used to identify individuals (Briand Peterson 1972). We have developed a similar coded system for identifying reticulated giraffe in Laikipia, Kenya (Shorrocks and Croft 2006) (Fig. 4.2). The 'neck pattern' is converted to a 'neck code', written for each individual. The neck pattern that is coded is the section that runs along the back of the neck, adjacent to the mane. The yellow lines, of the giraffe's pattern, that reach the mane are coded according to the angle they make with the mane, starting at the head end. Three types of line are recognized, right-angled lines (R), acute-angled lines (A) and obtuse-angled lines (O). Obviously, each giraffe has two codes, one for the right-hand side of the neck, the other for the left. The right-hand code for the giraffe in Figure 4.2 is ROAARAAROO. With three types of line (R, A and O) and ten

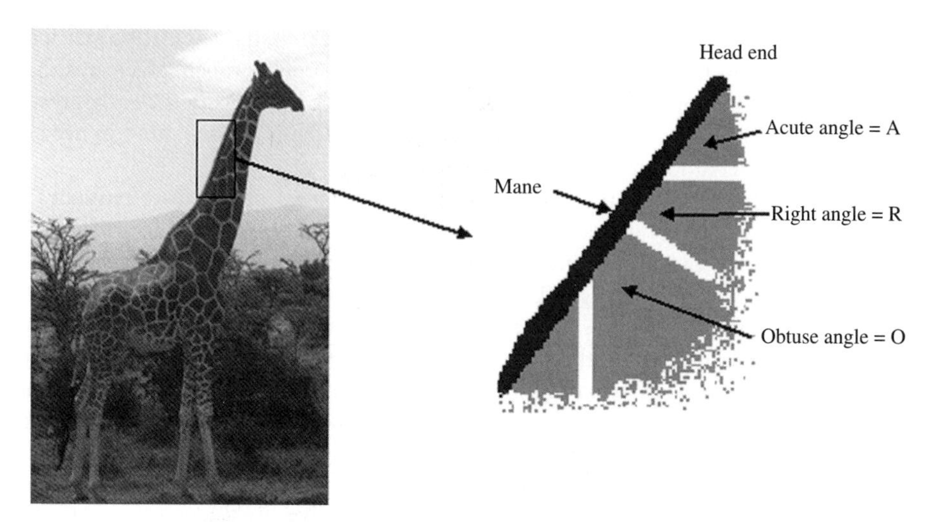

Fig. 4.2 Photograph of right-hand side of a reticulated giraffe showing neck pattern, and diagram illustrating the three types of line recorded. The giraffe in the photograph has the code ROAARAAROO (from Shorrocks and Croft 2006).

positions on the neck there are 3^{10} permutations. This equals 59049. With both sides of the neck observed it would be 3^{20} permutations or 3486784401. The advantage of this type of linear code is that it can be accumulated in a computer database, and new sightings checked against the stored codes, each day. This computerized searching is far more efficient than looking through photographs and allows hundreds of individuals to be recognized.

When natural markings are not available, individuals can be artificially marked. Putting numbered metal rings around legs is a very traditional way of marking birds, and the metal tags used by Schaller on lions, or by Birkett on *Acacia drepanolobium*, already mentioned, are similar techniques. Of course, a disadvantage of animals, over plants, is that they have to be caught before they can be marked. Birds are frequently caught in special nets, or ringed in the nest, small mammals are caught in live-traps, and large mammals are usually shot with an anaesthetic dart. One way of permanently marking large mammals (marks must be permanent otherwise you think there are more animals than there really are) is to cut notches in ears. In Nairobi N.P., the Kenyan Wildlife Service (Torchio and Manconi 2004) used a coded system of six notches (Fig. 4.3) which allowed them to mark up to 65 black rhinoceros, uniquely. Kruuk (1972), studying spotted hyaenas in the Ngorongoro Crater, Tanzania, used a similar system of ear notches to recognize up to 51 hyaenas. The KWS black rhino work was also involved in translocating surplus rhinos from Nairobi N.P., to a rhino sanctuary in northern Laikipia.

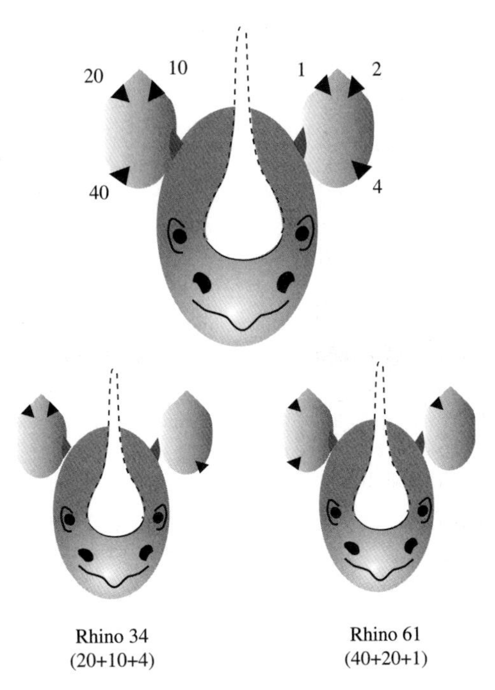

Fig. 4.3 Top: diagram of black rhinoceros head showing the position of of the six notches, and their assigned number, used by the Kenyan Wildlife Service. Bottom: two diagrams showing notch combinations used in two individuals (after Torchio and Manconi 2004).

These translocated animals were marked using a more hi-tech, but much more expensive, method that also allowed them to be easily located once released. After a rhino had been anaesthetized, using a dart, the tip of the horn was sawn off. A broad cavity was drilled into the base of the horn and a narrow hole drilled from the top of the horn into this basal cavity. A small radio transmitter (8 cm × 8 cm × 2 cm) is placed into this basal cavity, with its antenna projecting up through the narrow vertical hole. The openings in the horn are then sealed with resin. As a wholly insensitive appendage, a rhino's horn is the perfect receptacle for a sophisticated location-transmitting device. The embedded transmitter is pre-programmed to emit a unique signal, at specified time intervals, that can be picked up within a radius of seven kilometres. More usually, radio-transmitters are attached to collars which fit around the animals neck. Although older radio-collars were quite big, and only used on large mammals, modern radio-collars can be quite small. In Comoé N.P., Ivory Coast, kob antelopes were fitted with nylon radio-collars (total weight about 200 g) and located twice a day by triangulation (Fischer and Linsenmair 2001). In Queen Elizabeth N.P., Uganda, Hoffmann and Klingel (2001) were able to fit 2.5 g radio-collars to rodents (*Lemniscomys striatus*) that comprised only 5–7 per cent of their body weight, and tiny radio ear-tags have been used successfully on bushbuck (Dankwa-Wiredu and Euler 2002). The most hi-tech development is the attachment of satellite transmitters to radio-collars. In north-western Namibia, Lindeque and Lindeque (1991) attached 12 kg collars, containing both traditional VHF transmitters and UHF satellite transmitters, to seven elephants. A single individual within a family unit was collared so that in effect seven groups of elephants were tracked. Home ranges were found to vary between 2,302 km^2 and 10,738 km^2.

When animals are marked, released and recaptured (or otherwise re-examined) at a later date, an estimation of their population size/density can be made. If, (1) marked and unmarked animals behave in the same way (marks do not affect capture) and (2), the population is closed (no input or loss), then the ratio of **marked individuals to the total population size** is equal to the ratio of **marked recaptures to the total size of a second sample.** Let the unknown population size = P, the size of sample one = n_1, the size of sample two = n_2, and the number marked in sample two = m, then the equality stated above can be written

$$\frac{P}{n_1} = \frac{n_2}{m}$$

This is known as the Lincoln–Peterson estimate (Seber 1982). The main problem of this method of population estimation involves the second condition above; that the population remain closed between the two sampling events. This closure includes not only migration of individuals, but birth and death. One solution to this would be to take the second sample immediately after release of the first, marked, sample. However, this would inevitably mean that condition one was violated, because the marked individuals would be more likely to be recaptured. Because of this the two sampling events (marking and recapturing) are usually at least two consecutive days apart. To overcome the problem of input and output, you have to estimate the rate of gain or rate of loss. If either of these can be done, then a suitable adjustment can be made to provide a correct estimate of population size. To do this, at least one more sampling of the population must be done, and individuals caught on different days must be marked in a distinctive way. However, this is only a 'day' mark and individuals need not be identifiable. Details of these multiple mark/release/recapture methods will not be given here but they can be found in Seber (1982). Intriguingly, one of the first people to use these multiple recapture methods was Jackson (1933) who employed it for estimating populations of tsetse flies in East Africa. Most uses of this technique in savannah systems have employed the simple Lincoln-Peterson estimate, and the results must therefore be viewed with some caution. Kruuk (1972) marked 51 spotted hyaenas in the relatively small Ngorongoro Crater, northern Tanzania. By subsequently resighting these marked individuals (recaptures) he estimated that there were about 385 adult hyaena in the Crater. In the much larger area of the Serengeti ecosystem he marked 200 hyaenas, but obtained a much less precise estimate of about 3,000. In a more recent study, Thompson (1989) estimated the population of lovebirds (*Agapornis* sp.) living around Lake Naivasha, in Kenya. Mist-nets were used to capture the birds, which were marked with a leg ring. Because the marking of the lovebirds was quite time consuming, the two sampling occasions were quite separated, the first (n_1) between August and December 1985, and the second (n_2) between February and June 1986. Numbers caught, and ringed, on the first occasion (n_1) = 412, while n_2 = 255 and m = 15. This gave a population estimate of 7,004 birds.

Another method of estimating population size that uses marked, or naturally identifiable, animals is the 'removal' method. Animals are not, in practice, removed (because they would be replaced by new animals moving into the area), but simply identified. On each successive sampling occasion the number of 'new' unmarked (unknown) animals will decrease until all the animals have been marked. For example, Wittemyer (2001) observed elephants in the Samburu and Buffalo Springs National Reserves, Kenya. Each elephant within the study area was identified using sex, age and features unique to the individual. Photographs and drawings of these features were used to compile an identification dossier. Over a period of 21 consecutive months (from November 1997) population numbers were recorded (Fig. 4.4a). New individuals were recorded each month until the last three months when the numbers of identified elephants plateaus at 744—the estimated population of elephants using these two reserves. Of course once again, the method only works if the population is confined to the area in question. Otherwise you are estimating the population of an unknown area. Additionally, it is not usually necessary to continue observing the population until every animal has been recorded. Since the number of 'new' animals drops progressively, it is possible to extrapolate to 'no new' animals. Figure 4.4b shows this extrapolation, using only the trend from the 2nd to the 11th months data and a simple regression analysis. It predicts a total

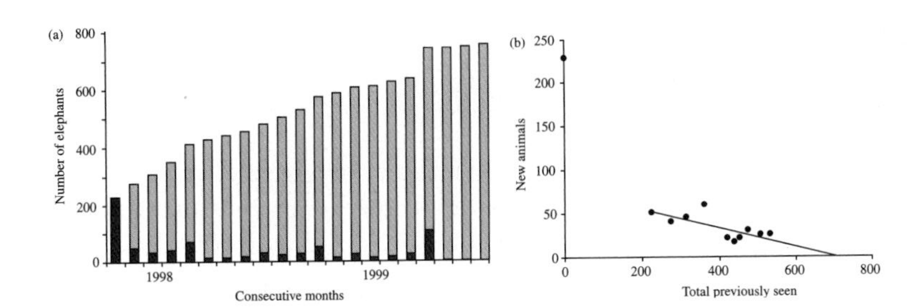

Fig. 4.4 Removal sampling of elephants in Samburu National Reserve in Kenya. (a) original data: black = newly identified, grey = previous total (from Wittemyer 2001). (b) population estimation using the regression method on data points 2 to 11.

population of about 740 elephants, not very different from the figure obtained after a further years counting.

Most population surveys in savannah habitats involve some type of transect. A line is travelled at a constant pace, by the observer, and all animals seen are counted. Sometimes a straight line transect is walked on foot, as was done by Dangerfield (1990) estimating termite (*Cubitermes sankurensis*) numbers in a miombo woodland savannah in Zimbabwe. More frequently a road transect is driven in a vehicle. This is often more convenient, and safer. The fact that the road transect is not a straight line doesn't really matter. The transect method is the same. In effect, a strip either side of the road is sampled. If the sample strip is a random sample of the study area, and if all animals within the strip are seen, then the estimation of density, or numbers, is relatively easy. For example, imagine that observers in a game reserve drive along its network of roads (20 km) observing all buffalo within 100 m either side of the road. They see 32 buffalo. The density (D) of the population would be $D = n/(2aL)$, where n = the number of animals (32), a is half the strip width (0.1 km), and L is the length of the transect (20 km). This would equal $D = 32/(0.2 \times 20) = 8.0$ buffalo per km^2. In practice, it is never so simple. The width of the strip will rarely be constant. For example, if the transect runs through dense vegetation the observed strip will be narrower. If the transect runs through open grass savannah, the observed strip might be wider. One answer to this problem is to record the sighting distance to each animal, and use this to estimate an average value for a. Population density is then estimated by

$$D = \frac{n}{2L}\left(\frac{1}{n}\sum\frac{1}{r_i}\right)$$

where n and L are as before, and r_i is the observed distance to each sighted animal i. For example, table 4.1 shows part of the data from a transect of white-eared kob (*Kobus kob leucotis*) in the Sudan. The transect length was 10 km.

The estimated density is therefore

$$D = \frac{12}{2(10)}\left(\frac{1}{12}\left[\frac{1}{0.15} + \frac{1}{0.20} + \frac{1}{0.16} + \cdots + \frac{1}{0.20}\right]\right) = 3.80 \text{ kob per km}^2$$

Although this estimate is one way to allow for variable strip width, it still assumes that all animals within the strip are seen. It assumes what is called a rectangular detection function (Fig. 4.5, line a). In practice the

Table 4.1 Sighting distances for white-eared kob, from a line transect in the Sudan.

Animal number	1	2	3	4	5	6	7	8	9	10	11	12
Sighting distance r_i	150	200	160	200	250	130	150	130	200	100	140	200

Modified from Krebs 1999.

probability of detection falls off the further an animal is from the line transect baseline. Line b in Figure 4.5 is the simplest such relationship. With this detection function, only half the animals within your transect strip would be seen, and in effect a is half the value you think it is. The constant a is therefore more than just the strip width. It estimates how wide the strip would be if every organism was seen and none were missed, and there are numerous ways of estimating it (Buckland *et al.* 2001). The shaded area in Figure 4.5 encloses the general zone for detection functions for wildlife populations. Eltringham *et al.* (1998) used a road transect (393 km) to estimate numbers of large mammals in the Mkomazi Game Reserve (2,850 km^2), northern Tanzania. Four surveys, each lasting three to four days, were undertaken during the 1996 dry season when the animals were more easily seen because of the sparse vegetation. They were able to estimate the numbers shown in Table 4.2. They should be regarded as minimum estimates.

The type of transect most associated with savannah is probably the aerial survey (Norton-Griffiths 1987). Rather than walking or driving the sampled strip, it is flown over, in a light aircraft, and the animals counted from the air. Sometimes, particularly if the animals are in groups, they are photographed and counted later. The technique was pioneered by Bernhard and Michael Grzimek (1960), who used it in the Serengeti ecosystem in the late 1950s. The flying method remains essentially unchanged today, although the statistical analysis surrounding it has become something of an industry. Only aircraft with high wings give an unobstructed downwards view. An observer sits in the back of the aircraft and, using markers placed on the window and wing struts to delimit a transect strip on the ground, counts the animals visible.

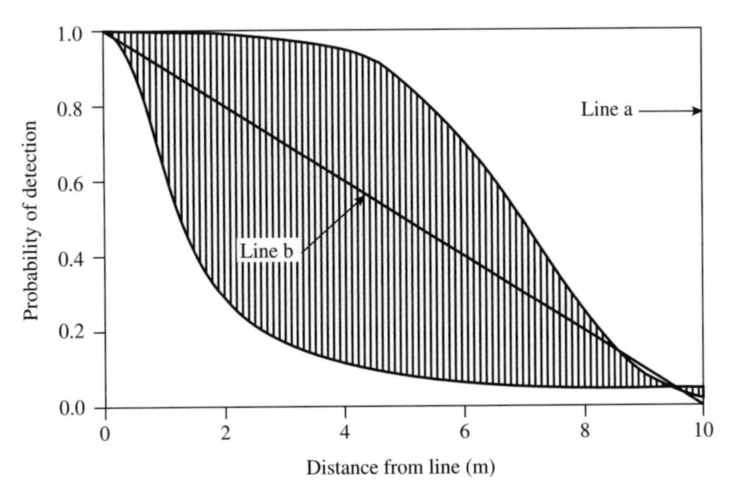

Fig. 4.5 Detection functions of a line transect survey (modified from Krebs 1999).

Table 4.2 Minimum estimates of the numbers of some large mammals in Mkomazi Game Reserve in the 1996 dry season.

Species	Numbers
eland	473
elephant	314
gerenuk	933
giraffe	979
Grant's gazelle	306
impala	3,564
dik dik	55,978
kongoni	840
lesser kudu	5,739
steinbuck	554
warthog	1,460
common zebra	1,438

From Eltringham *et al.* (1994).

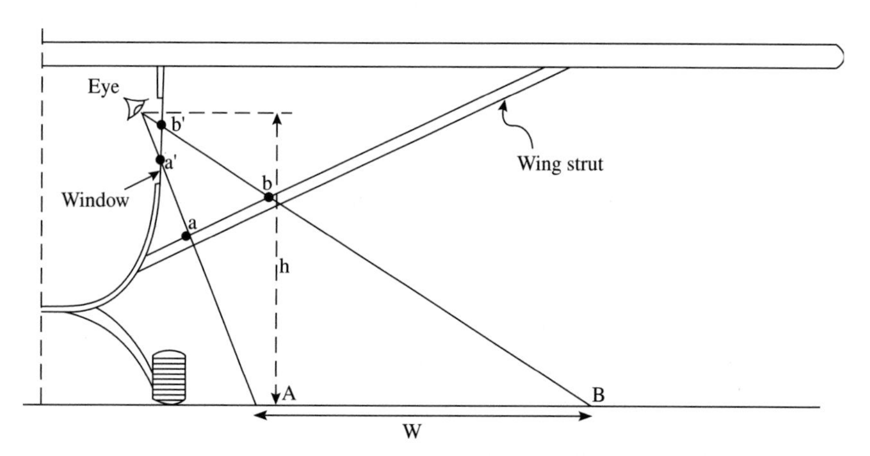

Fig. 4.6 Diagram of plane showing how the positions of window and strut markers are fixed while the aircraft is on the ground (after Norton-Griffiths 1978).

Figure 4.6 illustrates how these markers are positioned while the aircraft is on the ground. If W is the required strip width during the survey and H is the flying height chosen, then w (the strip width on the ground) is obtained from $w = W \times h/H$. This width, w, can then be marked on the hanger floor and the position of a′ and b′ (on the window) and a and b (on the wing strut) can be adjusted until they are in line with the margins of this width (A and B). The pilot simply has to fly at height H for the observer to 'see' a transect of the required width on the ground below. In East Africa such aerial surveys normally travel at 160 km/h, at a height of 100 m, giving a transect width of 150 m. In open terrain, and/or conspicuous animals,

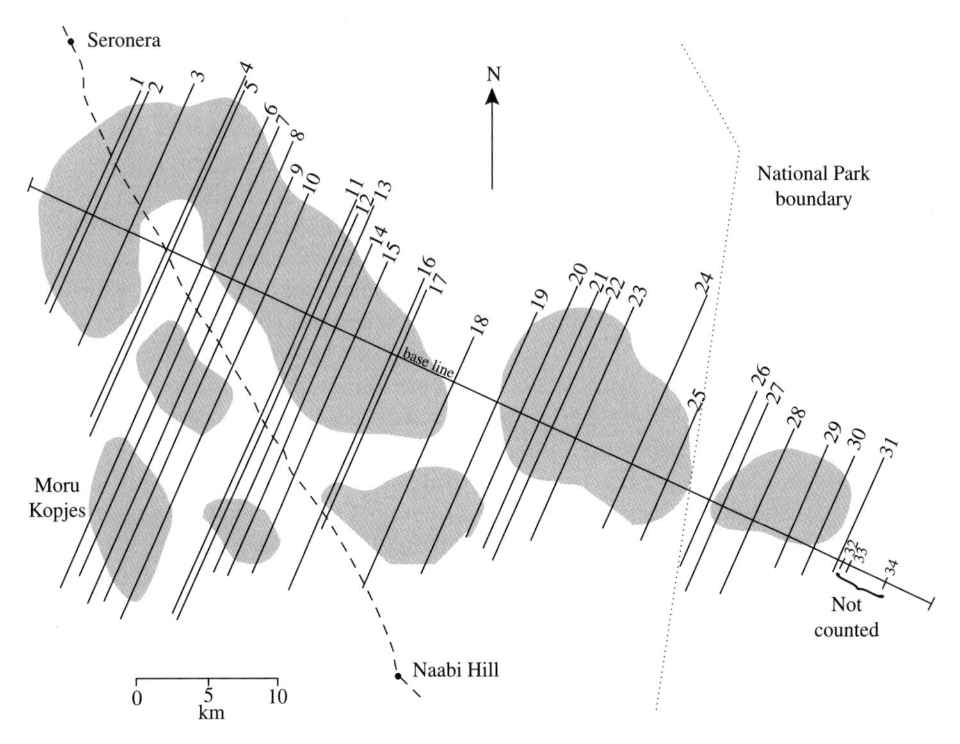

Fig. 4.7 Orientation of the base-line and of the random transects in the May 1971 sample count of the migrating Serengeti wildebeest (from Norton-Griffiths 1973).

the height can be greater and in more wooded terrain, and/or less visible animals, the height can be lower. Two observers can be employed, one on each side of the aircraft. Cameras can also be fitted to the underside of the aircraft and photographs taken of a transect strip directly below, rather than to each side. Of course, not all the savannah area being studied is usually flown over. This would require many hours of expensive flying time. The study area is sampled and the total population, or density, is estimated. This is a similar procedure to that employed in the road transect estimates above. A suitable baseline is usually fixed through the area to be sampled and random flight transects chosen at right-angles to this baseline. This procedure is well illustrated by the counting of the Serengeti migratory wildebeest, carried out by Norton-Griffiths (1973).

Figure 4.7 shows the migrating wildebeest herds (shaded areas), the survey baseline through these herds, and the randomly chosen transect flight paths (1 to 31) at right angles. These random flight transects were chosen before the flight, with the aid of a large-scale map, and in the event transects 32, 33, and 34 were not required because no herds were under these transects. In this particular example very large numbers of animals were involved and the observer operated a 35 mm camera, mounted on the door sill of the aircraft.

Along each transect, photographs were taken every ten seconds of flying time so that two levels of sampling were in fact employed. In all the 31 transects flown, the estimated number of wildebeest was 88,884, and the total population estimate (all the wildebeest in the shaded areas in Fig. 4.7) was 754,028 wildebeest, with 95 per cent confidence limits of ± 8.5 per cent. A slightly different type of aerial survey has been carried out in the Addo Elephant National Park, South Africa (Whitehouse, Hall-Martin, and Knight 2001). Annual helicopter based total counts have been conducted since 1978. Parallel strips are flown at a ground speed of approximately 80–120 kph, with strip width and flying height varying with vegetation and weather conditions. Unlike the light aircraft surveys, herds were circled to facilitate counting. Aerial surveys usually underestimate the populations on the ground (Caughley 1974). The proportion seen declines with flight speed and observer inexperience. In the Addo elephant count, above, it was estimated that about 7 per cent of adults were missed by the helicopter flights. However, for calves it was much higher, with about 48 per cent remaining undetected.

So far, estimating animal populations has involved counting animals. However, ecologists frequently count the signs left by animals rather than the animals themselves, such as footprints, feeding signs, and dung. In African savannahs, and particularly with elephants, counting dung is a frequently employed method of population estimation. An example is the survey of Jachmann (1991) carried out in the Nazinga Game Ranch, Burkina Faso, in western Africa. The Nazinga Game Ranch covers about 940 km^2. The vegetation is a tall-grass, tree and shrub, savannah with *Vitellaria paradoxa*, *Terminalia* spp., *Combretum* spp., *Acacia* spp. and *Detarium microcarpum* as the dominant trees. Jachmann estimated the elephant population by four methods, unfortunately not all in the same year: aerial counts of elephants (both total and sample), line transect foot surveys of elephants, vehicle road surveys of elephants and a dung count using quadrats. The 'dung' quadrats were positioned along the line transect used for the foot survey. Of course, simply counting the number of dung piles is not enough to estimate the elephants producing them. Jachmann had to estimate two other things: the rate of dung production and the rate of dung decay. Defecation rate was estimated by following elephants on foot for several hours. Dry season defecation rate was estimated at 14.1 droppings/elephant/day, while the wet-season rate was estimated at 27.2 droppings/elephant/day. By watching 31 droppings, over several weeks, the decomposition rate was estimated to be 0.59 per cent/day. Table 4.3 shows the estimates of population size obtained by the various methods. Using the total aerial survey as the best estimate, we might believe that the dung count gives a very good estimate of the elephant population at Nazinga.

In the next section we will look at what factors change the numbers of animals and plants, in savannah ecosystems. However, before doing that it is appropriate to pause and ask what these numbers, estimated by the techniques described above, can be used for. I think there are two main

Table 4.3 Elephant population estimates in the Nazinga Game Ranch, Burkina Faso.

Survey method	Year	Population estimate
Aerial (100% coverage)	1989	366
Aerial (6.1% coverage)	1988	610
Foot transect	1987	487
Foot transect	1988	306
Vehicle road transect	1988	293
Dung count	1987	353–396*

From Jachmann 1991.
* Two different assumptions about dung dynamics.

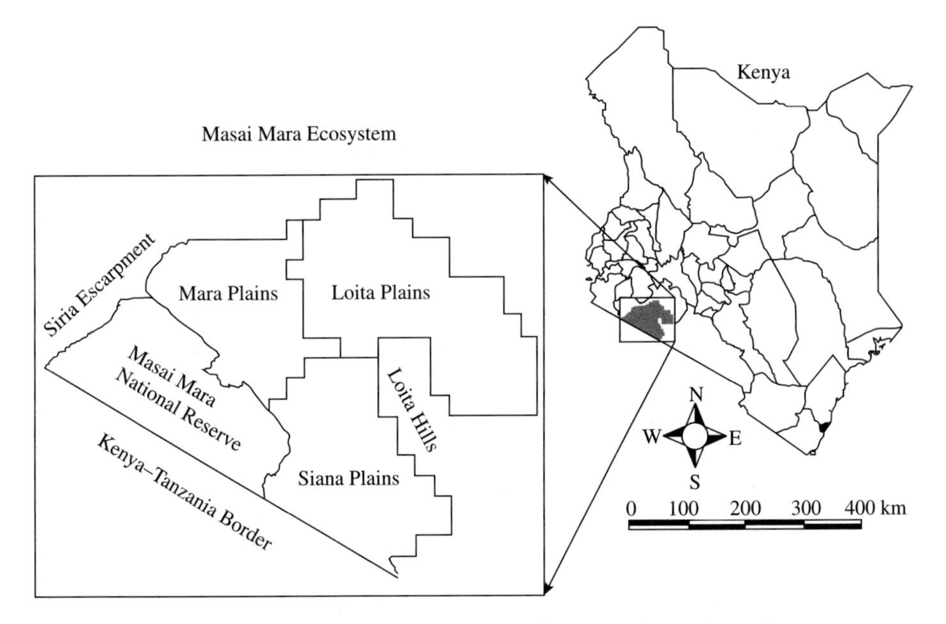

Fig. 4.8 Map showing the Masai Mara ecosystem (from Ottichilo *et al.* 2000).

uses. First, these numbers inform population models such as those described in the third section of this chapter. These models are an attempt to describe, and understand, the ecological processes that determine population numbers. Without the estimated numbers of real populations we would have nothing to compare our predictions to. Secondly, they alert ecologists, and conservationists, to changes in populations (trends) that we should know about. I will therefore end this first section, by describing a study that looked for such population trends, in savannah animals.

The Masai Mara ecosystem comprises the Masai Mara National Reserve and the adjoining group ranches (Fig. 4.8). The National Reserve is a formal conservation area owned by the Government of Kenya, while the group ranches are privately owned. The National Reserve and the Mara Plains are mainly *Themeda* grassland, the Loita Plains are mainly dwarf shrub and *Acacia drepanolobium* grassland and the Siana Plains are mainly *Croton* bush and other woody species interspersed with grassland. The main land uses are pastoralism, tourism and agriculture. Since 1977 the Department

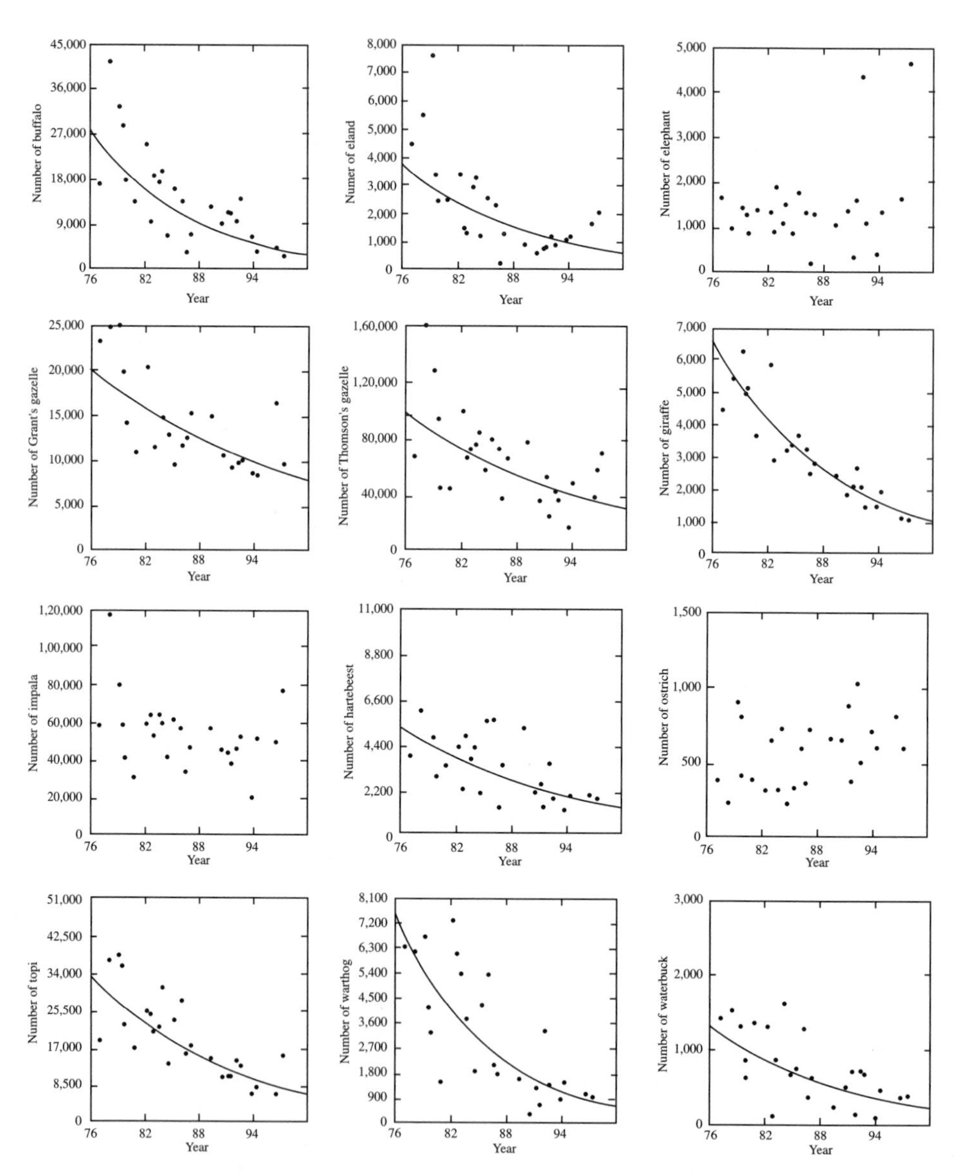

Fig. 4.9 Population trend for 12 wildlife species in the Masai Mara ecosystem between 1977 and 1997. Those graphs with trend lines are statistically significant (modified from Ottichilo *et al.* 2000).

of Resource Surveys and Remote Sensing have carried out systematic survey flights to estimate the wildlife populations. Ottichil *et al.* (2000) used census data spanning the years 1977 to 1997, to look for population trends in several species. These data are shown in Figure 4.9. Those graphs with log linear regression lines have significant trends ($P < 0.05$), all of them downwards.

Declines for individual species were: warthog 88 per cent, buffalo 82 per cent, giraffe 79 per cent, eland 76 per cent, waterbuck 76 per cent, topi 73 per cent, Cokes hartebeest 66 per cent, Thomson's gazelle 62 per cent, and Grant's gazelle 52 per cent. There were no significant downward trends in elephant, impala and ostrich. Livestock populations (cattle, sheep, goats, and donkeys), which were also examined, did not show any significant downward trend. The decline of most wildlife in the Masai Mara ecosystem, seen in Figure 4.9, could be due to a number of factors, acting individually or in combination. These include counting errors and biases, climatic effects, habitat changes, competition for forage, and poaching. Although difficult to assess, the authors conclude that counting errors and climate can be discounted. Counting errors tend to increase the variation around the population estimate rather than introduce a systematic effect, such as that observed here. Also the method of data collection did not change over the 20 years. There were severe droughts in this part of Africa during 1984, 1986 and 1993 which would have affected grass growth (Chapter 2) and potentially herbivore numbers, but any effect could not be detected in the data. A combination of the last three factors, which are known to be present, could be responsible. These possibilities will be revisited in later chapters.

What changes numbers?

The simple answer to this question is of course birth, death, immigration, and emigration. The movement of savannah animals can often cause dramatic local, and seasonal, changes in numbers. We will look at some examples of these animal movements in Chapter 5. However, population ecologists have often concentrated on birth and death, and I will examine these here. Birth and death are frequently age-specific, that is, the chance of an animal dying, or of a female giving birth, changes with their age. Such information can be summarized in a life-table, or their graphical counterpart (e.g. Figs. 4.10 and 4.12b). Unfortunately, with savannah animals, such detailed information is rarely available and the subsequent discussion of what influences population size is consequently less precise. For example, with mortality, the important distinction is often between what simply 'kills' and what 'regulates'. In the case of most savannah animals we only have lists of the former which, although very interesting, do not illuminate the causes of population regulation. I will therefore concentrate initially, on two studies where quite detailed information is available. I believe these

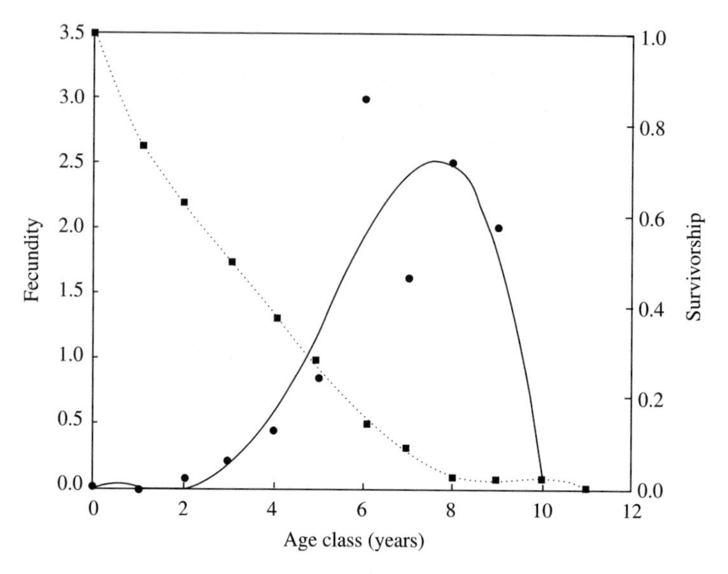

Fig. 4.10 Age-specific fecundity and survival in the African wild dog (*Lycaon pictus*) in the Selous Game Reserve, Tanzania (from Creel and Creel 2002).

build the solid framework for the less precise lists of mortality agents seen in the literature. These two studies involve African buffalo (*Syncerus caffer*) and wildebeest (*Connochaetes taurinus*), and both studies have been carried out by Tony Sinclair and his colleagues in the Serengeti ecosystem of East Africa. However, before looking at these two studies let me briefly return to this population problem of what kills and what regulates. The distinction is between 'density independent' mortality factors and 'density dependent' mortality factors. Both types of factor can change the numbers of animals in a population, but only density dependent factors will regulate numbers. With density independent mortality the probability of an individual dying remains the same irrespective of density. With density dependent mortality the probability of death changes with density, usually being higher in dense populations. Notice that this statement involves the probability, or chance, of dying, not the numbers dying. If ten buffalo are killed by lions in a population of 100 buffalo, and 100 are killed in a population of 1000, this is NOT an example of density dependence. For any individual, the chance of being eaten by a lion is 10 per cent, and the same in both populations.

The difference can be visualized by thinking of two different ways of changing the temperature of a room. If the room has a heating system controlled by a thermostat, this can be set at the required temperature. When the temperature of the room drops below this point the thermostat switches on the heating system, and the room temperature increases. When the temperature of the room rises above this point the thermostat switches off the heating system, and the room temperature decreases. The room temperature is regulated by this negative feedback mechanism, and maintains a more or less constant temperature. Alternatively, someone could periodically open a window to cool down the room if it got too hot (assuming the room is in a cool climate) but this would not be so precise, and if done without any consideration of the actual room temperature would produce a very erratic level of heating. Because many savannah populations show long-term stability, it is often assumed that they are regulated by density dependent factors.

African buffalo

Between 1965 and 1972, Sinclair (1974, 1977) studied buffalo populations in the Serengeti National Park, East Africa. During this period, census information suggests that buffalo were increasing in numbers, probably as a result of the removal of rinderpest (Chapter 5) (Fig. 4.11a). This, as we shall see later, was even more pronounced in the Serengeti wildebeest population (Fig. 4.15). There is also some suggestion that buffalo numbers started to level off in the early 1970s. What caused this? In chapter 2 we saw that there is a close relationship between annual rainfall and the biomass of green grass produced in East African savannah ecosystems (Figs. 2.1 and 2.11). In East Africa, for post rinderpest populations, there is a similar close relationship between buffalo density in an area, and mean annual rainfall (Fig. 4.11b). Variation about the regression line in Figure 4.11b is probably caused by the extent of permanent water in the dry season. This is particularly noticeable in the case of Lake Manyara N.P. which contains a number of permanently flowing springs as well as extensive swamps and lakeshore alkaline grasslands which increase the food supply to buffalo beyond that produced by rainfall. Figure 4.11b therefore suggests that buffalo are 'regulated', in most places, by rainfall acting through their grass food supply.

Sinclair measured recruitment as the number of yearlings (animals between one and two years old) per 100 adult females. During the period of study it fluctuated around an average of about 38 (32–43), suggesting that the observed increase in population was due to a change in mortality, rather than 'birth'. Mortality was, in adults, density dependent (the regression line in Fig. 4.12a) and age-specific (Fig. 4.12b). In the latter case, notice that it is juveniles and 'old' buffalo that have higher mortality rates. It is also revealing that the pattern of mortality in adults is rather similar between the sexes, and between the two study areas. This suggests similar mortality factors even though these two areas are over 100 km apart.

As stated at the beginning of this section, changes in population numbers are due to birth, death, immigration, and emigration. Understanding population change is therefore a matter of understanding the difference between potential fecundity and actual birth, and the difference between numbers at consecutive times in the life history (mortality). Within any generation, this can be visualized (Fig. 4.13) as a series of consecutive, declining numbers (N), with the differences between them either representing a reduction in fecundity ($N_0 - N_1$), or a reduction in individuals after birth (mortality).

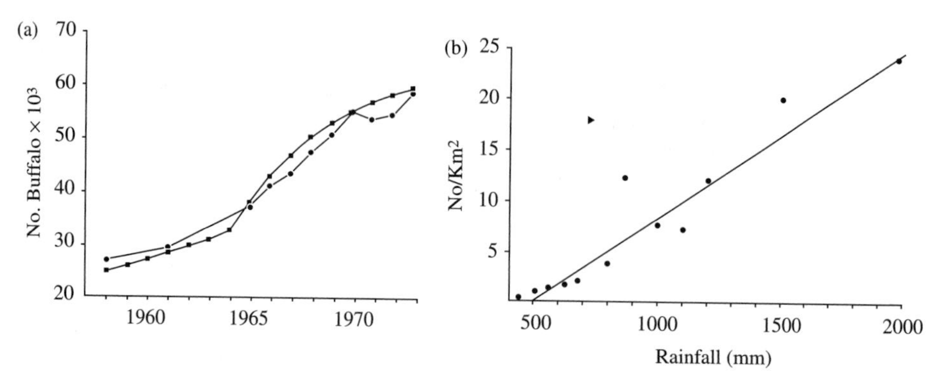

Fig. 4.11 Buffalo in East Africa. (a) (●) Population increase in the Serengeti between 1958 and 1973, (■) Population model of Serengeti buffalo using constant k_f and k_j, and density dependent k_a. Before 1964 an additional, constant, is added to juvenile mortality, representing mortality from rinderpest. (b) relationship between buffalo density and annual rainfall; regression lines excludes the Lake Manyara data point (▲) (both graphs modified from Sinclair 1977, with permission of Chicago University Press).

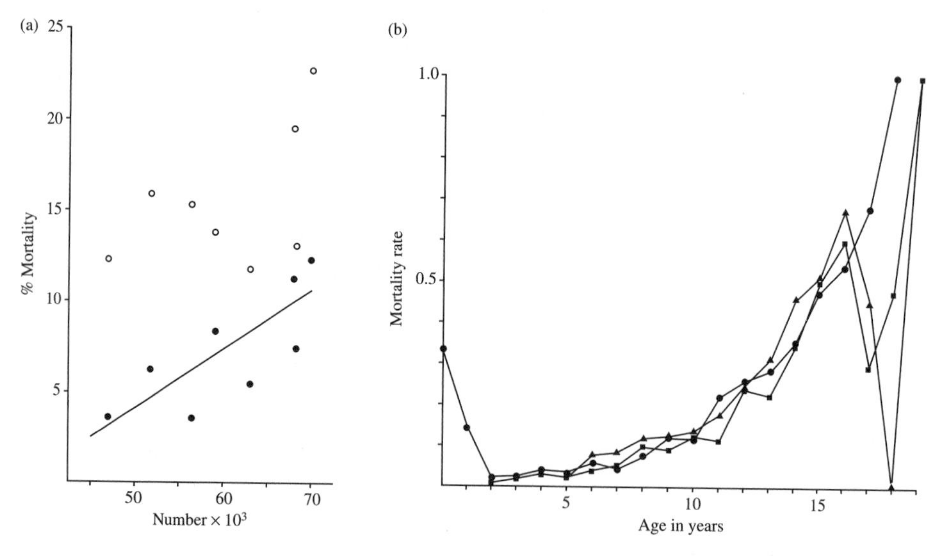

Fig. 4.12 Mortality in buffalo in the Serengeti. (a) mortality as a % of population size immediately preceding it in June. The regression line is for mortality of adults (closed circles). Open circles are juvenile mortality. (b) age specific mortality rate for female buffalo (circles), northern males (squares) and southern males (triangles). The northern study area was around Kogatendi, the southern study area around the Moru Kopjes (modified from Sinclair 1977, with permission of Chicago University Press).

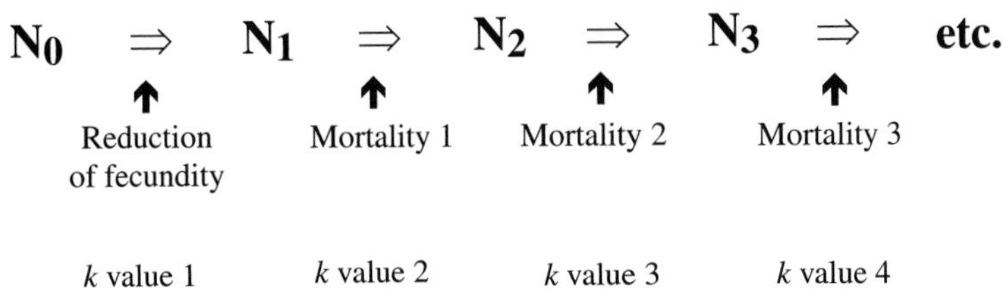

$N_0 \Rightarrow N_1 \Rightarrow N_2 \Rightarrow N_3 \Rightarrow$ etc.

Reduction of fecundity — Mortality 1 — Mortality 2 — Mortality 3

k value 1 — *k* value 2 — *k* value 3 — *k* value 4

Fig. 4.13 Schematic diagram showing the reduction in population numbers over one generation.

Reduction can be recorded in different ways. For example, mortality 1 could simply be written as the number of deaths. Simply using numbers has the advantage that total deaths can be simply obtained by summing m_1, m_2, m_3 etc. The disadvantage is that ten deaths in a population of 100 is more intense than ten deaths in 1000. Simple numbers are no guide to intensity, which is crucial in revealing the presence of density dependence. Mortality can also be represented by a mortality rate, which does reveal intensity (Fig. 4.12). This is simply the number dying out of the number available to die. Rates, however, cannot be simply added together to get an idea of the total mortality rate over the lifetime of the animal. A third way of representing mortality, or reduction, combines both the good qualities of these last two measurements. It is called 'killing power' (or *k* value) and is simply $\log_{10} N_x - \log_{10} N_{x-1}$ (Haldane 1949; Varley and Gradwell 1970). It can be summed to give total mortality and it is a rate (on a logarithmic scale). Notice *k* values can also be calculated if consecutive numbers change because of migration. In Sinclair's buffalo study, he was able to calculate *k* values for three reductions in population each year. These were a decrease in potential fecundity (birth) (k_f), mortality of calves under one year old (juvenile mortality)(k_j), and mortality of animals over one year old (adult mortality)(k_a).

The first of these was the difference between the number of calves born if all females reproduced and the observed number of pregnancies. By plotting these *k* values for each year, against log numbers before the reduction, it is possible to detect the presence of density dependence and therefore the potential causes of population regulation (Varley and Gradwell 1970). Sinclair (1974, 1977) did this for Serengeti buffalo (Fig. 4.14a). This analysis helps to focus the search for the causes of regulation. It does not, however, specifically identify the agents of reduction. For Serengeti buffalo, the only reduction that showed a significant positive slope with density was adult mortality, with a regression slope of $b = 0.2357$ ($p < 0.05$). The points on the juvenile mortality graph showed so much scatter that the positive slope of the regression line ($b = 0.1575$) was not statistically significant. The effect of this juvenile variation is transmitted to the graph of combined juvenile and adult mortality, which also has a non-significant regression slope. The regression line for loss of fertility ($b = 0.0449$) is very

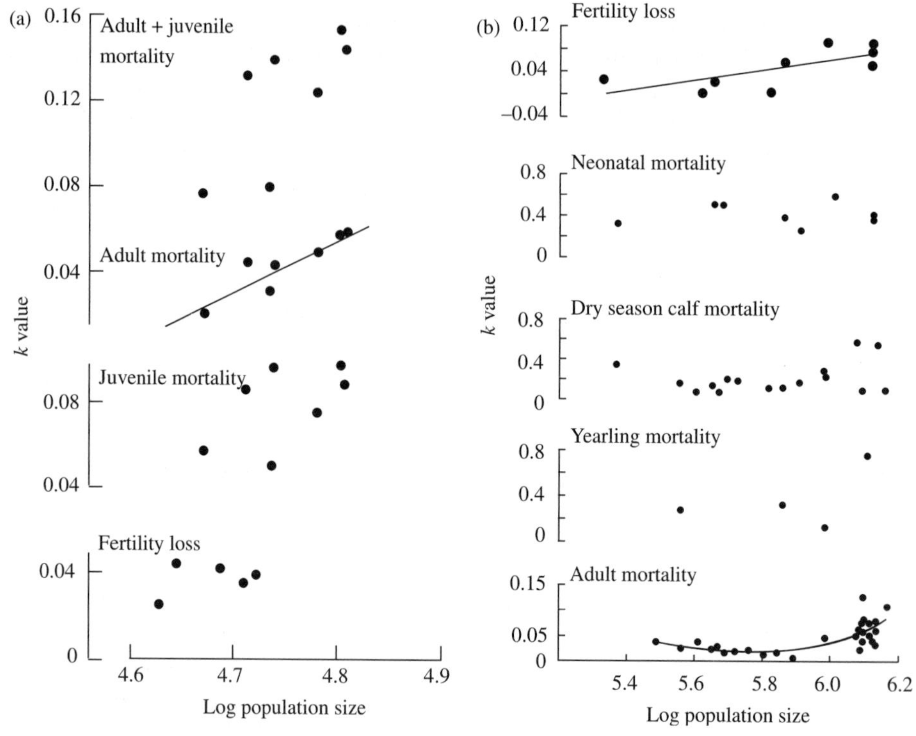

Fig. 4.14 *k* factor analysis for Serengeti buffalo (a), and wildebeest (b) (modified from Sinclair 1974 and Mduma, Sinclair, and Hilborn 1999).

close to zero, suggesting that this cause of reduction remains relatively constant with density. This analysis suggests, therefore, that adult mortality alone regulates Serengeti buffalo.

To examine this possibility further, Sinclair used the results of the *k* factor analysis to construct a simple, predictive, model of population growth. Starting with the estimated buffalo population from the census of 1965 he extrapolated population size, forward to 1976, and backwards to 1958. He used the density dependent regression line, in Figure 4.14a, for adult mortality, and simply used the average *k* for juvenile mortality and fertility loss. This *k* factor model gave a very good fit to the forward data predicting a buffalo population, in 1974, of 60,770. The estimated observed population was 61,134. Although there are population losses due to reduced fertility and juvenile mortality, it is the density dependent losses of adult mortality that regulates the buffalo population to a predicted equilibrium of around 63,500. Of course, young animals are very sensitive to a variety of causes of mortality and therefore, although juvenile mortality does not regulate the buffalo population, it does produce fluctuations in population size from year to year (average k_j is in fact greater than average k_a). A factor that reduces populations in this way is called a 'key factor', and is defined as the factor that contributes most to total population reduction (Royama 1996). In Serengeti buffalo, total reduction (K) $= k_f + k_j + k_a$. One way to identify a key factor is to plot successive values of K against each of the separate reductions, and when Sinclair did this, juvenile mortality (k_j) was the most important contributor to K. A simply way to express what is happening in these buffalo populations, from 1965 onwards, is to say that juvenile mortality causes fluctuations and that adult mortality compensates for these disturbances in a density dependent way.

The backward fit of the k factor model, to 1958, was not very good. The model predicted a much faster rate of increase than actually happened between 1958 and 1965. Sinclair surmised that this might be due to the occurrence of rinderpest in these early years. The data used in the k factor analysis, and the model, was collected in the years when rinderpest was no longer present in the Serengeti. We know that this introduced, exotic, disease caused repeated mortality in juvenile buffalo and wildebeest. There is no estimate of this mortality for buffalo but in wildebeest 1.73x more yearlings died when it was present. Sinclair extrapolated this wildebeest figure to buffalo and added it to the mean juvenile mortality. This new value of k_j was used in the k factor model, for the years before 1964. The new model's predictions are shown in Figure 4.11a. The new model fits the real census data quite well, suggesting that fluctuating, but sometimes quite high, juvenile mortality and density dependent adult mortality are the major determinants of buffalo population growth. But what causes this mortality, particularly the important density dependent adult mortality?

In Serengeti buffalo, as with most animals, the three main causes of mortality are predation, disease, and under nutrition. The major predators of buffalo in the Serengeti are lions (Schaller 1972), and this also appears the case for other areas such as Lake Manyara, Tanzania (Schaller 1972; Prins 1996), Kafue Park, Zambia (Mitchell, Skenton, and Uye 1965) and Kruger Park, South Africa (Pienar 1969). However, although the act of lion predation is spectacular, it is probably not an important source of mortality so far as population regulation is concerned. Schaller (1972) estimated that in the Serengeti, lions kill 2,468–2,961 buffalo each year. This represents only about 23–28 per cent of total adult mortality. Because of the high numbers of scavengers, calf mortality could not be estimated in the Serengeti. However, at Lake Manyara, where there are fewer scavengers, Sinclair (1977) estimated it at 11 per cent of calf kills. Bearing in mind that many predated buffalo are either old, diseased or undernourished it seems unlikely that predation is a major cause of mortality in the Serengeti.

What about parasites? About twenty trematode worms (flukes), six cestode (tape worms) and 28 species of nematode worm have been recorded from buffalo, along with three species of pentastomids (Sinclair 1977), although not all have been recorded from one area. Prevalence (per cent of the population infected) varies between both parasites and areas. For example, in Uganda one population of buffalo had 73 per cent of individuals infected by *Ashworthius lerouxi* (a nematode), while another population had zero prevalence (Woodford and Sachs 1973). In addition to these endoparasites there are several common ectoparasites. At least ten species of ticks have been found on buffalo in the Serengeti along with several species of biting flies, such as tsetse (*Glossina*) which is the vector for *Trypanosoma*. At least fifteen diseases have been reported for Serengeti buffalo including rinderpest, theileriases (a blood parasite that causes East Coast fever in cattle), babesiasis, anaplasmosis, and Allerton-type herpes virus (Sinclair 1974). Although these lists of parasites and diseases are interesting, their role in significant mortality is equivocal. The only exception to this statement is the exotic rinderpest. Most healthy individuals can acquire immunity to endemic diseases allowing them to 'fight' these parasites effectively. Young (in Sinclair 1974) found that in two captive animals, one in poor condition due to a lion injury, had a high infection of *Trypanosoma* while the other, in good condition, had a low infection. The major cause of poor condition in buffalo, and most other animals, is nutritional state and Scrimshaw, Taylor, and Jordon (1968) have reviewed the interaction between disease and nutrition. Their conclusions are clear. In a majority of cases, the nutritional state of the individual is crucial in determining death or survival from a disease. Sinclair (1974, 1977) therefore concluded that both predators and parasites, although causing the death of some individuals, were secondary causes of mortality. The main factor was under nutrition, as we suspected might be the case from Figure 4.11b.

Sinclair (1974, 1977) measured the percentages of crude protein in the diet of Serengeti buffalo, at different times of the year. The chemical analysis used the Kjeldahl method for nitrogen, multiplied by 6.25 to obtain the crude protein percentage. As the dry season progressed (July to October) the percentage of crude protein in the diet declined, for two reasons. The quality of the most nutritious component, grass leaf, changed from 12.7 per cent crude protein in March to 4.0 per cent in October, while the proportion of leaf (as opposed to the less nutritious stem) in the diet declined from 56 per cent to 13 per cent. Since the activity of the rumen's microflora is inhibited at crude protein levels below about 5 per cent (Chalmers 1961), this figure is approximately the minimum level of crude protein needed to maintain body weight. Below this level animals loose weight because they are using their own body reserves to compensate for the nutritional shortage. During the dry season, the diet of the Serengeti buffalo was below this minimum level. At the end of the dry season the buffalo's diet consisted of only 2.18 per cent crude protein. These trends were present in all age-groups and both sexes, suggesting a real shortage of good quality grass leaf, affecting all sections of the population. Sinclair also examined the 'condition' of some buffalo. When buffalo are undernourished, they use their body fat first and their bone-marrow fat second. Consequently, a decline in bone-marrow fat reflects a relatively severe state of undernourishment. Using this measure of condition, buffalo between two and ten years old were in good condition during the dry season while buffalo less than two years old, and greater than ten, showed depletion of bone-marrow fat during the dry season.

For the well-studied case of Serengeti buffalo therefore, we can conclude that the density independent juvenile mortality is mainly due to disease (rinderpest before 1964) and periodic undernourishment. The regulating density dependent mortality of adults is probably due to undernourishment in the dry season. This conclusion is supported by the *k* factor analysis (Fig. 4.14a), the *k* factor model (Fig. 4.11a), crude protein levels of grass, bone-marrow fat content, the relationship between rainfall and buffalo numbers (Fig. 4.11b) and the pattern of age-specific mortality (Fig. 4.12b). But what about other ungulates?

Wildebeest

After the removal of rinderpest, the population of Serengeti migratory wildebeest, like that of buffalo, increased dramatically (Fig. 4.15). There appear to have been three main phases of growth over the last forty years. Between 1960 and 1977 the population increased from about 0.25 to about 1.3 million. In the second phase, covering the sixteen years from1977 to 1992, the population entered an approximately stationary period with fluctuations between 1.1 and 1.4 million. Finally between 1993 and 1994 the population declined to 0.9 million following severe dry season mortality in the drought of 1993, and it has remained at that level. Figure 4.14b shows the results of the *k* factor analysis carried out by Mduma, Sinclair, and Hilborn (1999). The life stages for which they had data were fertility loss

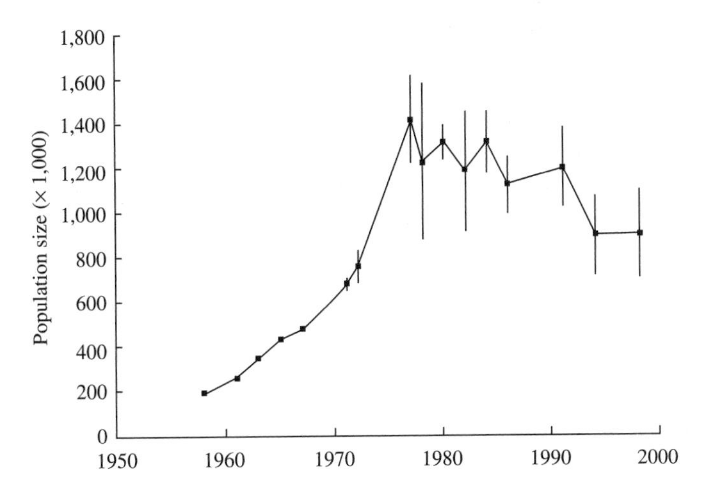

Fig. 4.15 Wildebeest population numbers in the Serengeti ecosystem from 1958 to 1998 (from Mduma, Sinclair, and Hilborn 1999).

(k_1), newborn (neonatal) calf mortality occurring in the wet season (March–June, ages 0–4 months)(k_2), dry season calf mortality (July–December, 5–10 months)(k_3), yearling mortality (ages 11–24 months)(k_4) and adult mortality (>24 months)(k_5). Calves, yearlings, and adults all tended to show higher mortality during the dry season.

The key factors were dry season calf mortality (k_3) and adult mortality (k_5). Significant density dependent mortality was found in adult mortality (k_5) and in fertility loss (k_1). Adult mortality showed a curvilinear density dependence. Between 1962 and 1972 (when the population was increasing), mortality was low and actually inversely density dependent ($b = -0.073, P = 0.0004$). From 1975 to 1994, when the population was stable or declining the linear slope was positive. In fact during this latter period there is clear evidence of delayed density dependence, since an even better relationship exists between k_5 and density the previous year ($b = 0.336, P = 0.086$). Just as with buffalo, wildebeest appeared to die from a number of causes, judged by examining carcasses. Predator deaths (26 per cent to 30 per cent), where attributed to lion (33.8 per cent), spotted hyaena (30.4 per cent), cheetah (3.3 per cent) and unidentified (33.3 per cent). Lions tended to kill middle-aged animals and hyaenas young and old. However, Mduma and his colleagues believe that 'predation played only a minor role in limiting the wildebeest population'. Only about 3 per cent of adult females died from predation so its effects on the number of births was small. The life stage that showed the most statistically significant, density dependent response was adult mortality, most of which occurred in the dry season. Adult mortality (k_5) was significantly, negatively, related to per capita food supply during the dry season ($b = -0.053$, $P = 0.0005$). What is more, per capita food supply during the dry season was also significantly, negatively, related to dry season calf mortality (k_3) ($b = -0.322$, $P = 0.002$) and not significantly related, but still showing a negative regression slope, to yearling mortality (k_4) ($b = -0.410, P = 0.063$), neonatal mortality (k_2) ($b = -0.017$, $P = 0.617$), and fertility loss (k_1) ($b = -0.100$, $P = 0.34$). The conclusion seems clear. For wildebeest, per capita food supply is crucial. Individuals die because they don't get enough food in the dry season and more importantly they die in a density dependent way. Predators kill wildebeest, but many of these individuals would have died from undernourishment anyway. If there were no predators it would simply take longer for the undernourished individuals to die.

The conclusion, therefore, from these two well-studied cases (buffalo and wildebeest) is that rainfall, acting through dry season food shortage is the most important factor affecting population regulation.

Population Models

In the previous section I looked at what regulates populations and affects their numbers. If the required detail is available, k factor analysis allows insight into these population processes. The analysis allows us to describe, in words, what we think might be happening—a verbal description, or verbal model. Another approach is to produce a mathematical description, or mathematical model, to help understand the processes affecting a population. These mathematical models are often more precise, and can be used to predict future numbers under various ecological scenarios. Also, despite what many field biologists may say, they are no more removed from reality than the verbal descriptions they replace. In fact we have already looked at one type of model when we examined buffalo populations—a k factor model. However, models can be built in different ways, and in this section I look at two quite different population models, one for the migrating wildebeest population of the Serengeti, and one for the plains zebra population of Laikipia, Kenya.

Wildebeest model

Mduma, Hilborn, and Sinclair (1998) used data, on the Serengeti wildebeest population, collected between 1961 and 1994. The data were (i) dry season rainfall measured from numerous stations throughout the Serengeti over the entire study (Fig. 2.10); (ii) 15 censuses of total wildebeest population size (Fig. 4.15); (iii) 24 estimates of yearling/adult ratio (Fig. 4.17b); (iv) 22 estimates of calf mortality (Fig. 4.16a); (v) eight estimates of dry season adult mortality (Fig. 4.16b). Pregnancy rate was estimated from autopsy, and analysis of hormone levels in faecal samples (Mduma 1996), and estimated to be 0.85, with very little variation between years. In any

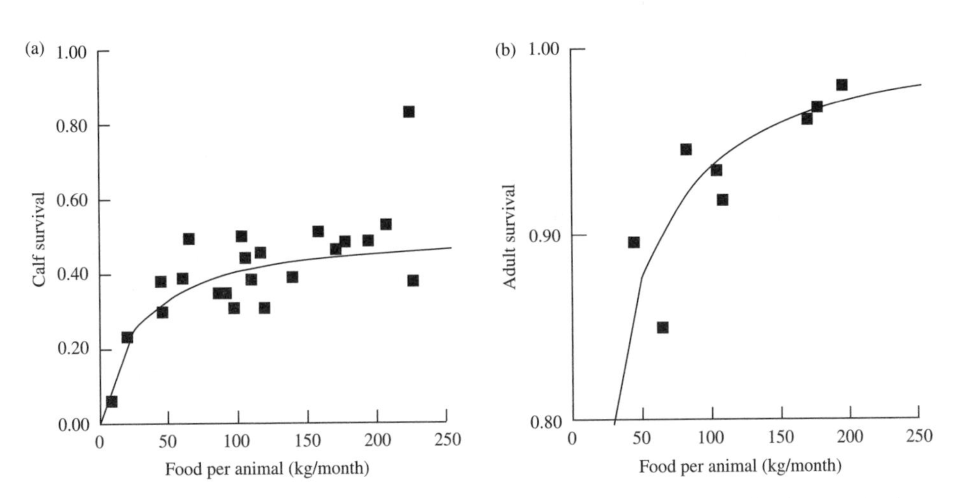

Fig. 4.16 Survival of (a) calves, and (b) adults plotted against estimated food per animal per month. The solid squares are the observed values and the lines are the rectangular hyperbolas (equation 3) fitted by the model. The shape of the fitted curves are different because the values of the equation constants, a and b, are different for calves and adults (from Mduma, Hilborn, and Sinclair 1998).

year, the relationship between rainfall and dry season grass production (Sinclair 1979) is described by the equation:

$$G = 1.25R \qquad \text{equation (1)}$$

where G is the amount of grass produced measured in kilograms per hectare per month (kg ha^{-1} month^{-1}), and R is the dry season (July–October) rainfall measured in millimetres and averaged over the northern Serengeti region. Remember, from the last section, that green food supply, in the dry season, is considered to be the limiting resource for wildebeest.

The amount of food per animal (F) is the total grass produced per month per hectare (from equation 1), times the number of hectares utilized in the dry season by the wildebeest (500,000), divided by the total number of wildebeest (T):

$$F = \frac{500{,}000 \cdot G}{T} \qquad \text{equation (2)}$$

Mduma and his colleagues divided the wildebeest population into three groups. Calves (up to one year), who have their own survival rate, yearlings (up to two years) who do not reproduce, and adults (over two years) who have a pregnancy rate of 0.85 and, along with the yearlings, an adult survival rate. However, survival rates are not constant. They are influenced by the per capita amount of food (F). The relationship between F and survival (s) was assumed to be similar for both calves and adults and described by a rectangular hyperbola, sometimes referred to as Holling's 'disc equation' (Holling 1959) and frequently used to described the relationship between food and herbivore response (Beddington *et al.* 1976; Crawley 1983). It is the simplest equation that gives zero survival when there is no food, and is at a maximum when food is unlimited (Fig. 4.16). The general form of this equation is:

$$s = \frac{a \cdot F}{b + F} \qquad \text{equation (3)}$$

The values of the constants a and b varied between calves and adults, so that the precise shape of the rectangular hyperbola differed between the two (Fig. 4.16). The initial conditions for the simulated population were those of the 1961 census. That is, 263,000 wildebeest, 10 per cent yearlings, 10 per cent two-year-olds, and 80 per cent three years or older. Each year from the number of 3+ individuals divided by 2 (assuming an equal sex ratio), $\times$ 0.85, we obtain the number of calves born. Using the rainfall for that year to calculate total grass and the number of wildebeest to get the food per individual, the calves surviving can be calculated. Using the adult survivorship curve, yearlings become 3 year-olds, and 3 year-olds become 4 year-olds, and so on. The only unknown quantities are the four constants in the two survivorship curves. Mduma and his colleagues used repeated computer simulations to fit different values of these four constants until the

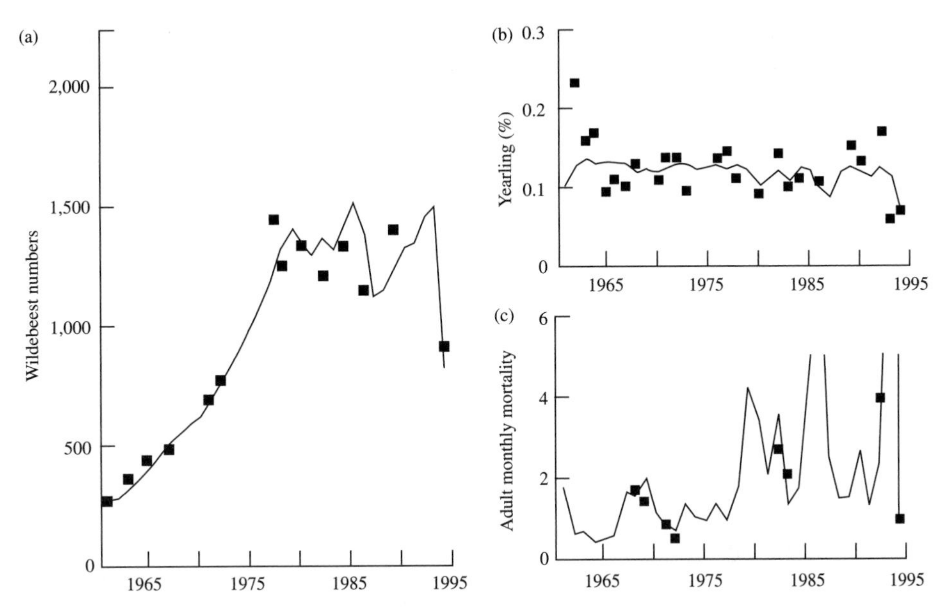

Fig. 4.17 Three sources of wildebeest data and the best fit population model. The solid squares are observations and the lines are the fitted model. (a) Total wildebeest population size, (b) percentage of the population that is yearling, (c) adult mortality rate per month of the dry season (from Mduma, Hilborn and Sinclair 1998).

best fit to three sources of data was obtained. These were (i) censused numbers of wildebeest, (ii) the percentage of yearlings, and (iii) the estimates of adult dry-season mortality (Fig. 4.17).

The survivorship curves for both calves and adults that give this best fit are shown in Figure 4.16. This best fit model, using the relatively simple equations 1 to 3, is a very good description of the real events unfolding in the Serengeti wildebeest population after 1961. It suggests that per capita food in the dry season, acting through mortality, is the dominant factor determining wildebeest numbers. Of course animals also die in the wet season, and animals will be killed by predators, but these other mortality factors, interesting though they may be, only add 'noise' to the population trajectory.

Zebra model

The Laikipia District of central Kenya (9,666 km^2) is comprised mostly of semi-arid savannah, divided into a mosaic of privately, publicly, or communally owned properties. Before the 1960s, wildlife was almost eradicated from Laikipia, largely by shooting for food. In 1977, Kenya introduced a national ban on the consumptive use of wildlife and subsequently most of the indigenous wildlife, including plains zebra (*Equus burchelli*), has returned. Georgiadis, Hack, and Turpin (2003) developed a simulation model of the dynamics of the plains zebra population in Laikipia. Monitoring of this population began in 1985, and has continued at intervals until 2001 (Fig. 4.18, black dots). Zebras in Laikipia do not migrate seasonally, although they do shift habitats in response to patchy rainfall. This annual rainfall, recorded, at five stations throughout the zebra range is also shown in Figure 4.18 (open circles). For the purposes of the model, the zebra population was divided into three age classes identifiable in the field: foals ($<$ 1 year old), sub-adults (1–3.4 years) and adults (3.4 years +). The division between sub-adult and adult is taken as 3.4 years because this is the youngest age of first conception in plains zebra. Sex ratio was assumed to be unity in foals and the sexes were treated separately in sub-adults and adults. Successive population numbers were generated by using 5 difference equations. For example, foal numbers in year x were calculated as equal to the number of surviving adult females in year $x-1$, that didn't give birth in year $x-2$ (females rarely give birth two years in succession), multiplied by the birth rate. Calculation of sub-adults (♂ and ♀) and adults (♂ and ♀) used similar equations, only using a survival rate rather than a birth rate, to link the two generations. Since Georgiadis, and his colleagues, lacked independent information about zebra carrying capacity (K) in Laikipia ($\cong$ to 'grass' in the wildebeest model) it was calculated using a power equation ($K = h{\cdot}R^i$) that allowed carrying capacity to increase, with rainfall, at a greater than linear rate (see Chapter 2). Both h and i were obtained each generation by an iterative best fit process. The five 'vital' rates (birth, survival of sub-adult males and females, and survival of adult males and females) were linked to carrying capacity, and density dependence, in a novel way. Each vital rate was made equal to a density independent component + a density dependent component. Survival of adult males, for example, was $s_m + \delta_m$, with s being determined as the minimal field observation, and δ

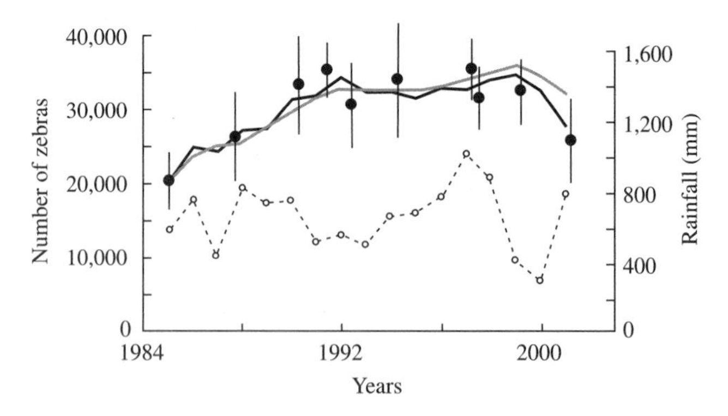

Fig. 4.18 Laikipia plains zebra population. Observed estimated population (black dots, vertical bars are standard errors). Best fit simulation model: rainfall mediated density dependent model (RMDD) (black line), rainfall dependent model (RD) (grey line). Fluctuations in annual rainfall (open circles) (from Georgiadis *et al.* 2003).

linked to K. This of course fits neatly with what we know about real populations (see the k factor analyses above); that they are influenced by both density independent and density dependent factors. The density dependent component of the vital rate (δ in the above survival example) was linked to carrying capacity in two different ways. One, called Rainfall Dependent (RD) was: $\delta = (c/K)$, while the other, called Rainfall-Mediated Density Dependence (RMDD) was: $\delta = (c \cdot N/K)$. In both equations, c is a constant defining the sensitivity to rainfall. Both models, using total annual rainfall, were able to simulate the observed zebra dynamics (Fig. 4.18), and generate total vital rates consistent with field observations. However, the RMDD model gave the best fit to the observed data.

Other species, in other areas

The conclusions from the two sections above seem clear. Savannah ungulates are regulated through their food supply (bottom-up regulation), determined by rainfall. However, these detailed studies involved only three species—wildebeest, buffalo, and plains zebra, all in East Africa. What about other species, in other areas?

One of the important pieces of evidence for the role of food in regulating ungulate populations is the presence of a correlation between rainfall and ungulate biomass. Figure 4.11b for buffalo was a good example. What is more, this type of association usually involves data points from many different savannah areas, suggesting that the relationship is not restricted to these three species, the Serengeti ecosystem, or even the east African region. Coe *et al.* (1976) first showed that the total biomass of large savannah herbivores was positively correlated to mean annual rainfall. They used data from 20 wildlife areas of southern and eastern Africa, in which rainfall varied from less than 200 to more than 1,100 mm per year. Since primary production is also positively related to mean annual rainfall in African savannah systems (see chapter 2), Coe *et al.* (1976) considered that the herbivore biomass/rainfall relationship reflected the effects of water availability on the herbivores food supply. This, of course, is the mechanism implicated in the wildebeest, buffalo and zebra case studies above.

However, the analysis reported by Coe *et al.* (1976) referred to large herbivore communities as a whole rather than to particular species. What is more, it was confined to what are sometimes called arid/eutrophic savannahs (Huntley 1982), characterized by high soil nutrients, relatively low rainfall and a low biomass of high-quality vegetation supporting a high total biomass of large herbivores (Chapter 2). Bell (1982) suggested that the herbivore biomass/rainfall relationship could be modified by soil nutrient status and that the positive correlation observed by Coe *et al.* (1976) only applied to savannahs with high soil nutrient status (e.g. soils of volcanic origin). Bell (1982) suggested that for savannahs with low nutrient soils the

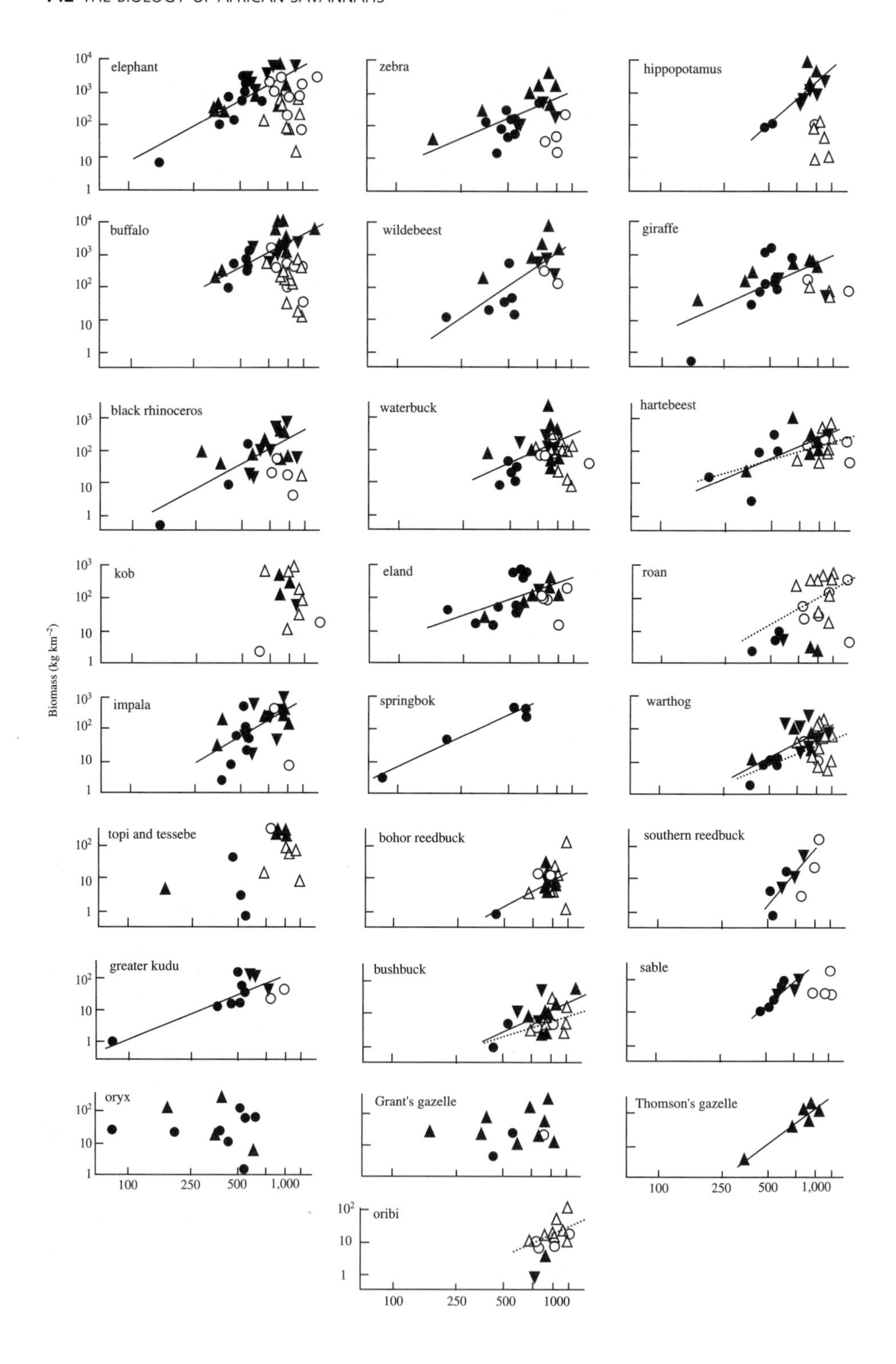

Biomass (kg km^{-2})
elephant
zebra
hippopotamus
buffalo
wildebeest
giraffe
black rhinoceros
waterbuck
hartebeest
kob
eland
roan
impala
springbok
warthog
topi and tessebe
bohor reedbuck
southern reedbuck
greater kudu
bushbuck
sable
oryx
Grant's gazelle
Thomson's gazelle
oribi
1
10
10^2
10^3
10^4
100
250
500
1,000
1000

relationship might even 'reverse' at high rainfall. Such savannahs would include the miombo woodlands of south- central Africa (Chapter 1) whose pre-Cambrian rocks frequently weather into highly leached soils with low supplies of nutrients. The biomass/rainfall relationship was therefore re-examined by East (1984) for individual species, with soil nutrient status being taken into consideration. These relationships are shown in Figure 4.19.

Three classes of soil nutrient status were distinguished following the broad classification used by Bell (1982): high (volcanics), medium (rift-valley sediments), and low (basement and basin sediments, granitic shields, and Kalahari sands). Areas of low soil nutrient status were further subdivided into those with mean annual rainfalls of less than or equal to 700 mm, and those greater than 700 mm. Least squares regression analysis, of log-transformed data, revealed significant relationships ($P \leq 0.05$) in arid/eutrophic savannahs for 19 out of 23 species for which sufficient data was available. Of the four remaining species, roan antelope had few data, topi/tsessebe had $P = 0.075$ and oryx and Grant's gazelle showed little association with rainfall. Notice that on soils of low nutrient status, some species (elephant, zebra, buffalo, giraffe, black rhinoceros, and eland) do show a decline in biomass at rainfall greater than about 700 mm (open symbols in Figure 4.19), although the regression for arid/eutrophic savannahs is still significant (solid line in Figure 4.19). Significant relationships between biomass and rainfall are also seen in several species on poor nutrient soils (hartebeest, roan, warthog, bushbuck and oribi) (Figure 4.19 dotted lines).

Although this survey of East (1984) is not a detailed examination of population processes, as were the studies on wildebeest, buffalo, and zebra, none the less the evidence for a bottom-up regulation of savannah herbivores seems quite clear. What is more, there are other field studies, not as detailed as those of wildebeest and buffalo in the Serengeti, that enforce this view obtained from the relationships in Figure 4.19. Studies on greater kudu populations in Kruger N.P., over the period 1974 to 1984 (Owen-Smith 1990), also reveal the controlling influence of annual rainfall on population dynamics, mainly through calf survival, but also through yearling and adult survival. This is particularly interesting because the greater kudu is predominantly a browser, rather than a grazer like wildebeest and buffalo. Mills *et al.* (1995) confirmed the strong positive influence of annual rainfall on changes in the abundance of kudu, as well as buffalo and waterbuck, in the central region of Kruger. Rainfall appears to determine primary production and this, in turn, regulates population size. This is a useful, general,

Fig. 4.19 Biomass/rainfall relationships for 25 herbivores species. Symbols indicate different nutrient status: high (▲), medium (▼), low with annual rainfall ≤700 mm (●), and low with annual rainfall >700 mm in southern and eastern Africa (O) and West Africa (△). Closed symbols are arid/eutrophic savannahs and open circles are moist/dystrophic savannahs. Statistically significant regressions ($P < 0.05$) are indicated for arid/eutrophic savannahs as solid lines and for all savannahs in areas of low nutrient status as dotted lines (from East 1984).

statement about the population regulation of savannah herbivores. It appears to apply to many African species, in many African savannahs. Of course there will be interesting and instructive exceptions. In the Serengeti ecosystem, populations of wildebeest and buffalo before 1960, were undoubtedly affected by rinderpest, an exotic disease. In many other parts of Africa the interaction between the rinderpest virus and wild and domestic populations of ruminants was the principal factor that kept these populations at reduced levels for many years (Sinclair 1979). In an outbreak in Lake Manyara in 1959, 250 buffalo (78 per cent of the population) died (Prins 1996), and in Kruger N.P. the buffalo population was only 20 individuals in 1902, while it had been common before the arrival of rinderpest (Stevenson Hamilton 1911). Of course rinderpest may even be exceptional as a disease. It was exotic, not endemic, and may have had more devastating results as a consequence. Many disease outbreaks seem to occur at the end of the dry season. This as been noted for rinderpest in the Serengeti (Sinclair 1979), rinderpest in Manyara (Prins 1996) and anthrax in Kruger N.P. (Pienaar 1961, 1967). As previously mentioned this may imply that disease is often removing individuals that are already dying through starvation.

In Lake Manyara N.P., both buffalo numbers (△ in Figure 4.11b) and lion numbers are exceptionally high. Lion predation on buffalo is also quite

Table 4.4 Reported predation rates on large herbivores of African savannah ecosystems. Serengeti and Ngorongoro are in central East Africa; Kalahari and Timbavati are in southern Africa.

Prey	Main predator	% Prey mortality	Study area	Reference
buffalo	lion	23–28	Serengeti	Sinclair and Norton-Griffiths 1982
wildebeest	lion	38–93	Kalahari	Mills 1984
wildebeest	lion	96	Timbavati	Hirst 1969
wildebeest	Lion	13–21	Serengeti	Sinclair and Norton-Griffiths 1982
wildebeest	lion and hyaena	15	Serengeti	Elliot *et al.* 1977
wildebeest	lion	13	Kruger	Mills and Shenk 1992
plains zebra	lion	90	Timbavati	Hirst 1969
plains zebra	lion and hyaena	59–74	Serengeti	Sinclair and Norton-Griffiths 1982
plains zebra	lion	11	Ngorongoro	Elliot *et al.* 1977
plains zebra	lion	14	Kruger	Mills and Shenk 1992
springbok	leopard	87	Kalahari	Mills 1984
gemsbok	several	84	Kalahari	Mills 1984
Th. gazelle	lion	8	Ngorongoro	Elliot *et al.* 1977
waterbuck	lion	82	Timbavati	Hirst 1969
kudu	lion	63	Timbavati	Hirst 1969
impala	leopard	55	Timbavati	Hirst 1969
giraffe	lion	34	Timbavati	Hirst 1969
warthog	lion	29	Timbavati	Hirst 1969

significant, amounting to some 89 per cent (Prins 1996, Table 5.4). Many other authors have reported high predation rates in savannah ungulates (Table 4.4), but it is difficult to know how to interpret these. These data come from carcasses that are found and identified as 'predator kills'. Predator kills are identified by seeing predators at a kill, seeing claw marks on kills without predators and even by hearing 'killing' noises the previous night. However, we know that even lions can obtain up to 15 per cent of their diet by scavenging (Schaller 1972) so that many of these 'predator kills' might be deaths from other causes. Predators also pick out individuals in poor condition and this implies that predators are often removing individuals that are already dying through starvation, and/or disease. In fact many authors, including some in Table 4.4, have questioned the role of predators in regulating savannah herbivores (Eloff 1984; Hirst 1969; Mills 1984; Pienar 1969; Schaller 1972). Equivocal support for this view is provided by a removal exercise carried out in Kruger N.P. In the late 1970s it was speculated that lion and spotted hyaena were responsible for the local decline of wildebeest and zebra populations (Smuts 1978). As a result the largest systematic culling operation in Kruger's history took place. Over a five-year period, 445 lions and 375 spotted hyaena were killed, but the reduction had no detectable influence on the population trends of the wildebeest and zebra (Whyte 1985). Unfortunately the cull had no detectable influence on the lion and hyaena population either so the proposed regulatory effect of these two predators was still unclear.

Mills and Shenk (1992) present some evidence for the 'impact' of lion predation on wildebeest and zebra populations in the south east of Kruger N.P. Using intensive observations, over a four year period, they estimated wildebeest, zebra and lion population size. They also estimated the number of 'killing' lions (females) and their individual annual kill rate (seven wildebeest and eight zebra). They used these two parameters to estimate the death rate of both prey species and, coupled with an estimate of prey fecundity, used these parameters in a simulation model. The model was then used to predict the predator numbers necessary to 'stabilize' the population, such that it showed little change over a four year period (like the wild prey populations). Unfortunately there is no regulation in this model (no density dependence) and the delicate balance observed relies on the model makers. If fecundity rose, or kill rate declined, by as little as 10 per cent the population of prey simply went into exponential increase (or into exponential decline if they moved in the other direction). The model demonstrates that predation, if very high, can have an impact on prey, but not a regulating effect. There is an important difference between predation being an 'important source of mortality' and 'predation regulating populations' of savannah herbivores.

Quite recently, two excellent analyses of long-term population counts for savannah ungulates have claimed to show that predators influence population numbers. One of these was in the Kruger ecosystem of South Africa (Owen-Smith and Mason 2005; Owen-Smith *et al.* 2005) and the other was in the Serengeti–Mara ecosystem of East Africa (Sinclair *et al.* 2003). In both cases there is certainly something very interesting happening demographically, but whether this is caused by predation is debatable. In the Serengeti–Mara, Sinclair *et al.* suggest that a 'natural experiment' (Diamond 1986) has taken place. They suggest that in an area of the northern Serengeti,

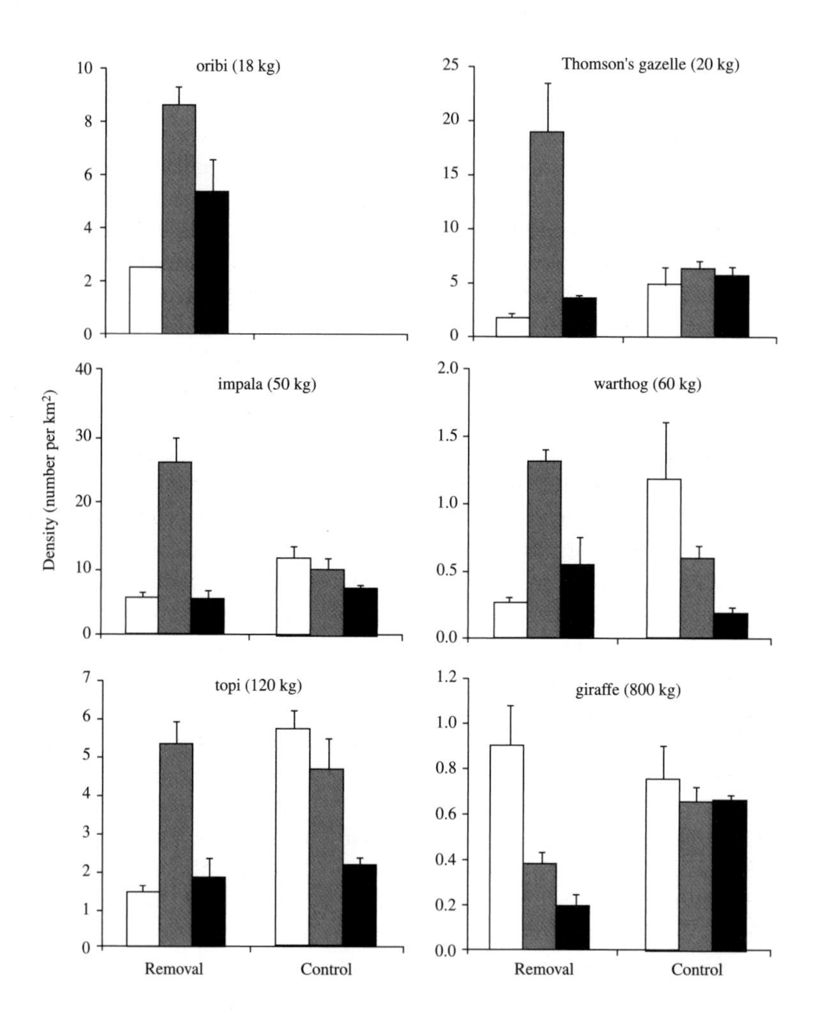

Fig. 4.20 Densities of six ungulates in a 'predator removal' area in northern Serengeti compared with an adjacent 'control' in the Mara Reserve. The period before predator removal (open bar) is for 1967–1980, the period of low predator numbers due to poaching (shaded bar) is for 1981–1987, and the period after predators returned (solid bar) included years from 1989 onwards. There is no control for oribi because they don't occur in the Mara. Error bars are one standard error (from Sinclair *et al.* 2003).

poaching and indiscriminate poisoning removed the majority of carnivores (lion, spotted hyaena, and jackals) for an eight-year period (1980–7), citing Sinclair (1995). However, there are no predator population numbers to confirm this, and what the Sinclair (1995) paper actually says is 'there is some circumstantial evidence that lions, which are the main predators in the area disappeared: roar counts were considerably lower in the study area in 1990 compared to other parts of the Serengeti (Packer 1990)'. Notice that only lions are mentioned and the comparison is with 'other parts' at this time, not the northern area prior to this date. In the contiguous Masai Mara Reserve predator populations are thought to have remained intact (Broten and Said 1995). During the period of 'predator removal', five smaller-bodied species (<150 kg), oribi (*Ourebia ourebi*), Thomson's gazelle (*Gazella thomsonii*), warthog (*Phacochoerus africanus*), impala (*Aepyceros melampus*) and topi (*Damaliscus lunatus jumela*), increased markedly in density in the northern Serengeti (no predators) relative to their population in the Mara (predators present)(Fig. 4.20). These northern Serengeti populations declined once predators returned to the area, after 1987. In contrast, the abundance of a large ungulate, the giraffe (*Giraffa camelopardalis*), did not increase during the predator removal period (Fig. 4.20). Clearly there is reason to think that something happened in the northern Serengeti during the period 1981 to 1987 that did not happen in the Mara. Was this due to a relaxation of predation? I think the evidence is equivocal. The evidence for reduced predators (perhaps only lions) is circumstantial. However, we do know that buffalo numbers in this northern area of the Serengeti were seriously reduced due to poaching about this time. Sometime between 1976 and 1982 approximately 85 per cent of buffalo (about 9,000 animals) in northern Serengeti were removed by poachers (Sinclair 1977; Dublin *et al.* 1990). This did not happen in the Mara Reserve. So would this have affected all the animals in Figure 4.20, except giraffe, through competition for grazing? We probably don't know but it's a possible scenario. Alternatively, the predation explanation for Figure 4.20 might be reinforced by Figure 4.21, also from Sinclair *et al.* (2003), which suggests that indeed only small species are subjected to high predation rates, while larger species are not. This may be because smaller prey have more predators (Chapter 5). Notice however that this mortality is for adults only and that we have already seen that mortality from predation often targets the young.

This issue of young versus adult mortality leads on to the second study in Kruger National Park (Owen-Smith and Mason 2005; Owen-Smith *et al.* 2005). Between 1965 and 1976, ungulate population numbers have been estimated using irregular aerial surveys, supported by ground counts. From 1977 to 1997 these surveys were carried out each year and from 1983 to 1996 population structure was also recorded. Where possible individuals were classified as adult (>2 years)(male or female), yearling (1–2 years) (male or female), and juvenile (<1 year). There are also extensive rainfall records for the period 1960 to 1997. About 80 per cent of the annual rainfall falls

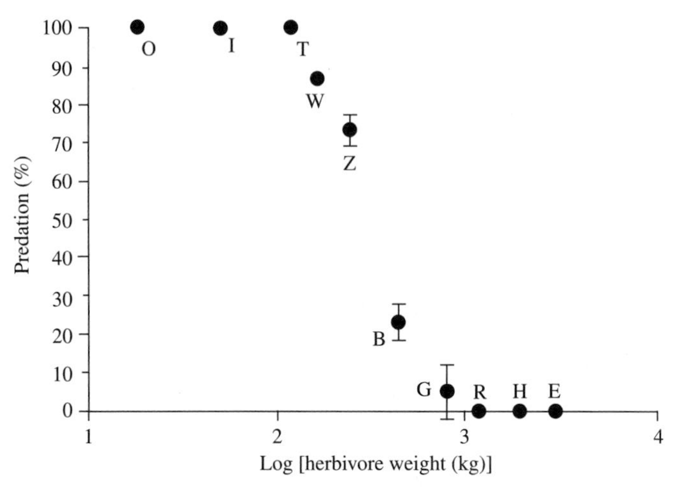

Fig. 4.21 The proportion of adult mortality accounted for by predation in 10 non-migratory ungulate populations in the Serengeti ecosystem. Error bars are 95% confidence limits. Species are: O = oribi, I = impala, T = topi, W = non-migratory wildebeest, Z = common zebra, B = African buffalo, G = Masai giraffe, R = black rhinoceros, H = hippopotamus and E = elephant (from Sinclair *et al.* 2003).

during the wet season (October to March) and only 20 per cent during the six months of the dry season. Figure 4.22 shows both rainfall and estimated numbers of some major ungulates over the period 1965 to 1996. Notice that for both rainfall and population numbers, figures are shown as smoothed 'running means'. For rainfall this is a five-year running average, while for numbers it is a three-year weighted running average, $N'_t = 0.25N_{t-1} + 0.5N_t + 0.25N_{t+1}$. At the start of this period average rainfall was low. In the 1970s to early 1980s it was high, but from about 1981 it started to drop and has remained low during the 1980s and 1990s. A feature of this latter period of low rainfall was very low dry-season rainfall. Following the 1981 drop in annual rainfall there appears to have been a change in population numbers across, what Owen-Smith and Mason (2005) called, a common 'break point' pivotal around 1987. While some species increased before this date and then persisted at high abundance afterwards (common zebra, wildebeest, impala and giraffe), others (buffalo, greater kudu, common waterbuck, warthog, sable antelope, tsessebe, eland, and roan antelope) declined after this point. In other words, some species ('stabilizing' species) did not appear to 'respond' to the low rainfall after 1981, while others ('declining' species) did. Notice that there appears to have been a five year lag in this 'response'. Owen-Smith and Mason (2005) estimated annual survival rates for juveniles, yearlings, and adults of all four stabilizing species and five of the declining species (buffalo, roan, and eland were not included).

For three species (giraffe, sable, and tsessebe) they could not estimate yearling survival. They estimated survival separately for the period before

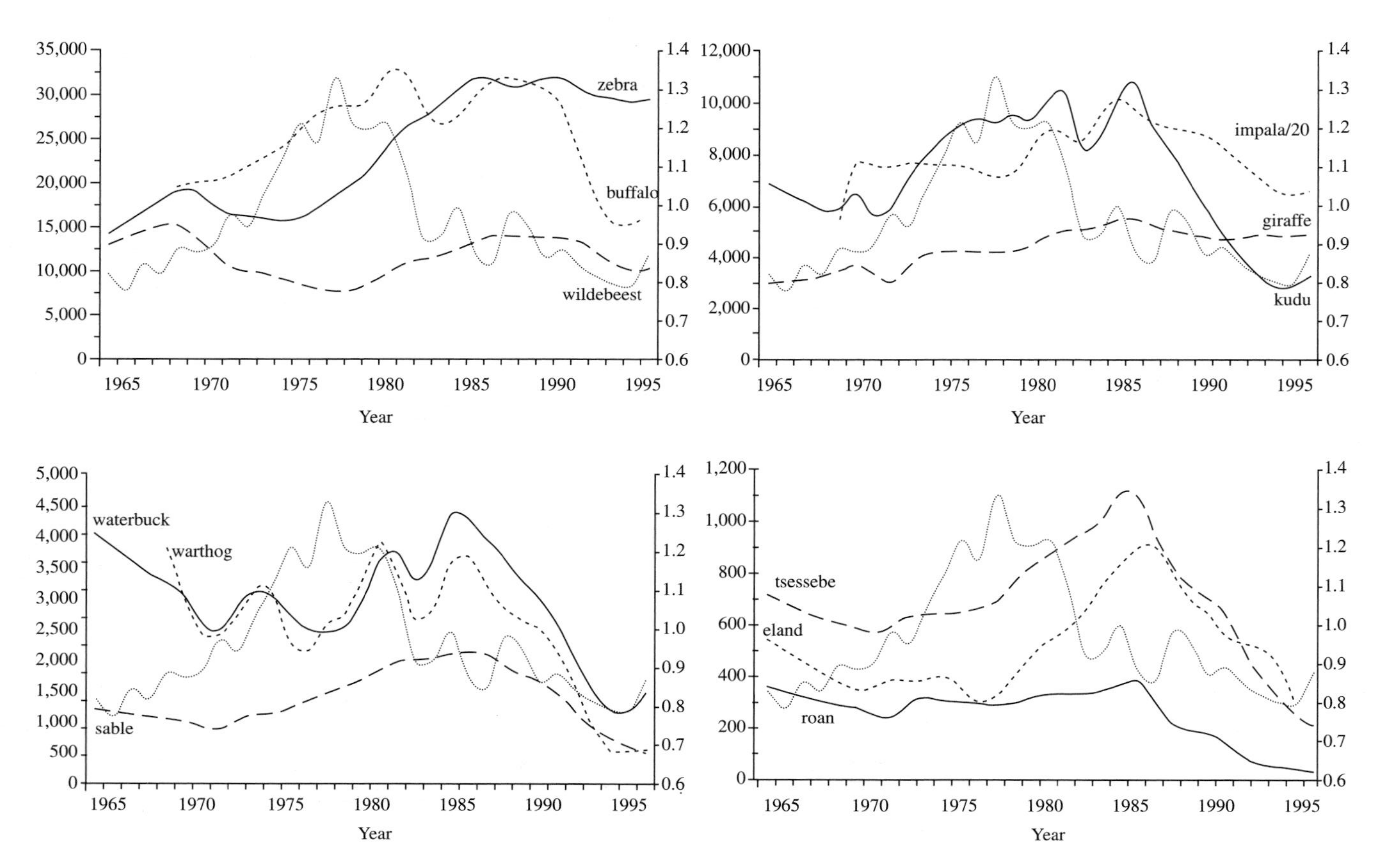

Fig. 4.22 Changes in population numbers of the major ungulates of Kruger National Park since 1965, in relation to mean rainfall (fine dotted line on each graph). All numbers are smoothed running means (see text). Impala numbers are divided by 20 (modified from Owen-Smith and Ogutu 2003).

Table 4.5 Number (out of total) of statistically significant differences between pre 1987 survival rates and lower post 1987 survival rates. Notice that not all yearling survivals are available for estimation.

	Juveniles	Yearlings	Adults
Stabilizing species	0/4	0/3	2/4
Declining species	1/5	1/2	4/5

Taken from Owen-Smith and Mason 2003.

1987, and for the period after, and tested for any indication of a significant difference between the two periods. All survival rates post 1987 (low rainfall) are lower than the survival rates pre 1987 (high rainfall) except for yearling wildebeest (significant, $P = 0.019$) and yearling waterbuck (not significant $P = 0.978$), which are higher. Therefore in 22/24 comparisons, the period after 1987 (lower rainfall) is associated with reduced survival. However, not all these reductions are statistically significant and table 4.5 summarizes the patterns that emerged. Owen-Smith and Mason (2005) draw attention to the fact that most statistically significant reductions in survival are associated with 'declining species', and adult life stages. The first observation can be interpreted as the cause of the decline in these species. With reduced rainfall all species showed lower survival but some, the declining species, showed much more significant reduction. So far, this explanation fits with the one that I have been suggesting in this chapter. That is, reduced rainfall, means reduced food, which in turn means reduced survival. However, Owen-Smith and Mason interpret the second observation (lower survival associated with adults) as indicating that it is predation, rather than malnutrition, that is the cause. It is predation that is producing the decline seen in so many of these Kruger populations after 1987. They suggest lions, whose kills are known to concentrate on the adult segment of medium sized ungulates (Pienaar 1969, Mills and Biggs 1993). As already stated, after 1987 rainfall, particularly dry season rainfall, was very low. The resultant lack of sufficiently good forage during the dry season could have reduced survival rates in adults either directly through malnutrition, or indirectly by making these adults more vulnerable to predation. With this in mind Owen-Smith *et al.* (2005) continued this analysis, and examined the relationship between survival and population density, rainfall and predation using linear multiple regression, fitted by standard least squares. They used the same nine species analysed in Owen-Smith and Mason (2003), plus roan antelope. Because density, through resource availability, will operate on adults and juveniles over an extended period, density dependence was assessed using a 3-year weighted average of the census totals, with the weighting centred on the year preceding that in which the survival estimate was made. For rainfall, both the total annual rainfall and that for the wet and dry season alone, preceding the count, were used. To assess the effect of what they called 'prior

Table 4.6 Best fit predictors of stage-specific survival rates, from alternative linear regression models.

Species	Stage	One predictor	Two predictors
Stabilizing species			
Zebra	Adults	Abundance	
	Yearlings	Annual rain	
	Juveniles	Prior rain	
Wildebeest	Adults	Prior rain	
	Juveniles	Prior rain	
Impala	Adults + yearlings	Abundance	
	Juveniles	Dry season rain	
Giraffe	Adults	Abundance	
	Yearlings + juveniles		Dry season rain + prior rain
Declining species			
Kudu	Adults	Annual rain	
	Yearlings	Annual rain	Annual rain + prior rain
	Juvenile		Dry + wet season rain
Waterbuck	Adults + yearlings		Dry season rain + abundance
	Juvenile	Dry season rain	
Warthog	Adult + yearling	Predation	
	Juvenile	Annual rain	
Sable	Adults + yearlings		Predation + dry season rain
	Juveniles	Annual rain	
Tsessebe	Adults + yearlings		Annual rain + predation
	Juveniles		Predation + dry season rain
Roan	Adults + yearlings	Predation	
	Juveniles	Annual rain	

From Owen-Smith *et al.* 2003.

rainfall', that might affect the vegetation cover and its composition, a 4-year running average prior to the survival estimate was used.

No direct information on predator abundance was available, because these species are not readily visible from the air during a census. They therefore devised a 'proxy' index to represent this measure. All 10 ungulate species, plus buffalo, are taken by lions (Pienar 1969) and all these were regarded as potential prey. The annual total counts for each prey species were transformed into a 'carcass biomass' through multiplying by their estimated mortality (to get the number of carcasses, and then by the species biomass (to get the estimated amount of food available for lions). Since uneaten carcasses are rarely encountered in Kruger, this annual 'carcass biomass' was regarded as the total food available for lions and therefore a proxy for lion numbers. The initial regression analysis included 'numbers' and annual rainfall as the two potential predictors. Once the relationship between survival rate and these predictors (if any) were established, either predation (carcass biomass) or 'prior rainfall' was added to the regression model. Table 4.6 summarizes the best fit predictors.

Table 4.7 Periodic estimates of the total population size for the five major carnivores of the Serengeti ecosystem, over a 25-year period from the late 1960s.

Species	Year	Numbers
spotted Hyaena	1967	2,207 ± 120
	1977	3,306 ± 432
	1986	5,214 ± 828
	1991	9,500
lion	Mid 1970s	2,000–2,400
	1991	2,800
leopard	Mid 1970s	800–1,200
	1991	840
cheetah	Mid 1970s	220–500
	1991	600
wild dog	Mid 1970s	150–300
	1991	101

Even if you believe that 'carcass biomass' = predation, the evidence for this being an important predictor of survival rate (and therefore population change) is not great, although it is interesting that the five cases where 'predation' appears in Table 4.6 are all in the declining species, and four of them affect adults. However, some measure of rainfall seems to dominate these best fit predictors, suggesting that vegetation is influential in determining ungulate survival once again. Both the Serengeti and Kruger analyses of population change are important attempts to look for patterns that could be interpreted as predation. At the moment the evidence is intriguing, but circumstantial. We clearly require more detailed studies of the factors influencing the population dynamics of savannah ungulates.

Finally, we turn to predator populations, and what regulates them. The problem is that we have few good estimates of predator population numbers over time. Even in the Serengeti and Kruger National Parks, where most of the long-term surveys of ungulate populations have been carried out, we have no detailed series of estimates for carnivores (as seen in the last two analyses). In the Serengeti, for example, we have occasional estimates, that usually coincide with individuals doing short-term research on carnivores, and these are shown in Table 4.7. These estimates allow us to view the changes that have occurred, in predator composition (Figure 4.23), over this 26-year period, and this is helpful, but they do not allow any detailed understanding of how these predator populations are controlled. Spotted hyaena numbers have clearly increased from the 1960s, and lions may also have increased. This may be a response to the increase in wildebeest numbers, after the removal of rinderpest. There is also a suspicion that spotted hyaena have interacted unfavourably with wild dogs, a topic I will look at in Chapter 5.

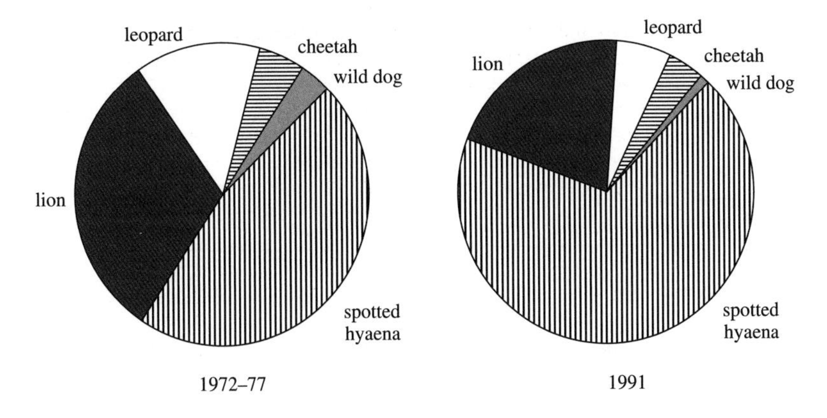

Fig. 4.23 Pie diagrams showing the change in predator composition in the Serengeti National Park over a 25 year period from the mid 1960s.

East (1984), in his survey of biomass/rainfall relationships for herbivores, also looked at the relationship between the biomass of five large savannah carnivores (lion, spotted hyaena, cheetah, leopard, and wild dog), and the biomass of their preferred size class of prey. There was a significant positive relationship for four of the carnivores, and these are shown in Figure 4.24. Like the Serengeti lions and hyaenas, this implies that more prey means more carnivores, suggesting again a bottom-up regulation of savannah animals.

The scenario that emerges for population control in savannah mammals is therefore one of a series of positive relations depicted by the boxes below.

That is, more rain produces more grass (and other vegetation), which produces more herbivores, which produces more carnivores. Endemic disease may periodically cause episodes of mortality, in both herbivores and carnivores, but probably mainly kills individuals in poor nutritional condition at the end of the dry season. Predation may sometimes cause high mortality, but this may also be born mainly by nutritionally weak or old individuals. Perhaps there is top-down regulation in Kruger and perhaps there is top-down regulation of small prey in Serengeti. I have indicated this by a dotted arrow. However, the evidence is not, at the moment, compelling.

Notice that in Figure 4.24 there is no graph for wild dog (*Lycaon pictus*). This is because there was no significant relationship between dog biomass and their prey biomass. The populations dynamics of this predator are however quite interesting, and well studied. They involve the special problems of small populations, and their vulnerability, and competitive interactions with other carnivores. This leads us to consider populations of two

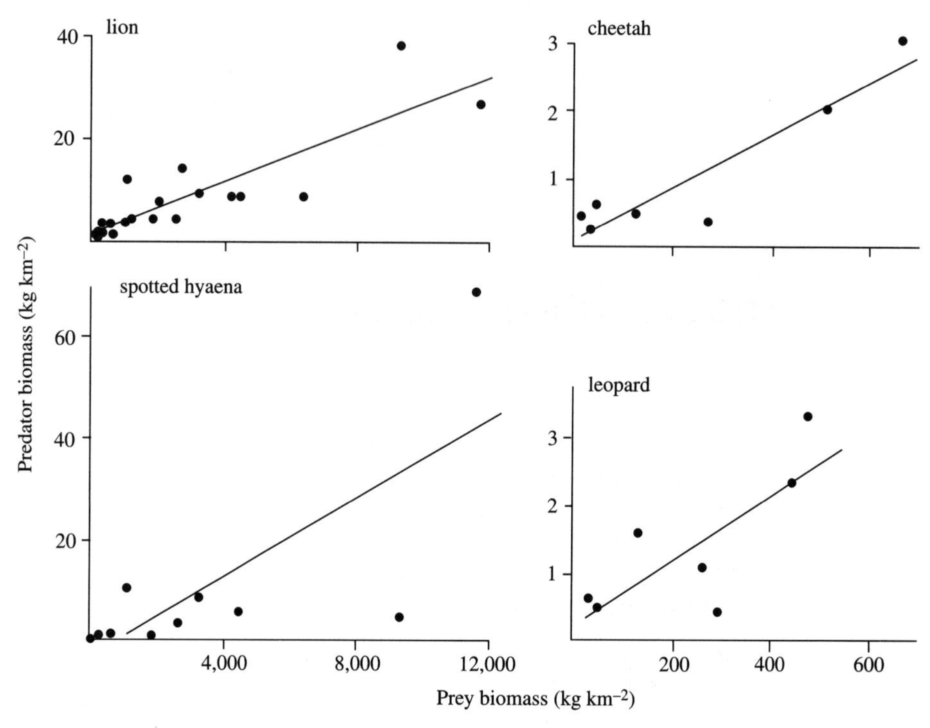

Fig. 4.24 Relationship between the biomass of large carnivore species and the biomass of their preferred size class of prey (lion and spotted hyaena: topi and hartebeest to buffalo) (cheetah and leopard: Thomson's gazelle to impala and kob). All relationships are statistically significant as judged by linear regression (solid lines) (leopard $P = 0.05$, rest $P = 0.01$) (from East 1984).

interacting species more directly, and is the subject of the next chapter. However, one final comment on single species populations. This chapter has dealt with natural factors that influence birth and death. There are also human imposed factors that act via tourism, hunting, poaching, habitat removal, and fire. These are mostly dealt with in Chapter 6, although fire will appear in Chapter 5.

5 Species interactions

Interactions between species have been given many different, often confusing names. A simple and unambiguous method of classification, that I have always found useful, is to use the 'effect' that individuals of one species have upon the population growth of another species and *vice versa*. We ask the question: in the presence of species A (+) does species B

(1) increase its numbers (+)
(2) not change its numbers (0)
(3) decrease its numbers (−)

relative to when species A is absent (−)? The same question is asked of species A in the presence of species B. The answers can be conveniently summarized in a 3 × 3 table (Table 5.1).

Because of the symmetry in the table, there are actually only six types of interaction. These are frequently called neutralism (0 0), commensalism (+ 0), predator/prey, parasite/host, herbivore/plant interactions (+ −), amensalism (0 −), competition (− −) and mutualism (+ +). The neutral interaction may well be very common in many ecosystems, but does not contribute to the dynamics of the system, and would be regarded by most people as uninteresting. Commensalism could be regarded as an extreme example of mutualism in which the beneficial effect of one species on the other is rather weak and undetectable. Amensalism could also be regarded as an extreme example of competition, in which the adverse effect of one species on the other is much greater than the reciprocal effect. This chapter will therefore examine three types of interaction between savannah species: 'predator/prey' (+ −), competition (− −) and mutualism (++). However, before proceeding it is worth drawing attention to a rather confusing interaction that is sometimes observed. An ecological situation can arise where two species appear to show the reciprocal negative (− −) effects associated with interspecific competition, but this is in fact the result of predation by a third species. This is called apparent competition.

Table 5.1 A simple classification of species interactions.

	Effect of A on B		
	+	0	−
Effect of B on A			
+	+ +	+ 0	+ −
0	0 +	0 0	0 −
−	− +	− 0	− −

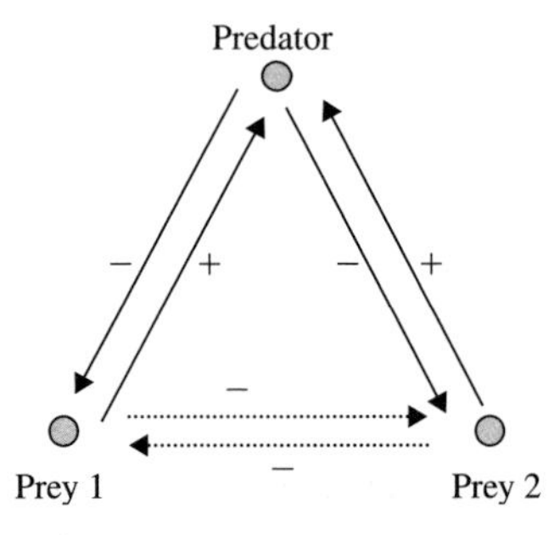

Fig. 5.1 Diagram showing apparent competition between two prey species. The solid lines indicate direct interactions between prey and predator, while the dotted lines show indirect (apparent) interactions between the two prey species.

Consider the situation shown in Figure 5.1. A single species of predator attacks two species of prey. The predator–prey interactions are of a − + type, and therefore both species are adversely affected by the predator, and the predator is positively affected by both species of prey. This means that the positive effect that prey 1 has on the predator will, in turn, increase the negative effect upon prey 2, and *vice versa*. The overall consequence of this is that there will appear to be a − − interaction between prey 1 and prey 2, even if they are not competitors for any essential and limiting resource.

Predator–prey type interactions (+ −)

This type of interaction actually encompasses three interactions that may appear rather different. They certainly tend to be studied by different ecologists. They all have in common the fact that one species (carnivore, parasite, or herbivore) consumes the body, or part of the body, of another species (prey, host, or plant). In the case of the predator, the prey is killed. With the parasite the host is sometimes killed, but may recover from the infection. With herbivores, the whole plant is usually not killed or consumed (frequently just leaves and fruit are eaten) and the 'prey' regrows those parts that have been consumed. Exceptions to this general rule might be elephants who can sometimes kill trees, and rodents that consume whole plants in the form of seeds. I will look at each of these + − interactions, and detail some of their special features, which have relevance to savannah ecosystems.

Herbivores and plants

Interaction between grasses and bovids (antelopes and buffalo), in African savannahs, has probably taken place since at least the late Pliocene, some five million years ago (Gradstein *et al.* 2004). Both grazer and grass are inextricably linked, and the adaptive radiation of both groups occurred simultaneously in the ensuing Pliocene and Pleistocene geological periods. Bovids evolved grinding teeth and complex digestive systems to deal with grasses (Chapter 3), and grasses evolved defences to combat grazing, such as the deposition of silica. Grasses also have a growth form that 'allows' continuous grazing (Chapter 2), and show 'compensatory growth' as a response to grazing pressure (McNaughton 1979b). This compensating growth, stimulated by grazing, is the product of several mechanisms, including enhanced photosynthetic capacity, more efficient use of light (due to reduced leaf shading), reduced leaf aging, nutrient recycling and the stimulating effects of herbivore saliva (McNaughton 1979a). This leads to the idea that moderate grazing will actually stimulate above ground net primary productivity, above that of ungrazed areas, although heavy overgrazing would still depress productivity (McNaughton 1979b). This model of grazing is illustrated in Figure 5.2. This predicts that there will be an optimum level of defoliation at which there is a balance between residual leaf area, and photosynthesis per unit of leaf area, which maximizes net productivity. A quantification of this predicted pattern was obtained in the northern Serengeti and Kenya Mara

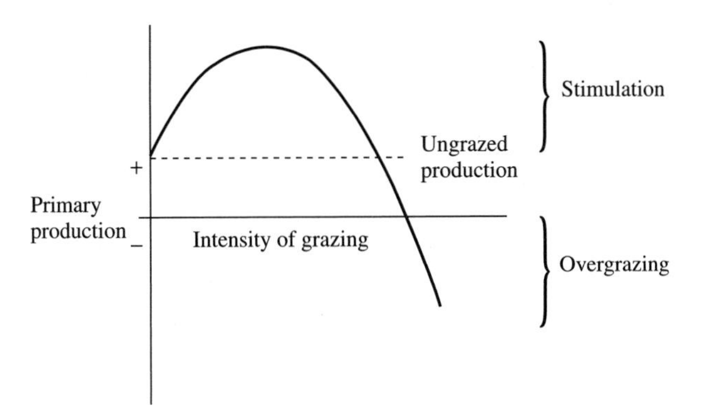

Fig. 5.2 McNaughton's grazing model, predicting the level of primary production of green grass, versus the intensity of grazing. The level of primary production in ungrazed areas is shown as a dashed line, which coincides with zero grazing (modified from McNaughton 1979b).

during August and September 1974 (dry-season), and January 1975 (wet-season) (McNaughton 1979b). During the periods that measurements were made, wildebeest were the principal grazers, and the results are shown in Figure 5.3. During the dry-season (Fig. 5.3a) the above-ground green biomass declined in control stands were grazing was prevented. This was because ungrazed grasses dried out, subsequent to flowering, despite the fact that this was an unusually wet dry-season. Intermediate levels of grazing resulted in more primary production and overgrazing resulted in negative values. The wet season grazing system (Fig. 5.3b) appears slightly different, although a similar optimization curve is produced. As in the dry season, moderate grazing promotes productivity. The fact that the curve is asymmetrical, and skewed to the right, suggests that wet season savannahs are stimulated more by light grazing, and are more resistant to overgrazing, than grasslands in the dry-season. In addition to stimulating growth, wet-season grazing also suppresses flowering, because the grass diverts nutrients from flower production to leaf production. This model of grazing dynamics, leads to the intriguing idea that herbivores manipulate their food supply, to enhance the quality of food available on later visits. McNaughton's contention (1984) is that the herbivores (in this specific example, wildebeest) can produce 'grazing lawns' that are kept in a state of high above ground productivity. This was an idea originally put forward by Vesey-FritzGerald (1974) when he looked at the interaction between buffalo and various grasses in Arusha National Park, Tanzania. In *Cynodon dactylon* grassland he noted that optimum utilization resulted in a short grass lawn being maintained by the buffalo. More recently, by monitoring exclosed vegetation plots over a five-year period, when wildebeest numbers were high, Belsky (1992) showed that grazing was essential for the grazing system to persist. In the absence of grazing, the 'grazing lawns' were rapidly transformed into rough pasture with dense foliage, in which the short-statured species favoured by the grazers disappeared. In fact McNaughton (1979b) had also noticed this

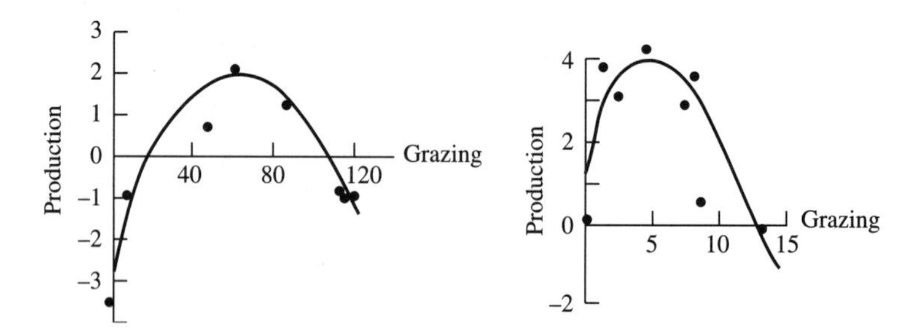

Fig. 5.3 Relationship between grazing and net above-ground primary production in the Serengeti–Mara ecosystem, East Africa. (a) Dry-season: production is measured in g/m^2/day subsequent to the grazing period, grazing = g/m^2 initial green biomass eaten by wildebeest. (b) Wet season: production = g/m^2/day, grazing = g/m^2/day green biomass consumed (modified from McNaughton 1979b).

effect on species composition. He re-examined areas in southern Serengeti that had originally been fenced over ten years before, and Table 5.2 shows how preventing grazing had changed the frequency of four species of grass. Inside the exclosures grass species diversity had declined due to the dominance of a few taller-growing species that invest heavily in stems (such as *Pennisetum*). Inside the exclosures, grass height was much greater, there was an accumulation of dry foliage, and a doubling of the grass biomass invested in stems. Not only does grazing appear to maintain the grass in a 'young' leafy state, with species preferred by grazers, but it also helps to reduce the risk of fire. Herbivores therefore are confronted by a shifting mosaic of available forage. After grazing on one patch for a while they will sequentially move on to new patches, only to return to the original grazing lawn when it has regrown new leaf. This proposed sequence has been called 'cyclic grazing' (Drent and van der Wal 1999). McNaughton (1979) noted that Serengeti wildebeest tended to revisit grassland patches at regular intervals, harvesting the high-quality fresh growth resulting from a previous grazing bout. In Lake Manyara National Park, northern Tanzania, Prins (1996) collected a large data set on the visitation interval of buffalo to patches of the grass *Cynodon dactylon* (Fig. 5.4). Patches (black areas in Fig. 5.4) are, on average, about 20 ha in area, and are more or less uniform stands of this grass. In Manyara, buffalo rely to a large extent on *C. dactylon* for their nutritional needs. Over a six-month period buffalo returned, on average, after 4.5 days (histogram in Fig. 5.4). Clipping experiments, followed by subsequent regrowth suggested that the ideal return time would be about five days (Prins 1996). However, the median may be a better indicator of the typical return time for a buffalo, and this is only three days. Prins suggests that buffalo are forced to return earlier than ideal because elephants would otherwise take the new grass.

Not all herbivore/plant interactions, in savannahs, involve grazers. A smaller, but significant, number are browsers. However, woody plants produce defensive compounds of far greater diversity and toxicity than grasses. In savannahs, condensed tannins in particular have a deterrent effect against browsing ruminants. Figure 5.5 shows how this affected browse preference for three herbivore species in Kruger National Park, in southern

Table 5.2 Frequency (%) of four grass species inside and outside exclosures in the southern Serengeti.

	Inside exclosure	Outside exclosure
Andropogon greenwayi	0	56
Sporobolus marginatus	0	20
Pennisetum ultramineum	72	5
P. mezianum	26	3

Data from McNaughton 1979b.

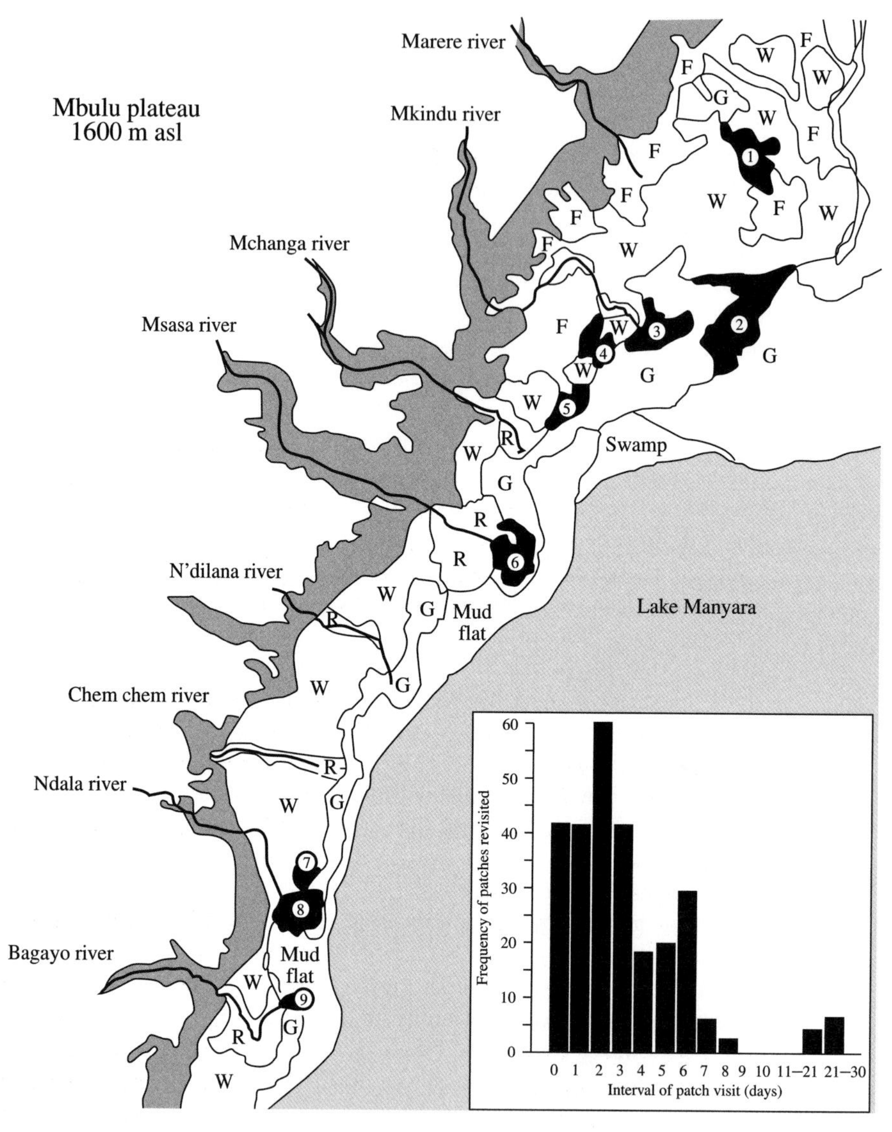

Fig. 5.4 Visitation intervals of buffalo utilizing patches of *Cynodon dactylon* in Lake Manyara National Park. *Cynodon* patches are indicated in black. The grey area is escarpment and the vegetation codes are R = riverine bush, F = forest, W = woodland savannah, G = grassland savannah (modified from Drent and van der Wal 1999, after Prins 1996).

Africa (du Toit 2003). In fact du Toit points out that avoidance of condensed tannin has a stronger effect on feeding preference than selection for leaf nitrogen or phosphorus. What is more, ruminants can accurately regulate their intake of specific toxins (du Toit *et al.* 1991) to match their detoxification abilities. This could explain why many browsers take many

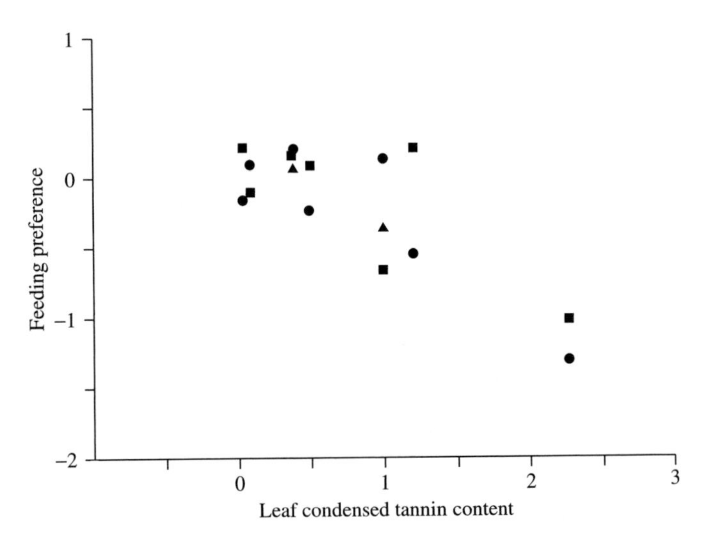

Fig. 5.5 Feeding preference of kudu (■), impala (●), and steenbok (▲) in central Kruger (Figure modified and reprinted from *The Krugar Experience* by Johan T. DuToit, *et al.*, eds. © 2003 Island Press. Reproduced by permission of Island Press, Washington, D.C.).

small meals from a wide range of plants, and plant parts. Steenbok in central Kruger, for example, feed on the bright yellow and toxic fruits of *Solanum panduraeforme* but rarely take more than one fruit from a single bush, even if it is laden with fruit (du Toit 2003). Browsers also affect the trees they browse, particularly young trees. An example from Laikipia, in central Kenya, is shown in Figure 5.6. *Acacia drepanolobium* trees were tagged (Chapter 4), and the damage inflicted by browsers recorded over a three-year period. The principal browsers were elephant, black rhinoceros and reticulated giraffe, and the damage inflicted by each species could be recognized. Elephants either pushed over trees, or broke the main stems and left the bark hanging in strips. Rhinos made a clean cut of the main stem, and giraffe ate the leaves and growing tips. Damage was placed into two categories: tree killed, or tree 'reversed' into a lower height category. What is clear is that browsing damages this *Acacia* to quite an extent. Giraffe had little effect, but black rhinoceros and elephants kill, and reverse, many trees particularly young saplings.

Of course, savannah ecologists and wildlife managers have frequently been worried by the 'elephant problem' (for example, Croze 1974a and 1974b), coupled with an often observed trend in savannah vegetation from more wooded, to more open grassland savannah (Tanzania: Vesey-Fitzgerald 1973, Kenya: Leuthold and Sale 1973, Uganda: Laws 1970, Zambia: Caughley 1976). However, the problem is more complex than it seems at first sight. Elephants actually prefer to eat green grass, fire may play a pivotal role and the woodland to grassland savannah trend may in fact be

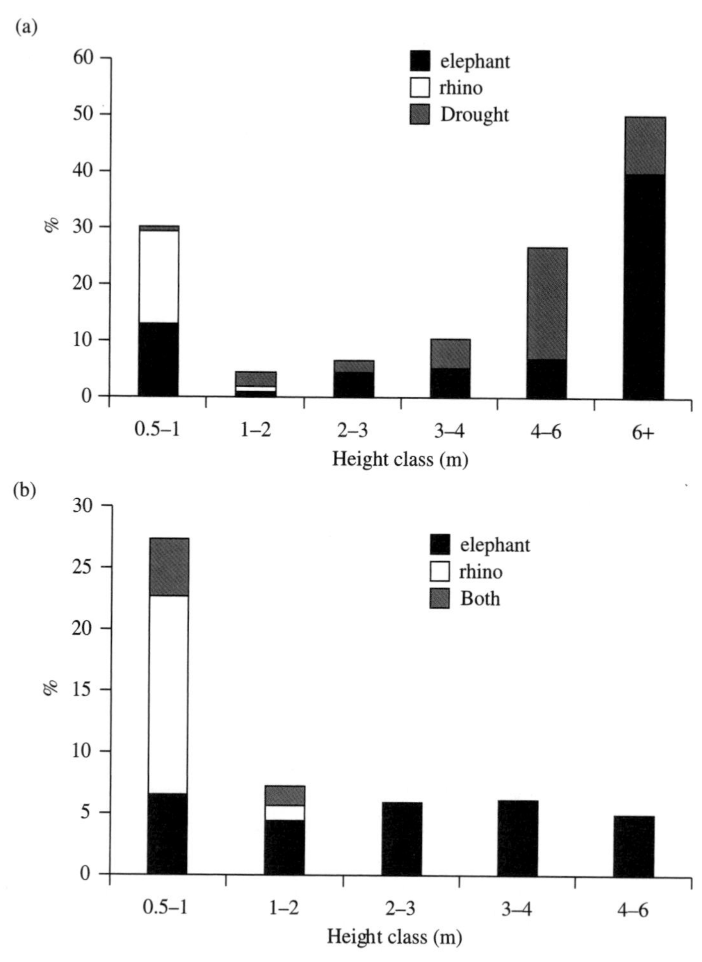

Fig. 5.6 Herbivore damage on trees. The percentage of *Acacia drepanolobium* (a) killed and (b) reversed in each height class over a three year period (from Birkett and Stevens-Wood 2005).

cyclical and quite natural. The interaction of these three factors (elephants, trees, and fire) have been particularly well described for the Serengeti–Mara ecosystem (Norton-Griffiths 1979; Dublin *et al.* 1990; Dublin 1995).

The reconstruction of the vegetation dynamics of the Serengeti–Mara ecosystem involved the collection of historical information as well as more recent observations, and the construction of a simple mathematical model. In the early 1900s, Swahili slave traders, and European explorers, hunters and naturalists describe the Serengeti–Mara as open grassland with occasional *Acacia* trees, rather like much of the Mara Reserve today. The recently introduced rinderpest had reduced ruminant numbers to low levels, and the elephant population had suffered from heavy ivory poaching during the

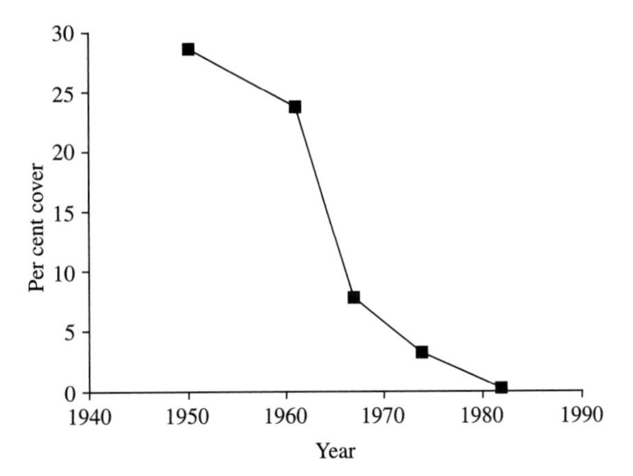

Fig. 5.7 Mean percentage cover in Acacia woodland savannah in the Masai Mara Reserve. Percentage cover values are derived form five sets of aerial photographs taken between 1950 and 1982 (from Dublin 1995).

previous decade (Spinage 1973). Human numbers were also low, and consequently bush fires were infrequent. Tree recruitment was probably good and during the next thirty to fifty years dense woodland savannahs became established and heavy infestations of the tsetse fly became a serious problem (Ford and Clifford 1968; Ford 1971). It was at this time that the colonial administration started to introduce measures to protect these 'pristine' woodland savannahs, not realizing that less than fifty years earlier they had been open grassland savannahs.

By the late 1950s, and early 1960s, these woodland savannahs had started to decline. Ungulate populations had not yet recovered from the rinderpest epidemic and, as a consequence, a large standing crop of dry grass resulted in widespread bush fires. These were started by Masai to improve grazing pasture and clear tsetse infested bush, by hunters, and by the Park authorities. These fires cleared large areas of bush and attracted ungulates to 'grazing lawns'. Elephant numbers also increased, not so much because of population growth but because increasing human activity around the Park area forced elephants into the Serengeti–Mara, increasing the local density. Both fire and elephants had a detrimental effect upon tree recruitment. An analysis of aerial photographs (Dublin 1995) from the Masai Mara shows a steady loss of *Acacia* cover (Fig. 5.7).

Between the 1960s and 1980s two important changes occurred in the Serengeti–Mara ecosystem. First, with the eradication of rinderpest, the numbers of wildebeest increased dramatically (Chapter 4). More of the green grass was removed and consequently there was less dry grass available to fuel fires. The incidence of fires dropped dramatically and by the 1980s it was as low as 5 per cent in the Mara. Second, in April 1977, the border between Tanzania and Kenya was closed, and remained closed until 1986.

This closed the main tourist route between the Mara and the northern Serengeti and tourist numbers to the Serengeti dropped from 70,000 in 1976 to 10,000 in 1977. One consequence of this was a drop in the operating budget of the Serengeti which was linked to income from visitors. As a result anti-poaching patrols dropped to 60 per cent of that prior to the border closure. Two animals immediately suffered: black rhinoceros and elephant. Rhinos were effectively removed from the Serengeti part of the ecosystem (Fig. 6.19a), and are still absent (although occasional animals visit from the Ngorongoro Crater). The Serengeti elephant population declined by 81 per cent from 2,460 in 1970, to 467 in 1986. Of these, over 1,500 were killed by poachers, while the remaining 400 to 500 sought safe refuge in the Masai Mara, where tourist numbers, and anti-poaching patrols, were still high.

To help understand this intriguing interaction between trees, elephants and fire, Dublin *et al.* (1990) constructed a 'tree-population' model to describe and visualize these events. Animals and fire have three different effects on tree seedlings, which these authors describe as 'killing' (complete removal), 'reversing' (removal to ground level with the possibility of resprouting) and 'inhibiting' (reduced, and kept in the height-class below 1 m). Trees were placed into five height-classes, the last being the adult class. The model computes the tree recruitment rate as the ratio of trees entering the adult height-class to the number of adults dying each year. It uses fourteen simple equations, informed by field data, to describe the transfer of individual trees between height-classes (the dynamics of the tree population). These 'transfers' are affected by elephant damage, fire damage and the effect of wildebeest trampling and the browsing of other antelopes. My depiction of their results is shown in Figure 5.8 along with my

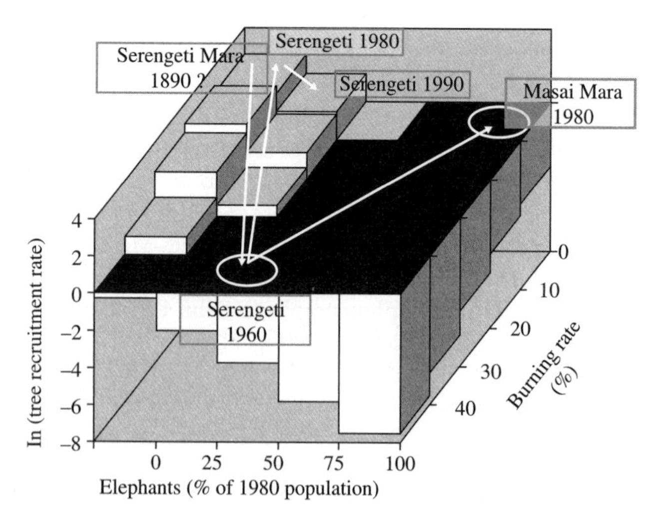

Fig. 5.8 Tree recruitment, as a function of elephant numbers and fire incidence, as predicted by the population model of Dublin *et al.* 1990.

interpretation of events in the Serengeti–Mara ecosystem. Tree recruitment is shown on the vertical axis and the important point is that all outcomes above zero (the black surface) are positive for tree recruitment, while all outcomes below are negative for tree recruitment. It is also important to remember that this graph describes recruitment not how the system will 'look' at a particular time. The 'look' will follow the pattern of recruitment by many years. For example, around 1900 tree recruitment was high but there were few trees. Over the next few decades tree cover increased. One interesting feature of these predictions is that an 'ecosystem' can move to two divergent 'equilibria', as represented by the Mara and the Serengeti post 1980. With the reduction of fires and fewer elephants trees have recovered in the northern Serengeti while across the Tanzanian/Kenyan border the higher incidence of elephants have kept the grassland savannahs open.

In Chapter 1 we saw how savannah biomes in different parts of the world, although showing similarities, contained rather different ecosystems. Within African savannahs a variety of broad types exist, sandwiched between deserts and tropical forests. In the latter part of Chapter 2 we saw how within one of these broad types (the Serengeti ecosystem) there were areas containing different soils, rainfall, and vegetation. Even the southern 'grasslands' were not uniform, with 'short', 'intermediate' and 'long' having their own growth forms, and species. Now we find that even within a localized area of savannah there can be yet other vegetation mosaics, produced by plant defences, grazing and fire. Savannah plants, and their associated animals, clearly inhabit a mosaic environment, heterogeneous on many different scales.

Predators and prey

Just like plants and herbivores, bovids and their carnivores have probably been interacting in African savannahs for at least five million years. Carnivores have developed physical (claws) and behavioural (hunting techniques) attributes that facilitate capture, and prey have developed physical (long thin legs) and behavioural (herding) attributes for escaping capture. In this section I will look at some of the characteristics of these ecological interactions. One of the predictions of ecological theory is that predator and prey will fluctuate numerically in a regular manner. In the original Lotka-Volterra model these predator-prey cycles are rather fragile, with neutral stability, but they have been replaced with models that have 'stable limit cycles' as part of their behaviour (see Taylor 1984, chapter 10). Possibly one of the best known, and most debated, examples in vertebrates is the 10 year cycle of snowshoe hares (*Lepus americanus*) and lynx (*Felis lynx*) in Canada for which more than a 100 years of population estimates are available from fur-trapping records (Elton and Nicholson 1942; Krebs *et al.* 1995). The original explanation is that predator numbers drive numbers of prey, but it is just as feasible that predator numbers simply follow prey, who in turn follow their plant food, which in turn follows some climatic variable such as rainfall. In this case, if a plant-herbivore-predator system is driven by a climatic variable that doesn't cycle, or fluctuate in any regular manner, there may be no cycles at all. For most African savannahs we do not have the extensive population data that would allow us to search for this 'characteristic' feature of predator-prey interactions. Long runs of herbivore population data are limited and long runs of predator population data are non-existent (Chapter 4).

However, large herbivore populations in Kruger National Park (KNP), South Africa, have reportedly shown quasi-cyclic fluctuations extending back to the 1920s (Stevenson-Hamilton 1947; Owen-Smith and Ogutu 2003) and these have recently been analysed by Ogutu and Owen-Smith (2005). The background to these data are detailed in Chapter 4, and here I look only at the reported oscillations. Previous authors (Tyson 1986; Owen-Smith and Ogutu 2003) have reported an approximately eighteen-year cycle in rainfall and therefore Ogutu and Owen-Smith analysed this data also. They analysed the population data for twelve ungulate species (those in Fig. 4.22), and four of these (buffalo, eland, giraffe and common zebra) are shown in Figure 5.9. They used two kinds of correlation. Autocorrelations within each species (Fig. 5.9a), and cross-correlation between each species and rainfall (Fig. 5.9b). If populations show regular cycles of abundance, the autocorrelation between 'peaks' will produce positive autocorrelations (with the lag indicating the cycle length), and those between 'peaks' and 'troughs' negative autocorrelations (with the lag indicating half the cycle length). The half-cycle length indicated for those species in Figure 5.9a are

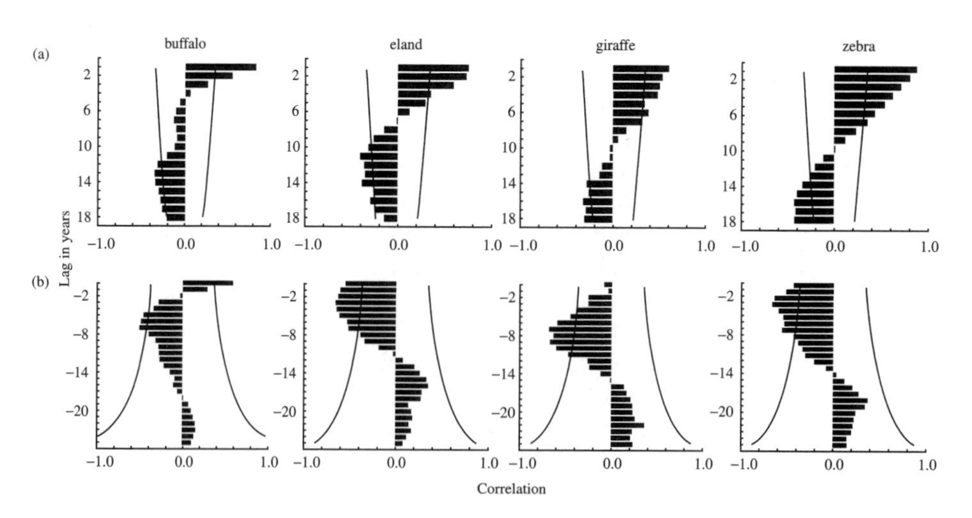

Fig. 5.9 Correlograms for four Kruger ungulate species. (a) autocorrelations with previous population estimates, (b) cross-correlations between population estimates and rainfall. Lines are 95% tolerance limits (modified from Ogutu and Owen-Smith 2005).

approximately 12–14 years for buffalo, 11–12 years for eland, 16–18 years for giraffe and 16–18 years for zebra. The range of such half-cycle lengths was from 16–18 (giraffe and zebra) to 8 (warthog), and there is some indication that those species with the longer generation time had the longer population cycles. Notice however that all the suggested cycle lengths are less than the observed cycle in rainfall. There is some 'effect' of rainfall seen in the cross correlations (Fig. 5.9b), with most species showing a positive correlation with the rainfall of sixteen to twenty years earlier. It is therefore difficult to see these fluctuations in ungulate population size as a simple, direct, response to a cycling environment (rainfall) and the link with generation time may well indicate an influence of predators.

Of course, predators in African savannahs do not only catch one type of prey, as in classical predator-prey models and the Canadian lynx and snowshoe hare system. They take a variety of prey items, as indicated for the larger Kruger carnivores in Figure 5.10. Notice that the two largest prey categories (100–350 kg and > 350 kg) are only taken by spotted hyaena and lion, the largest predators. The interaction between predator and prey is constrained by their physical size, and small predators usually catch smaller prey than large predators (of course small predators can catch the smaller young of larger predators). This means that 'larger' prey may 'escape' the predation pressures imposed on smaller prey. Indeed this was an explanation put forward by Sinclair *et al.* (2003) to explain the results of their northern Serengeti–Mara analysis, outlined in Chapter 4 (Fig. 4.21). In this ecosystem, 28 species of ungulate are prey for ten species of large carnivore.

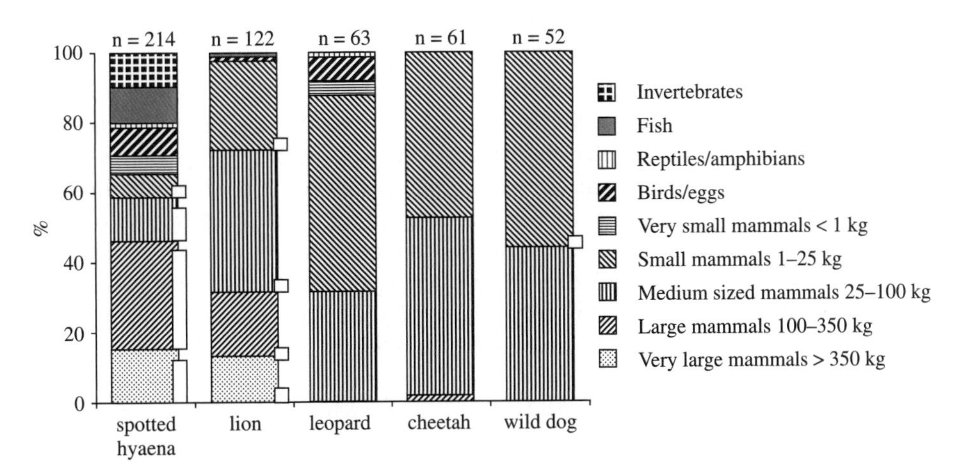

Fig. 5.10 Percentage occurrence of different prey categories in the diets of large carnivores in Kruger National Park (n = the number of food items eaten). The narrow histogram on the side of a species' histogram shows the proportion of items scavenged (from Mills and Funston 2003).

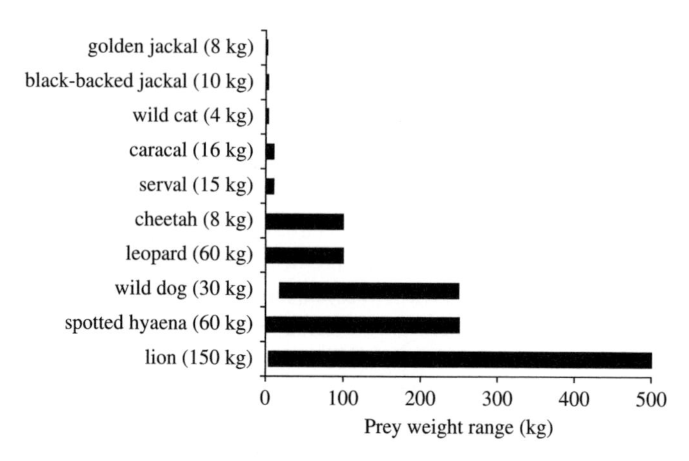

Fig. 5.11 The range of weights of mammal prey consumed by carnivores of different sizes in the Serengeti ecosystem (from Sinclair *et al.* 2003).

As in the Kruger ecosystem, each predator species shows a characteristic pattern in the size of their prey, and the diet range of smaller carnivores is nested within that of larger carnivore species (Fig. 5.11). Of course within the larger range each predator has its preferred choice of prey. Thus the lion's preferred prey (38 per cent of animals consumed) is the wildebeest and common zebra (170–250 kg), but 44 per cent of the diet is made up of smaller prey (2–45 kg). Buffalo and giraffe comprise 5–15 per cent. Notice however, that in terms of biomass consumed, rather than individuals, these three categories are 63 per cent, 14 per cent and 15–24 per cent respectively (Sinclair *et al.* 2003). Each carnivore therefore includes prey outside their preferred size range but is inefficient at catching it. Lions are less efficient at catching gazelles than catching wildebeest, but they take gazelles if the opportunity arises. Therefore, despite their smaller size, gazelles are less vulnerable to lion predation than are wildebeest. Because of the ranges shown in Figure 5.11, ungulates with smaller bodies suffer predation from many more predators than do larger ungulates (Fig. 5.12). Small species of antelope, such as oribi, are prey to five species of cats, two canids, hyaena, and many smaller carnivores in the Serengeti. Medium sized antelopes like wildebeest are prey to only three cats, wild dog and hyaena, while larger ungulates like buffalo and giraffe are predated by lion only. Of course this does not necessarily mean that oribi suffer more predation than wildebeest. Lion may be very efficient at catching wildebeest. How, and when, do the large carnivores of the African savannah catch their prey? Table 5.3 compares the details for the five largest predators.

In addition to the restrictions of body size, already mentioned, predators clearly take different species in different areas. This is often because predators simply take those prey that are most abundant in the local area.

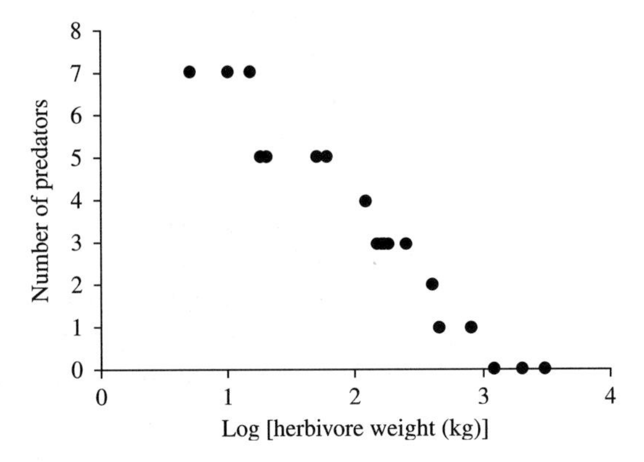

Fig. 5.12 The range of weights of mammal prey consumed by carnivores of different sizes in the Serengeti ecosystem (from Sinclair *et al.* 2003).

However, local habitat can also restrict predation. In Kafue National Park, in western Zambia, puku (*Kobus vardonii*) and lechwe (*Kobus leche*) are abundant, but they are hunted with great difficulty by lions, who avoid the swampy habitats used by these prey. Seasonal habitat changes can also influence predation. In this same area, buffalo form 38.4 per cent of lion kills in the dry season, when vegetation density is reduced by bush fires, while in the wet season buffalo form only 18.1 per cent of lion kills. Probably one of the most intriguing instances of habitat structure influencing predation is that observed by Prins (1996). He studied buffalo predation, by lions, in Manyara National Park, Tanzania (see Figure 5.4 for a map of the area). By noting where buffalo carcasses, that were the result of lion kills, were found (kills are not moved) he was able to estimate how 'risky' different habitat types were for buffalo. *Sporobolus spicatus* grasslands were risky, bare mudflats were relatively safe. However, to understand the relationship between lions and buffalo it is necessary to look at combinations of vegetation type rather than individual vegetation patches in isolation. Lions try to stalk their prey unobserved. They therefore require vegetation cover. In Manyara, buffalo spend most of the day in the open grassland, where it is difficult for predators to approach. The best strategy for lions is to observe their potential prey from cover, in the ecotone between grasslands and woodlands. Here they can look for the best position to start their attack when buffalo move in their direction.

The frequency distribution of kills in Figure 5.13 shows that this border between closed and open vegetation (an ecotone) is a 'killing zone' for buffalo. Of course one of the reasons that buffalo are 'safe' in the open grasslands and mudflats is that they can be vigilant and this leads us to consider the defences against predation that prey employ. Many savannah ungulates live in groups. There are benefits and costs to such behaviour (see Krause and Ruxton 2002 for a complete review), here I will briefly examine some

Table 5.3 Predation data for the large African savannah carnivores.

	cheetah	leopard	lion	spotted hyaena	wild dog
Hunting time of day	Day	Mainly night	Mainly night	Night and dawn	Day
Number of animals hunting	1	1	1 to 5	1 to 3 for gnu and gazelles 4 to 20 for zebra	Whole pack 2 to19
Method of hunting	Stalk then long fast sprint	Stalk then short sprint	Stalk then short sprint	Long distance pursuit	Long distance pursuit
Distance from prey when chase starts	10 to 70 m	5 to 20 m	10 to 50 m	20 to 100 m	50 to 200 m
Speed of pursuit	Up to 87 km per hour	Up to 60 km per hour	50 to 60 km per hour	Up to 65 km per hour	Up to 70 km per hour
Distance of pursuit	200 to 300 m	Up to 50 m	Up to 200 m	0.2 to 3.0 km	0.5 to 2.5 km
Hunting success rate	37 to 70%	5%	15 to 30%	35%	50 to 70%
Commonest prey species in East Africa	Thomson's and Grant's gazelle, impala	impala, Thomson's gazelle, dik dik, reedbuck	zebra, wildebeest, buffalo, Thomson's gazelle, warthog	wildebeest, Thomson's gazelle, zebra	Thomson's gazelle, wildebeest, zebra, Grant's gazelle
Commonest prey species in southern Africa	springbok, kudu (calves),warthog, impala, puku	impala, bushbuck, waterbuck, warthog, nyala	zebra, wildebeest, buffalo, gemsbok, springbok	wildebeest, buffalo, impala, springbok kudu, zebra, giraffe	impala, kudu, reedbuck, roan, duiker
% kills lost to other carnivores	10 to 12%	5 to 10%	≈ 0% Serengeti to 20% Savuti	5% Serengeti 20% Ngorongoro	50%
% of diet obtained by scavenging	≈ 0%	5 to 10%	17 to 47%	33%	3%

Data from Bertram 1979, Bothma and Walker 1999.

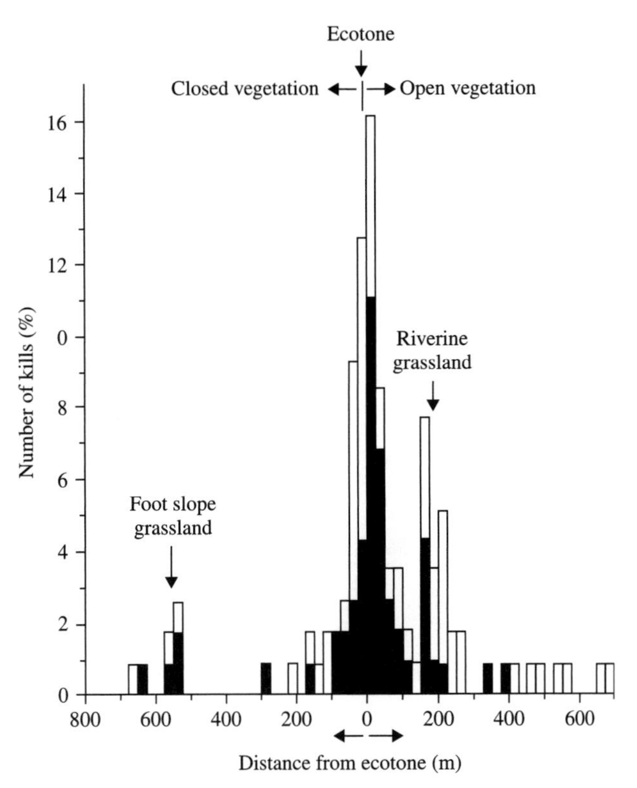

Fig. 5.13 Distribution of buffalo probably and certainly killed by lion (solid bars) and those possibly killed by lions (open bars) in relation to the distance from the ecotone between open grasslands and woodlands. The small peak on the left indicates a second, minor, ecotone between foot slope grasslands and woodlands. The peak at the right indicates the riverine delta grasslands where lions can easily hide in the many gullies (from Prins 1996).

of the stated anti-predator benefits. These basically fall into three categories: increased vigilance ('many eyes' theory), predator confusion, and dilution of risk. Increased vigilance, with increased group size, is a consequence of more eyes looking for predators coupled with the transmission of information about detected predators throughout the group. As a result, an individual in a group doesn't need to detect a predator itself in order for it to be aware of the potential attack. The 'many eyes' idea suggests therefore that as group size increases an individuals 'share' of the group vigilance can decline, leaving more time for other activities such a foraging.

Figure 5.14 shows some data for impala in Nairobi National Park (Shorrocks and Cokayne 2005). Selected individuals within both breeding, and bachelor, herds were observed and the amount of time devoted to eight behaviours noted. These were foraging, vigilance (head up), moving, grooming, allogrooming, lying down, and urinating. Some of these

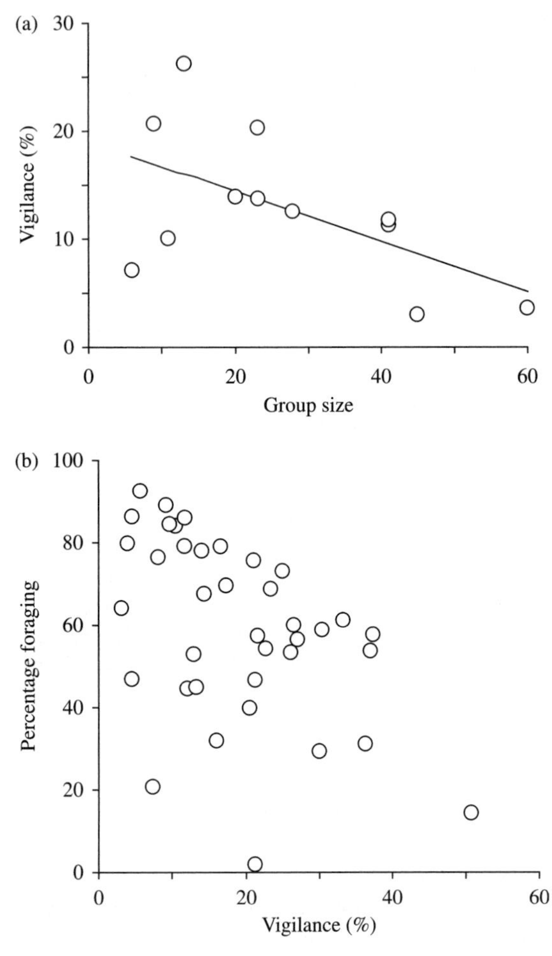

Fig. 5.14 Vigilance in impala. (a) Relationship between group size and % time vigilant for breeding females ($y = 19.15 - 0.2x$, $F = 4.74$, $P = 0.05$). (b) Correlation between % time vigilant and % time foraging for 38 individuals in both breeding and bachelor herds (from Shorrocks and Cokayne 2005).

behaviours are not mutually exclusive (e.g. an individual can move and be vigilant). Figure 5.14a shows the data for randomly selected females within breeding herds. Selected individuals spent less time being vigilant in larger groups, and overall individuals spent more time foraging when they were less vigilant (Figure 5.14b). Interestingly, it may be unsafe for individuals to 'cheat' and spend all their time foraging. FitzGibbon (1989) found that cheetahs hunting Thomson's gazelle preferentially attacked low-vigilance individuals. The second anti-predator effect of living in groups, predator confusion, is thought to arise because of an inability of the predator to single out and attack individual prey within a group. They are distracted by too

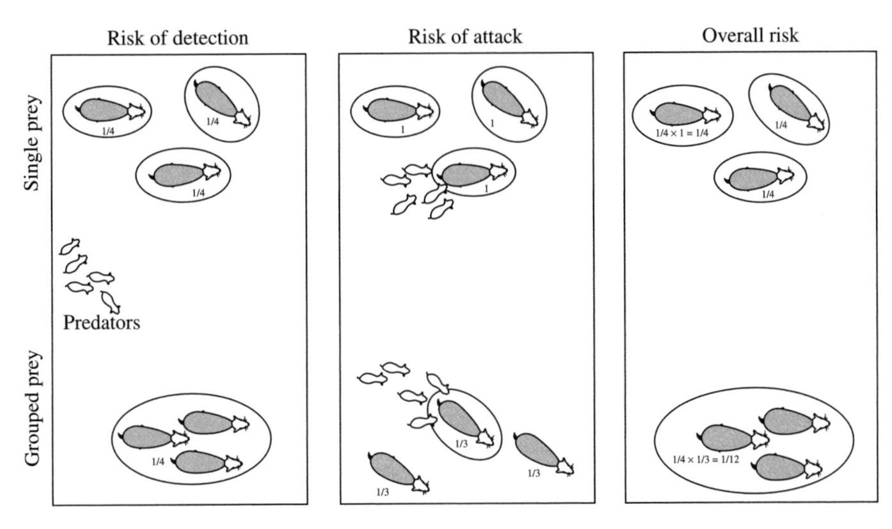

Fig. 5.15 The suggested anti-predator 'dilution effect' of being in a group. In the 'search phase' (left-hand box) there are four targets each with a detection probability of ¼. In the 'attack phase' (middle box) the risk of attack is ⅓ for the three individuals in the discovered group, while the risk of attack is 1 if you are a discovered lone individual. The right-hand box gives the overall probabilities of being attacked for an individual in a group or alone (modified after Krause and Ruxton 2002).

many targets. It's an appealing idea, but empirical support for the confusion effect is limited.

The 'dilution effect' basically says that there is safety in numbers. Included within this category of group benefits are two separate probabilities: a lower risk of detection, and a lower risk of capture when detected. The hypothetical example in Figure 5.15 shows the probability of detection during the search phase of the predator, and the probability of being selected as a target by the predator during the attack phase. The combination of the two probabilities shows how the overall probability of being captured by a predator could be reduced by being in a group. However for this to work, the predator must only take one prey during an attack, and the probabilities must be 'similar' to those in Figure 5.15. For example, if we changed the 'risk of detection' for the group to ½, perhaps because it was more easily seen from a distance (the risk for individuals would now be ⅙), the overall group probability would now rise to $\frac{1}{2} \times \frac{1}{3} = \frac{1}{6}$, the same as the overall individual probability ($\frac{1}{6} \times 1 = \frac{1}{6}$) and the 'safety in numbers' would disappear. In fact, the only data on this kind of risk assessment, in savannah ungulates, would suggest that the situation is less straight forward than the simple theoretical example in Figure 5.15. Creel and Creel (2002) calculated the 'risk of death' for an individual wildebeest, when the predator was the wild dog (*Lycaon pictus*), in the Selous Game Reserve, Tanzania. This measure of 'overall risk' took into account all the processes by which herd size might affect an individual's vulnerability to predation. These included 'risk of detection', the decision to hunt or not, the success of hunts, and the final 'risk of attack'. Notice that not all these are depicted in Figure 5.15. Their results are shown in Figure 5.16, and it is important to remember that these are probabilities obtained from observations made in the field, not a theoretical assessment. The 'risk of death' does in fact change with group size. However, the lowest individual 'risk of death' is obtained when herd size is intermediate (around 40), not at the largest group size actually observed in Selous Game Reserve. Larger groups of wildebeest are more easily detected than smaller groups and this cancelled out any benefit of reduced 'risk of attack'. This advantage, for individuals in groups of intermediate size, leads us to consider the distribution of group sizes naturally observed in savannah ungulates.

The distributions of group (herd) size have been recorded for several African ungulates (Sinclair 1977; Wirtz and Lörscher 1982; Creel and Creel 2002). They usually show a right-skewed frequency distribution (Figure 5.17a and b). Small groups are very common, medium sized groups and large groups less so. However, if we re-plot such distribution data, in terms of the group size experienced by each individual, we find that most individuals in fact choose to live in groups of intermediate size (Fig. 5.17b and d). In other words, large groups are uncommon, small groups are common but

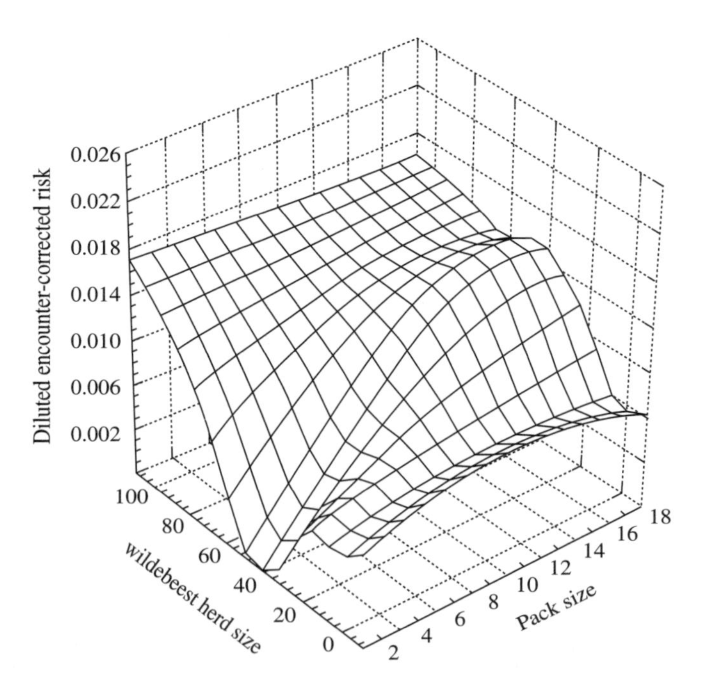

Fig. 5.16 The 'risk of death' for an individual wildebeest as a function of wildebeest group size and wild dog pack size (from Creel and Creel 2002).

comprise a relatively small number of individuals. Being in a larger group might confer better anti-predator advantages (higher vigilance, lower risk of attack), but being in too large a group has additional costs, such as increased competition for resources and, from the wildebeest example, higher risk of detection. There appears to be a medium group size that maximizes the benefits, and minimizes the costs. Of course these benefits and costs will vary from species to species, so we might expect group size to vary for savannah ungulates with different ecologies.

In a paper that has influenced African ecologists for over thirty years, Jarman (1974) compiled information on 75 species of ungulates, and compared their body size, diet, group size, habitat preference, and antipredator behaviour. His conclusions were based entirely on narrative description. Using more quantitative, statistical, methods this data has been re-examined by Brashares *et al.* (2000) and their conclusions, relevant to group size and

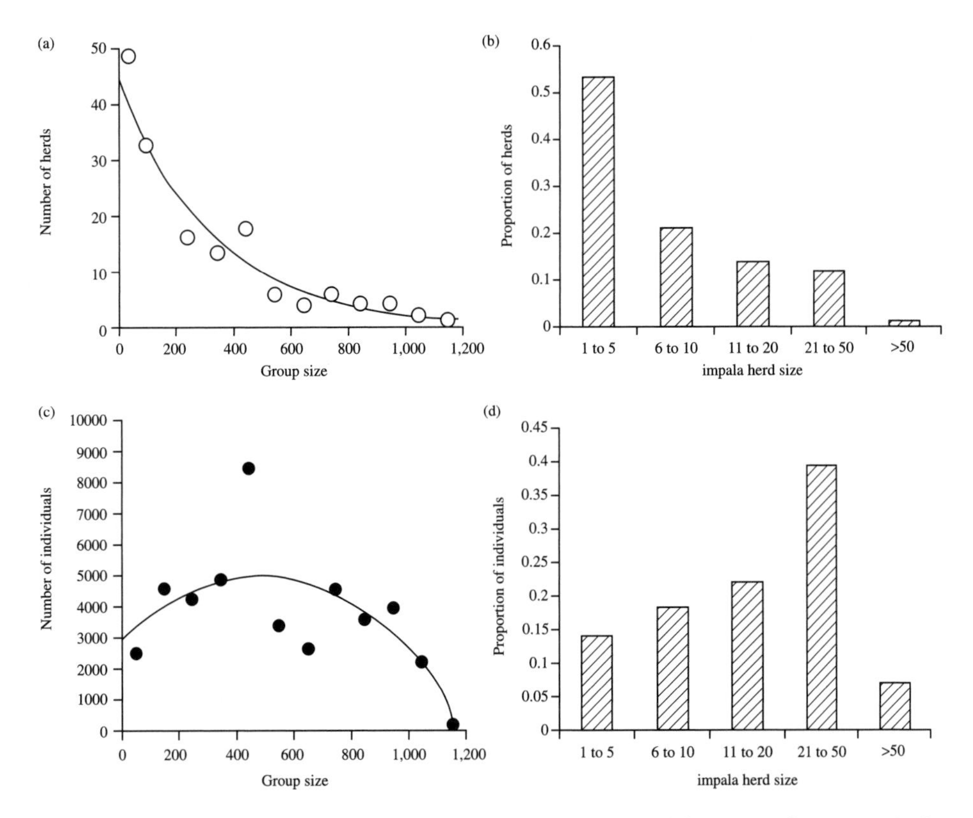

Fig. 5.17 Group (herd) size in two African savannah ungulates. (a) frequency of group size in the African buffalo, in Serengeti National Park (c) same data but showing the group size experienced by individuals (data from Sinclair 1977). (b) frequency of group size in impala, in Selous Game Reserve, (c) same data but showing the group size experienced by individuals (data from Creel and Creel 2002).

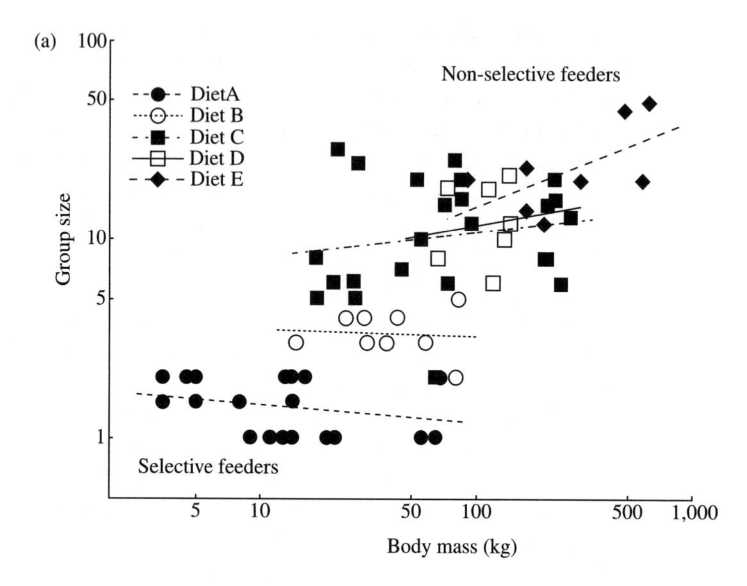

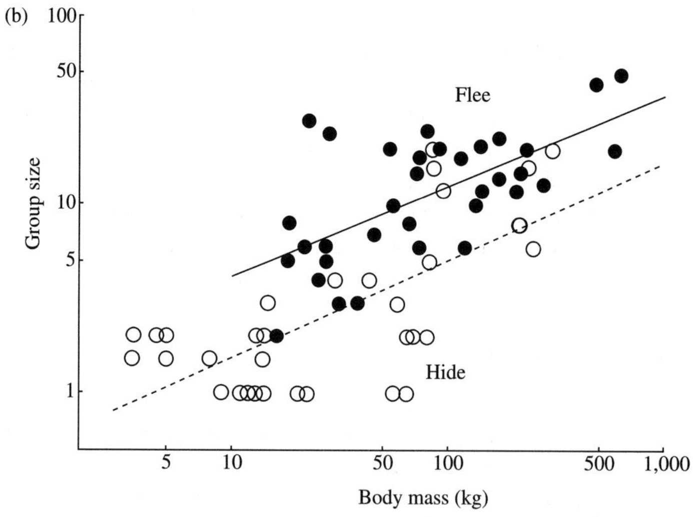

Fig. 5.18 (a) Group size versus body mass ($\log_{10}$ axes) for 75 species, or subspecies, of African antelope separated by diet (Jarman 1974). A: selective browsers, B: selective feeders on grass or browse, C: feeders on a range of grasses and browse, D: feed unselectively on grasses, E: feed nonselectively on a wide range of grasses and browse. Lines are least square regressions. (b) Group size versus body mass ($\log_{10}$ axes) for 75 species, or subspecies, of African antelope separated by anti-predator response (Jarman 1974). Antipredator behaviour is divided into two general categories, antelope that flee to avoid predation (●) and those that hide (O). Lines are least square regressions (from Brashares *et al.* 2000).

antipredator behaviour, are shown in Figure 5.18. Jarman's (1974) conclusion that group size and body size varied predictably with feeding style was confirmed (Figure 5.18a). Non-selective roughage-feeding antelopes are larger, and occur in larger groups, than selective feeders. Antelopes that flee when faced with a predator are more likely to occur in larger groups than those that avoid detection by freezing or hiding (Fig. 5.18b). The original narrative explanation (Jarman 1974) is still the 'best story' to explain these inter-related ungulate traits (body mass, group size, diet and antipredator behaviour) (Figure 5.18).

Small bodied species (such as steenbok and dik dik) have small mouths and narrow muzzles that facilitate their specialization on the most nutritious parts of plants (diets A and B). The clumped dispersion, and limited availability, of these high-quality plant parts, coupled with the relatively higher metabolic demands of small bodied mammals (Fig. 5.25), results in competition for food and adoption of territorial spacing behaviour. These species are therefore likely to live alone or in pairs, in closed habitats, in which 'freezing' or 'hiding' is the best predatory defence. At the other end of the size spectrum, large-bodied species (such as eland and buffalo) feed less selectively on course grasses (diets D and E). In part this is because they lack the morphology and dexterity to feed selectively. The widespread supply of coarse grasses results in little competition for food (but see later in this chapter) and therefore little selection for spacing behaviour. Because of the low nutritive value and seasonal availability of course grasses, there would be little selection for spacing behaviour. These species would have to forage over much greater distances, in more open habitats. This would favour formation of large groups, and the anti-predator tactics of group vigilance, and fleeing when attacked.

Parasites and hosts

We know from serological studies that many African mammals contract diseases of various kinds. In other words, when tested, they are found to contain antibodies to various pathogens. For example, at least eleven viruses, five bacteria and four protozoa are recorded from free-ranging populations of wild dogs (*Lycaon pictus*) (Woodroffe, Ginsberg and Macdonald, 1997). Adult spotted hyaena (*Crocuta crocuta*), from the Serengeti, have antibodies to at least ten pathogens, four of which (rabies, canine herpes, canine parvovirus and canine adenovirus) are identical to those listed for wild dog (Mills and Hofer 1998). With the cats there is recorded exposure of, cheetah to anthrax and feline coronavirus, of lions to canine distemper virus, and lions and leopards to feline immunodeficiency type viruses. Many of these diseases also affect domestic animals, which appear to serve as a reservoir for the pathogens, from which they can be transmitted to wildlife populations.

However, this type of serological data does not tell us how severe the effects of the pathogens might be. A high seroprevalence (many individuals having antibodies) may simply indicate that most animals become infected early in life, but that the resulting disease is mild and most animals recover and become immune. Alternatively, the same high seroprevalence might indicate that the population has recently been exposed to a highly virulent disease, and that only those individuals that survived (and are therefore sero$^+$) are now present. However, although we do not know how severely these pathogens affect most savannah animals, we do know that some pathogens cause high mortality in some species. Rabies (a rhabdovirus), for example, is an endemic, multispecies disease in many areas of sub-Saharan Africa, with sporadic epidemic cycles. It has been diagnosed in 33 carnivore species and 23 herbivore species. In1989, a pack of wild dogs, living at Aitong outside the Masai Mara Reserve in southern Kenya, was decimated by rabies. The following year, at least one wild dog died of rabies in the nearby Serengeti N.P., and in 1991 all the wild dog packs under study in the Serengeti ecosystem disappeared (Fig. 5.28). Rabies was suspected, and is also known to have killed wild dogs in Namibia and Zimbabwe. An epidemic of rabies is known to have halved a population of Ethiopian wolves in the Bale Mountains (Sillero-Zubri *et al.* 1996). In the Aitong pack the disease spread rapidly (rabies is transmitted mainly by biting). The time from the recording of the first infected dog to the death of the last of the 21 dogs that died was less than two months (Kat *et al.* 1995). Animals, such as wild dogs, living at low density, in groups, may therefore be at particular risk, since disease seems to have caused local population declines in other areas. For example, sightings of wild dogs declined dramatically after an outbreak of anthrax (which is known to kill wild dogs) in ungulates in the Luangwa Valley, Zambia, and population declines of wild dogs in Zimbabwe in the 1980s coincided with an epidemic of rabies in jackals. One of the most dramatic disease outbreaks, in a carnivore population, occurred in the Serengeti National Park in the early 1990s. By July 1994, an epidemic of canine distemper virus (CDV) had affected 20 per cent–30 per cent of the 3,000 lions in the Park, and in a monitored population of 250 lions, 87 died or disappeared. CDV was originally thought to be mainly a disease of canids. However, blood samples from Serengeti lions in 1985 suggest that the population may have been initially exposed to CDV around 1980 and that the virus has been increasing its pathogenicity in felids since that time. In addition, during this lion epidemic, several spotted hyaena cubs, below the age of six moths, died from CDV, and a molecular analysis of the virus isolated from hyaenas and lions indicated that they were more closely related to each other, than to CDV found in the nearest domestic dog population (Mills and Hofer 1998). Cases of hyaenas killed by CDV have also been reported from South Africa, Namibia, Zambia, Malawi, and Ethiopia.

Of course the most dramatic African wildlife epidemic did not directly affect carnivores, although it almost certainly affected them indirectly (see Chapter 4). This was the rinderpest epidemic that spread throughout sub-Saharan Africa, from about 1890. Rinderpest is a highly contagious viral disease of cattle, domestic buffalo, and some species of wildlife. The name is German, meaning cattle-plague. It is characterized by fever and high mortality. Rinderpest virus is a single-stranded RNA virus in the family Paramyxoviridae, genus *Morbillivirus*. It is immunologically related to canine distemper virus, human measles virus, peste des petits ruminants virus, and marine mammal morbilliviruses.

Most of the historical spread of rinderpest is likely to have occurred in the Indian subcontinent, the Near East, and Europe, following the domestication of ungulates. Since ungulate species are at low density in the Sahara desert they would probably not have supported a continuous infection of the disease. The desert would therefore have acted as a barrier to the spread of rinderpest into Africa. The pandemic that occurred at the end of the nineteenth century was initiated by the accidental introduction of a few infected cattle into the Horn of Africa in 1890. This caused a major pandemic which spread mainly via the movement of cattle along trade routes. It moved westward and southward and reached southern Africa in 1896. This initial '1890 epidemic' is thought to have killed 80 per cent to 90 per cent of all cattle in sub-Saharan Africa. Unfortunately, wild ungulates are infected by contact with cattle, drinking at the same water etc, and table 5.4 shows the main recorded epidemics of rinderpest in East Africa and the wildlife species affected. More recently, another rinderpest outbreak that raged across much of Africa in 1982–4 is estimated to have cost at least US$500 million in domestic livestock losses. The impact that rinderpest had on wildlife populations is seen in the dramatic six fold increase in the wildebeest population of the Serengeti ecosystem, following its disappearance after a cattle vaccination campaign in the early 1960s (Chapter 4). This was followed by an increase in predator populations, particularly the spotted hyaena.

Another disease that has had severe effects on wildlife populations is anthrax (*Bacillus anthracis*). After a susceptible host ingests spores, the bacilli enter the bloodstream, replicate exponentially, and produce fatal septicaemia. In the Kruger ecosystem in southern Africa, epidemics occur in

Table 5.4 The main recorded epidemics of rinderpest in East Africa and the species of wildlife affected.

Year	Species affected
1890	Most ungulate species
1897	hartebeest and kudu
1913–1921	eland and giraffe, then buffalo, bushbuck and reedbuck
1929	buffalo, bushbuck, and warthog, then eland and waterbuck
1931	buffalo, giraffe and wildebeest
1937–1941	buffalo, eland and giraffe, then buffalo, eland and kudu
1949	eland, and then wildebeest
1960	eland, kudu, warthog, buffalo, bushbuck, giraffe, impala and oryx

Modified from Simon 1962.

episodes of 6–20 years. These episodes tend to occur during dry periods, and in seasons when over abundant populations of herbivores are congregating around stagnant surface water. Although most mammal species in Kruger are susceptible to anthrax, kudu appear particularly vulnerable, with maggots of blowflies playing a pivotal role in the epidemic. Blowflies feed on infected carcasses. Infected droplets produced by these blowflies, contaminate leaves, which are browsed by kudu that are infected and die, so repeating the cycle. Vultures and hyaenas also act as disseminators of the disease. Although lions in Kruger have been known to develop acute septicaemia, carnivores are generally more resistant to anthrax, only developing a localized buccal form of the disease, in the membranes of the head. For example, during a serious outbreak in Etosha National Park, Namibia, lions, spotted and brown hyaenas, and black-backed jackals all fed from infected carcasses but showed no signs of the disease themselves. Similarly, during an epidemic in the Luangwa Valley in 1987, one area of only 80 km^2 yielded the carcasses of 101 hippos, 60 buffalo, and 20 elephants, along with puku, kudu and other ungulates, but only one spotted hyaena and two leopards carcasses.

One of the complications in assessing how severe the effects of these pathogens might be, is that their effects depend upon the condition of the host. Animals that are stressed may be more susceptible to disease, and the subsequent effects of a disease might be more severe. For example, canine distemper appears to be more pathogenic to wild dogs kept in captivity, than in free-ranging populations. Another example is provided by Sinclair (1977). Sarcoptic mange, in the Serengeti, affects both buffalo and wildebeest calves. However, while it was commonly seen in the severe dry season of 1968 and 1969, it was not observed in the mild seasons of 1971 and 1972. In fact Sinclair (1974) suggests that since disease (and predators) kill mainly animals in poor condition, it is agents such as malnutrition that kills many of these animals, rather than disease. Pathogens and predators may simply hasten the process (Chapter 4). Climatic effects also influence disease. In the Kruger ecosystem, epidemic foot and mouth disease, and anthrax are typically associated with the dry season, when animals congregate at water holes, providing ideal conditions for disease transmission. Rift Valley Fever, African horse sickness, theileriosis and other tick-borne diseases are associated with the wet season when these disease vectors (carriers) are more abundant.

Competitive interactions (− −)

First some definitions, and observations. There are both intraspecific (between individuals of the same species)(see Chapter 4) and interspecific (between individuals of different species) competitive interactions. This section of Chapter 5 will look at interspecific competition between African savannah species. An important distinction can be made between *interference* competition, and *exploitation* competition, although elements of both can often be found in the competitive interactions between species. With exploitation competition, individuals interact indirectly, via a common resource that is limiting, or in short supply (e.g. the grazing ungulates below). With interference competition, individuals interact directly, damaging each other in the process of obtaining a resource that might, or might not, be in short supply (e.g. the kleptoparasitism of carcasses by predators below). However, the essence of interspecific competition is that individuals of one species suffer a reduction in fecundity, survival or growth as a result of resource exploitation, or interference, by individuals of another species. Interspecific competition is frequently highly asymmetric. The consequences are often much greater for one of the species involved (e.g. wild dog in the kleptoparasitism example below). Finally, in discussions of interspecific competition one frequently encounters the term *ecological niche.* This term was first used by Joseph Grinnell in 1917 to simply mean the range of physical conditions under which a species could survive and reproduce. It thus had only physical dimensions. Now however, we include resource dimensions, such as seed size, grass species, grass height, prey size, foraging height, and so on. Only part of each niche dimension is used by a species, not simply all that between two limits. Most ecologists visualize the use as being described by a 'utilization function', often represented in models as a normal curve. For example, a seed-eating bird might take all seeds between small and large, but prefer (and therefore use more) seeds of intermediate size. Figure 5.19 shows a visualization of this idea. Usually a

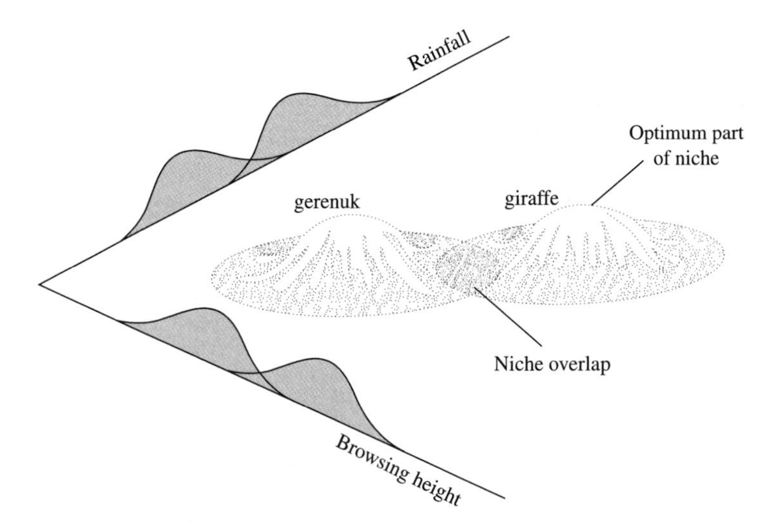

Fig. 5.19 A visualization of the ecological niche. Two niches are shown, each with two dimensions. Because both types can be important in real niches, one physical dimension (rainfall) and one resource dimension (foraging height) are illustrated. To help place this idea into a savannah context, gerenuk and giraffe have been used as plausible examples. Gerenuk prefer more arid savannahs, and browse lower down in trees and shrubs, than giraffe. The diagram is modified from Shorrocks (1978).

species has a larger ecological niche (*fundamental niche*) in the absence of competitors than it has in their presence (*realized niche*).

When I discussed predator/prey interactions I stated how important the Lotka Volterra model had been in forming ecologists' views on those interactions and their population consequences. This is even more so for the Lotka Volterra model of two species competition. From this theoretical model, and early laboratory studies, arose the widely held view that interspecific competition led either to competitive exclusion of one species, or coexistence. The Lotka Volterra model predicts that coexistence will only occur if intraspecific competition is greater than interspecific competition, for both species. This condition is achieved if the two species occupy different niches. Then, individuals of each species will meet individuals of their own species more frequently than they meet individuals of the other species. One of the weaknesses of this 'solution' to the coexistence problem is that the Lotka Volterra model does not predict how dissimilar the two niches have to be. Later models, exemplified by MacArthur and Levins (1967) and May (1973), attempted to rectify this and define the limits to similarity that would allow coexistence. These models, using normally distributed resource utilization curves, and competition between three species along one niche dimension, suggested the simple solution that $d/w > 1$ would lead to stable coexistence (where d = the distance between adjacent resource utilization curve peaks, and w = the standard deviation of these curves). However, Abrams (1983) found that alternative resource utilization curves, and competition in several dimensions would often lead to lower values of d/w being compatible with coexistence. There may be no general, precise, solution to this problem, and savannah ecologists may have to measure specific resource utilization curves and dimensions.

In theory, interspecific competition is obviously important—excluding some species and determining which species coexist with others. But how important are these theoretical effects in the real world? There is no doubt that competition *sometimes* affects community structure. Equally, no one now believes that it is always of overriding importance in each, and every case. If other agents (e.g. lions, rinderpest, and rainfall) keep numbers low, so that competition is negligible, then competition cannot always be a potent force in structuring communities.

One way of answering this question is by examining the results of field manipulation experiments, in which one species is removed from, or added to, an ecosystem and the response of the other species observed. Schoener (1983) looked at 164 studies, in the ecological literature. Unfortunately, published field manipulations are not representative of the world's ecosystems. In the terrestrial studies, for example, most of those found by Schoener were carried out on temperate systems. Conclusions may therefore be biased. Nevertheless, he found that 89 per cent (terrestrial), 91 per cent (freshwater) and 94 per cent (marine) of world-wide studies demonstrated the existence of interspecific competition. When he looked at single species, rather than single studies, he found that 76 per cent showed competitive effects sometimes, and 57 per cent showed competitive effects always. Connell (1983) reviewed just 72 studies (215 species and 527 different experiments) of field manipulations. He found evidence of competition in all the studies, in more than 50 per cent of the species and in about 40 per cent of the experiments. These two literature surveys suggest that interspecific competition might be rather important in structuring all ecosystems. But what about tropical savannahs? There is no indication from these two surveys that different ecosystems behave very differently, so we might assume that savannah ecosystems have competitive effects of this order of magnitude. What is more, these surveys suggest that vertebrate studies showed even higher values than the mean percentages given above, while many insect studies gave lower values. Competitive interactions may therefore be particularly important in the savannah's large mammal community.

How have savannah ecologists attempted to study interspecific competition? There have been some 'natural' experiments with savannah ungulates, and I will examine some of these below. However, savannah ecologists (along with many other ecologists) have more frequently looked at 'community patterns', that might result from interspecific competition. Uncovering niche differences is one such method that I will examine below. Other community patterns will be examined in Chapter 6. However, if several species coexist, and use the habitat in slightly different ways (have slightly different niches), is this evidence of competition? Many ecologists think it is, and believe these patterns are the result of past competitive interactions (the 'ghost of competition past' Connell 1980). However, as Begon *et al.* (1996) say "The trouble is . . . there is no proof". The danger with using pattern to infer process (and this includes the statistical techniques of correlation and regression) is that usually a pattern can be caused by more than one causative agent. A 'competition pattern' could be present as a result of non-competitive influences. Giraffe and kudu may well forage at different heights (see Fig. 5.27), as a result of past competition. They may also do this simply because they are different sizes. A result of past phylogenetic pathways that have nothing to do with competition. We have to distinguish between species traits being the specific result of competition (and therefore indicating past competition), and simply adaptations to a general mode of life. Resource partitioning and niche studies provide a start to examining interspecific competition, but savannah ecologists need to collect additional data to strengthen any competitive interpretation. This could include, for example, evidence of present resource limitation, observed niche shifts when one species is absent (a natural experiment), and looking for alternative explanations. It is against this general competition background that you should interpret the savannah examples below.

The structure of savannahs: competition between trees and grass

A central question in savannah ecology concerns the mechanism allowing the long-term coexistence of trees and grasses. I have already touched upon this topic in Chapter 2 when the central role of rainfall on 'tree-grass dynamics' was briefly examined (see Fig. 2.2). From this graph it is clear that mean annual precipitation (MAP) sets an upper limit to tree cover. However, within this rainfall limit there are other limiting factors at work. These are fire (also indicated in Fig. 2.2) and herbivory, and we have also looked at how these can limit tree recruitment in this chapter (Fig. 5.8). Notice that rainfall, fire and herbivory all appear to 'limit' trees, rather than 'promote' grass. The implication in Figure 2.2 is that the regression line shows the limit imposed on tree cover, by MAP, but that for any particular MAP, tree cover can be lower because of fire or herbivory. But what role does interspecific competition play in this? Traditionally, many savannah ecologists believed that trees and grasses competed for water in the soil. Following from the Lotka Volterra model, this led to the line of argument, that if trees and grasses compete for essential resources, yet they coexist over a wide range of savannahs, they must in some way have separate ecological niches. For example, Walter (1971) suggested that there was a separation of 'root niches' with trees having sole access to deeper soil water and grasses having superior access to surface soil water. Notice that this implies that it is small saplings, with surface root systems, that will be most affected by interspecific (actually inter life form) competition. Of course, young trees will also be more vulnerable to browsing, and fire damage. Biologists working in different African savannahs have claimed both field evidence 'for' (Helsa *et al.* 1985; Knoop and Walker 1985), and 'against' this root niche proposal (Belsky 1990; LeRoux *et al.* 1995; Seghieri 1995; Mordelet *et al.* 1997), although those working in eastern and southern Africa seem to favour the root niche idea, while those working in the humid savannahs of western Africa seem to find positive evidence lacking. There may be a difference between the two types of regional savannah. However, there may also be a difference of interpretation. Seghieri (1995) studied three sites, in a savannah in northern Cameroon. Root profiles consistent with the root niche idea were found and yet the author failed to be convinced of the usefulness of the root niche idea. Figure 5.20 shows three root profiles, for three woody species, in three different soils. Importantly, previous study had shown that the two deeper rooted woody species (*Acacia seyal* and *A. hockii*) actively grew over a longer period than the rainy season while *Lannea humilis* grew only during the rainy season.

Of course coexistence between trees and grasses, resulting from a long-term stable equilibrium of the classic Lotka Volterra kind, may be an illusion. We saw earlier in this chapter, with the elephant–fire–tree dynamics of the Serengeti–Mara system, that tree cover may well go up and down over time, because of long-term changes in tree recruitment. These demographic events can lead to 'non-equilibrium' grass/tree coexistence through climatic variation and/or disturbance which limits successful tree seedling germination, establishment and transition to mature size classes. Sankaran *et al.* (2004) reviewed both the classic equilibrium (competition) and

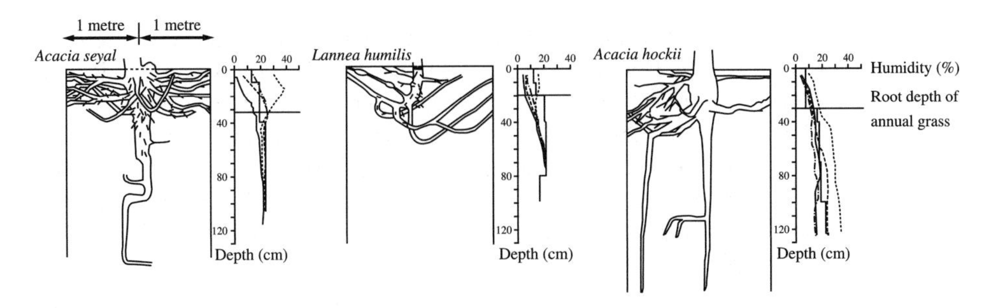

Fig. 5.20 Root profiles of three woody species from northern Cameroon. The humidity profile, to the right of each root profile, shows the moisture content of the soil (modified from Seghieri 1995).

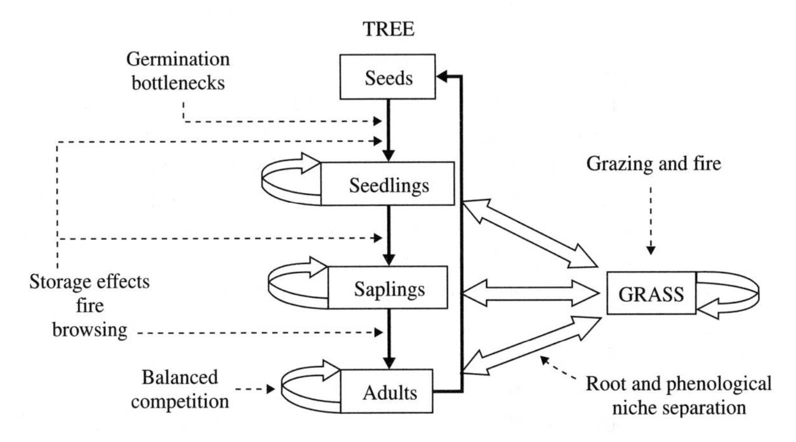

Fig. 5.21 Framework showing the demographic and competitive explanations for grass–tree coexistence in savannahs. Solid arrows show transitions between tree life-history components and open arrows represent competitive effects, both intra and interspecific (modified from Sankaraan *et al.* 2004).

non-equilibrium (demographic) proposals, and attempted to integrate the two views (Fig. 5.21). Competition-based proposals include the root niche separation already mentioned, 'phenological niche separation' and 'balanced competition'. Niche separation by phenology is based on the fact that savannah trees are able to store water and nutrients, and therefore achieve full leaf expansion quickly with the onset of rain (Scholes and Archer 1997). They also tend to retain leaves for several weeks at the end of the growing wet season, after grasses have started to senesce. The idea is that trees have exclusive access to resources early, and late, in the growing season. They have niche separation in time. With balanced competition, coexistence is achieved because the superior competitor (trees) is limited by intraspecific competition before it can exclude the inferior competitor, grasses (Scholes and Archer 1997). In contrast, proponents of the demographic view maintain that the critical problem for savannah trees is demographic and not competitive in nature.

Tree recruitment, and transition into the adult life stage, can be limited by fire, drought, and browsing. For example, variation in tree seedling establishment over time caused by random variation in rainfall, coupled with low adult tree mortality, leads to tree recruitment being pulsed in time. In effect the recruitment potential of trees is stored (*sensu* Warner and Chesson 1985) in the adult population and released in good years. In arid savannahs, tree recruitment can also be enhanced by the localized deposition of tree seeds in herbivore dung. Understandably, tests of these various explanations have usually been site specific (Sankaran *et al.* 2004), and inevitably all are supported in some sites and none are supported in all. Figure 5.21 is probably an accurate reflection of what happens between grasses and trees, but the detectable strengths of the arrows in Figure 5.21 will vary considerably. What I have tried to show throughout this book is that although savannahs are a recognizable biome, local variation in soils, rainfall, fire, herbivory, and predation impose subtle variations on the savannah theme. This is why in Figure 2.2 there is a clear upper limit to tree cover, imposed by MAP, but an amazing range of site specific variations beneath it.

Resource competition: niche differentiation among herbivores

Eltringham (1974) described an event in Queen Elizabeth National Park, Uganda, that could be viewed as a 'removal experiment' indicating interspecific competition. In 1957 a high density of hippopotamus was artificially reduced to very low numbers and then maintained at this level until 1967. Buffalo which were originally low in numbers, increased six fold by 1968. Eltringham believed that this was because vegetation cover, and therefore buffalo food, increased threefold when the hippopotamus were removed. Other areas of the Park, unaffected by hippopotamus, did not show such increases in vegetation, or buffalo numbers. This type of field manipulation experiment, that might throw light on the importance of interspecific competition in structuring savannah communities, is rare. Instead, savannah ecologists have looked for natural experiments (Diamond 1986) of the kind mentioned in Chapter 4. Several reintroductions into protected areas have been made that can be used to suggest the presence, or absence, of interspecific competition between savannah ungulates. In Kruger National Park, white rhinoceros (a grazer) was successfully reintroduced (Pienar 1963; Penzhorn 1971; Novellie and Knight 1994) even though it is approximately the same weight as hippopotamus, also a grazer, suggesting they don't compete for their grass resource. On the other hand, the oribi reintroduction in 1962 was not successful (Pienar 1963), perhaps because of competition with klipspringer and steenbok, which are a similar size. However, the diets of these two species contain more browse than that of oribi (Chapter 3). In the Bontebok National Park, in southern Africa, common reedbuck did not establish itself possibly because of interspecific competition from the similarly sized blesbok (both are predominantly grazers that take the occasional browse)(Novellie and Knight 1994). Unfortunately these reintroductions, and others mentioned by Prins and Olff (1998), can fail for a variety of reasons that may have nothing to do with interspecific competition. We usually do not have enough information to be sure.

The migrating wildebeest of the Serengeti ecosystem provide two natural experiments, where one species increases its numbers, allowing savannah ecologists to observe the effects on other potential competitors. The first 'experiment' (Sinclair and Norton-Griffiths 1982) was a singular event. During the 1960s, following the removal of rinderpest, the Serengeti wildebeest population underwent a dramatic increase in numbers (Chapter 4).

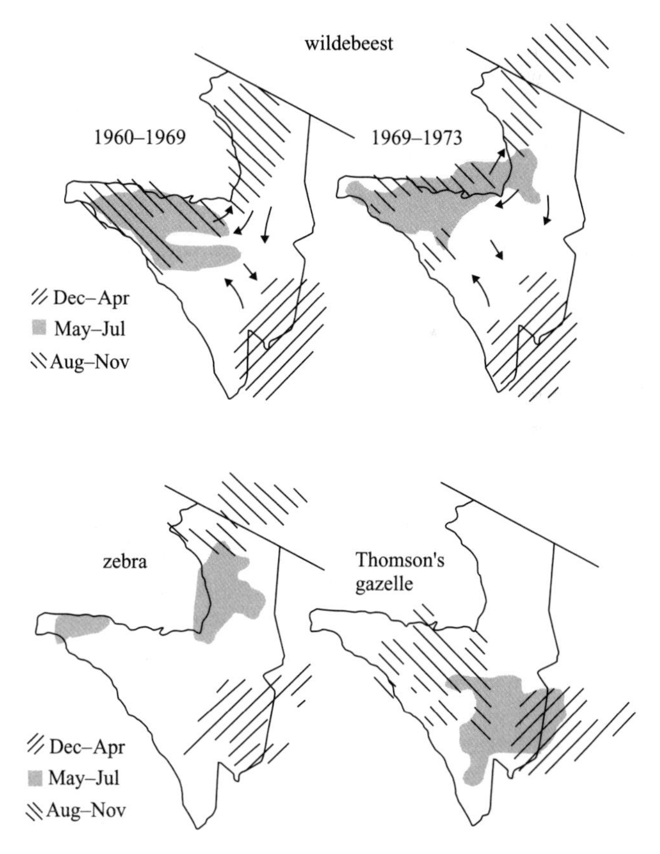

Fig. 5.22 Summaries of the ungulate migrations in the Serengeti–Mara ecosystem. The top maps show the wildebeest migration for two separate time periods, 1960 to 1969 and 1969 to 1973. The botton graphs show the zebra and Thomson's gazelle migration between August 1969 and August 1972 (from Maddock 1979, based on data in Pennycuik 1973).

The second 'experiment' (Sinclair 1985) occurs repeatedly every year as huge numbers of wildebeest arrive in the northern part of the Serengeti, in Tanzania, and the Masai Mara in Kenya. Since both 'experiments' involve the migrating wildebeest, and its two fellow travellers, it is convenient to pause and briefly describe this amazing seasonal event.

Observations on the migrations of the Serengeti wildebeest date back to the mid-1950s (Pearsall 1957; Swynnerton 1958; Grzimek and Grzimek 1960b; Talbot and Talbot 1963) with more detailed data analysis provided by Pennycuick (1975)(Fig. 5.22). Wildebeest spend the wet season (December to April) on the grass plains in the southeast of the ecosystem. At the start of the dry season they move northwest, into the western corridor (May to July), finally moving north as the dry season advances (August to November). They finally return south as the new wet season approaches. The timing of the movement, to and from the plains, depends upon the rainfall. In dry years they move north earlier, and move further north. There has also been a long-term change in the pattern of migration (Fig. 5.22). After 1969 the wildebeest migration penetrated further north into the Masai Mara. This is thought to be a consequence of the increase in wildebeest numbers in the 1960s. Zebra and Thomson's gazelle show a similar pattern of movement, although Thomson's gazelle tend to remain longer on the plains and never move beyond the western corridor. At the end of the wet season some zebra move west, but many move directly north (Fig. 5.22). The three migrating species have similar but slightly different diets (overlapping but not identical niches) and in looking for evidence of present day competition this should be born in mind. Wildebeest and Thomson's gazelle are ruminants, zebra are not (Chapter 3). Wildebeest take green grass, Thomson's gazelle green grass and herbs and zebra can take dry grass (Chapter 3). I will now return to the two 'natural experiments'.

Sinclair and Norton-Griffiths (1982) examined the results of the 1960s wildebeest population increase on the other two species of migrating ungulate, common zebra and Thomson's gazelle. Since there is evidence for intraspecific competition for green grass in Serengeti wildebeest (Chapter 4), it seems reasonable to assume there may also be interspecific competition for green grass between wildebeest and other grazing ungulates. If this was so, the dramatic increase in wildebeest during the 1960s should have had a demonstrable, long-term, effect upon the species that take part in the seasonal migration. Figure 5.23 shows the numbers of all three species over the relevant time period. There was no significant change in the zebra population over the twenty years. Perhaps this is not surprising since they have a food niche that is probably sufficiently different from wildebeest for them not to compete. Being hind-gut fermenters, zebra can process large amounts of dry grass quickly, an option not available to the slower, but more efficient, ruminant digestive system (Chapter 3). However, there has been a decline in Thomson's gazelle. There is a significant difference ($P < 0.02$) between the estimates of 1971 and 1980, although not

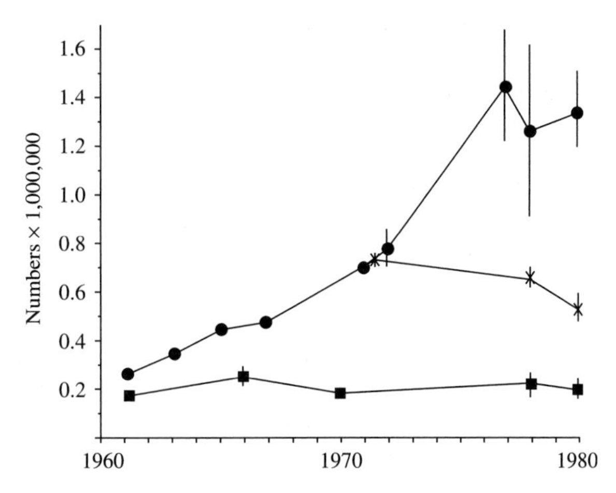

Fig. 5.23 Population estimates of wildebeest (●), Thomson's gazelle (X), and zebra (■) in the Serengeti. Vertical lines are one standard error (modified after Sinclair and Norton-Griffiths 1982).

between 1971–1978, and 1978–1980. Since there is some overlap in the diet of wildebeest and gazelle this decline may be viewed as a consequence of interspecific competition. However, interpretation of Figure 5.23 is made more complicated by the possible presence of a mutualistic interaction between the three species, which I will outline in the last section of this chapter.

In a second paper, Sinclair (1985) examined the effect on other ungulates that wildebeest have when they arrive in the northern Serengeti and Masai Mara. Observations were made from a vehicle while travelling through the Mara Reserve, and all animals within 250 m either side of the vehicle were recorded (Chapter 4). Numbers counted for wildebeest, and the eight potential competitors, are shown in Table 5.5. The increase in wildebeest, and zebra, in the July and August dry season counts, is clearly seen. The other migrating species, Thomson's gazelle, does not get this far north during the seasonal movement (Fig. 5.22) but there is a resident population in the Masai Mara. Interestingly, all seven 'resident species' show a decrease in observed numbers once the wildebeest arrive. This could reflect a movement of individuals in response to the arrival of vast numbers of migrating grazers.

If present-day interspecific competition is present we might expect some displacement of resident ungulate species from certain areas, when the very large numbers of wildebeest arrive. During the vehicle transects, Sinclair looked at this possibility by recording each ungulates use of (1) habitats (grassland savannah, open woodland savannah, thickets or riverine), (2) grass species (*Themeda triandra*, *Eragrostis tenuifolia*, *Hyparrhenia filipendula*, *Pennisetum mezianum* or *Cynodon dactylon*), (3) grass greenness

Table 5.5 Numbers of individuals counted during the Mara transects reported in Sinclair 1985. In the text all but the first two species are referred to as 'residents'.

Species	June 1982	July 1982	August 1983
wildebeest	443	55,003	224,936
zebra	2,208	23,173	28,191
Thomson's gazelle	8,290	4,867	9,629
grant's gazelle	695	455	443
topi	5,184	4,173	6,370
Coke's hartebeest	1,339	1,030	447
impala	3,709	3,177	3,971
waterbuck	135	53	292
warthog	299	120	237

(green, half green, green trace or dry), and (4) grass height (10, 25, 50, 75 or 100 cm). Of course there is some overlap in these categories. For example, grass species are correlated with habitat. The preferred habitat of wildebeest was short, green, *Themeda* grassland savannah. In June before the wildebeest arrived, all species (with the exception of impala who preferred open woodland savannah) had 30 per cent of their population in this 'wildebeest habitat'. Surprisingly, this proportion remained similar in July when 50 per cent of the newly arrived wildebeest population were using this habitat. Only at the height of the dry season (August) did use of this habitat drop markedly for zebra, topi, and Coke's hartebeest, although it actually increased for Thomson's gazelle. Sinclair also calculated per cent overlap between pairs of species. Overlap with wildebeest was high and did not generally decrease in the dry season. However, there was a decrease in overlap in 'habitat' use (17 pairs) and 'grass height' use (18 pairs). Overlap for 'grass species' was the same as that for 'habitats' and overlap for 'grass greeness' remained high (average 96 per cent) and constant for all species. However, there was a high degree of overlap between all species. So is competition not important in structuring these ungulate assemblages? Certainly the evidence for present day competition is equivocal. But what about its ghost? These ungulates might not be competing now because past competition has produced niche separation that allows them to coexist now.

There is a long history of looking for niche separation in savannah systems. These include separation by habitat (Lamprey 1963; Bell 1970; Jarman 1972; Ferrar and Walker 1974; Sinclair 1977; Ben-Shahar and Skinner 1988; Ben-Shahar 1990), by plant food species (Field 1968; Jarman 1971; McNaughton and Georgiadis 1986; Prins *et al.* 2006), and by plant parts eaten (Gwynne and Bell 1968; Bell 1970; Sinclair 1977). A case can certainly be made for niche separation. Some ungulates are ruminants (wildebeest), and some are not (zebra), thus allowing both the efficient processing of green and dry grass (Chapter 3). Some species are browsers on trees (giraffe), some are browsers on herbs (Grant's gazelle), some are grazers on

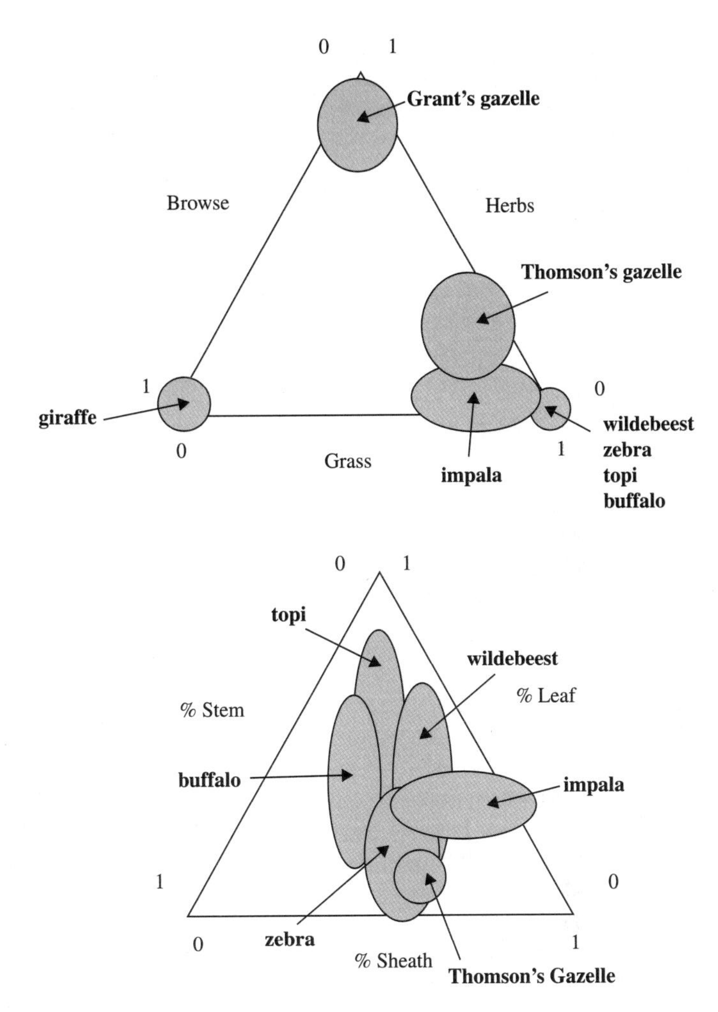

Fig. 5.24 Triangular graphs showing the % composition of ungulate diet in the Serengeti system. The top graph shows separation into grass, herbs and browse (trees and shrubs). The bottom graph shows separation of the grazers into three grass components (chapter 2) (data from Sinclair and Norton-Griffiths 1979).

grass (wildebeest), and others are mixed feeders (impala). Within grazers, some are large bodied with wide mouths (buffalo) while some species are small bodied, with narrow mouths (Thomson's gazelle), allowing them to be more selective feeders. Figure 5.24 shows how this dissection of resources can be used to visualize niche separation between potential ungulate competitors in the Serengeti ecosystem. In the top graph there is a separation into woody browse, herbs and grass, while in the bottom graph the separation is a finer one into 'parts' of the grass resource (Chapter 2). The resources used by the major herbivores do therefore appear to differ. This is partly related to body size. In general, among related species with similar digestive capabilities, smaller species require better-quality food because they have a higher, relative, metabolic rate, but larger species require larger absolute quantities of food (Figure 5.25a). The relationship between energy turnover and body size is: energy turnover = $K + W^{0.75}$, where K is a constant and W = body weight. The relationship between body weight and protein turnover is very similar: protein turnover = $K + W^{0.74}$. The significance of this is that the maintenance requirement of protein per unit of body weight increases with decreasing body size, in the same way as the requirement for energy increases (Fig. 5.25b).

Given a series of animals, such as ruminants, with comparable digestive systems, the smaller animals require a diet with a higher proportion of protein and soluble carbohydrates, at the expense of fibre. These differences have tended to produce at one end of the spectrum, small ungulates with small narrow mouths adapted for carefully selecting discrete, high-quality food items. At the other end of the spectrum, large species with mouths adapted for rapid ingestion of large quantities of undifferentiated items, possibly of low quality (Fig. 5.26). Notice that this size/specialist ranking also corresponds to the order of movement in the Serengeti migration, as the dry season approaches. In the Serengeti it also corresponds to the order

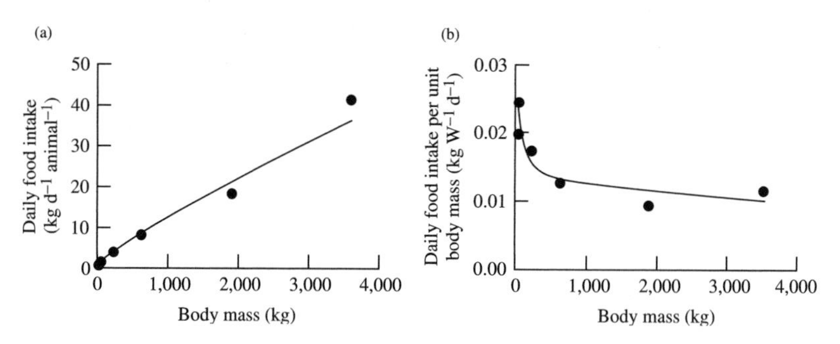

Fig. 5.25 The relationship between body mass and (a) daily food intake per animal and (b) daily food intake per unit body mass for 6 African grazers. The species are, smallest to largest, Thomson's gazelle, impala, wildebeest, buffalo, hippopotamus and elephant (data from Delany and Happold 1979, graphs from Prins and Olff 1998).

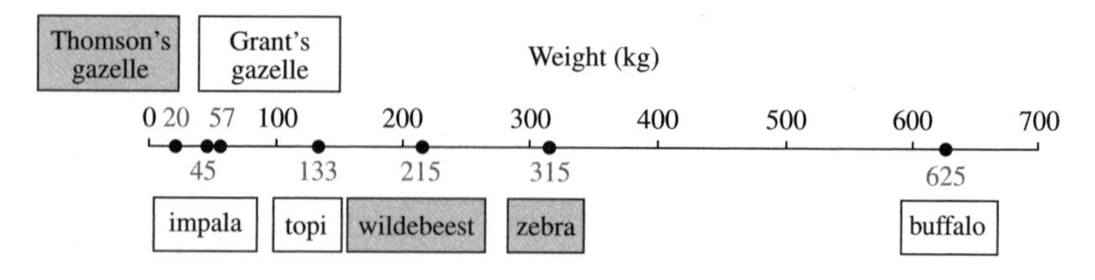

Fig. 5.26 Scale of body weights for a selection of African savannah ungulates. The three species involved in the seasonal Serengeti migration are shaded.

of the lesser movement down the catena (slopes) as the dry season approaches, including buffalo, topi and impala in the correct order.

This type of niche separation is not just confined to grazers. There is also evidence that browsers may have different niches, thus avoiding interspecific competition. Figure 5.27a shows the popular natural history view of this resource separation. Du Toit (1990) observed the height at which four species of ungulate were browsing, in Kruger National Park, South Africa (Fig. 5.27b). Giraffe allocated almost 90 per cent of feeding time to feeding above the height ranges of kudu, impala and steenbok. Kudu allocated 33 per cent of their feeding time to the height range 1.2–1.7 m, which was little used by giraffe and impala, and beyond the reach of steenbok. Among kudu, impala and steenbok there was a common pattern of increased mean feeding height during the dry season. The implication from Figure 5.27b is that feeding height stratification separates these species on an important niche dimension, thus promoting coexistence. It should be remembered however, that mature trees grow from seedlings and saplings. Excessive browsing by the smaller species on these young trees might have long-term consequences for the mature tree population. Browsing height stratification between giraffe, and kudu + impala + steenbok, does not exclude the possibility of resource competition.

The search for niche differences has, more recently, led people into more complex forms of multivariate statistical analysis, such as correspondence analysis. Examples, from the northern Transvaal (ten ungulates including impala, sable, roan, and waterbuck) can be found in Ben-Shahar and Skinner (1988) and, from southern Mozambique (three small ungulates,

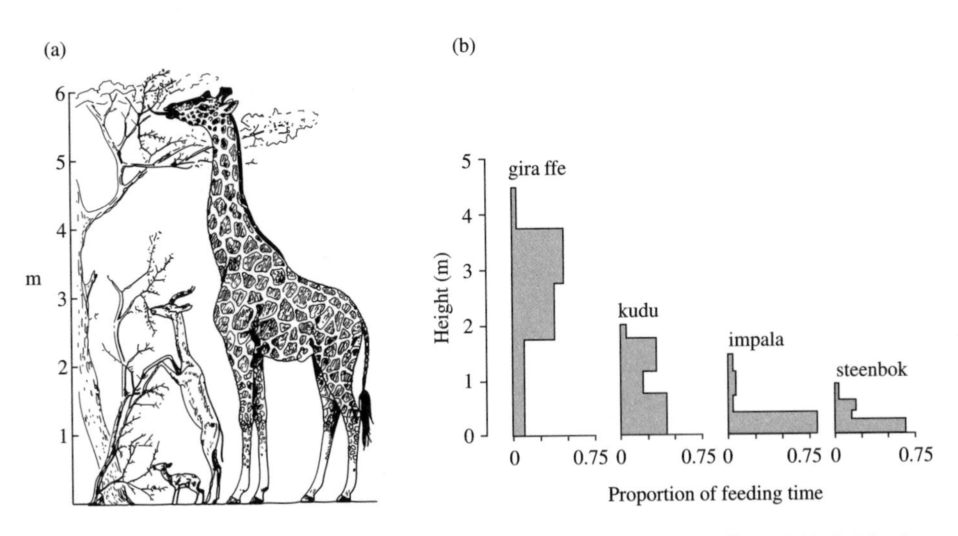

Fig. 5.27 Niche separation of savannah browsers. (a) the popular view (from Orbis Publications 19*xx*), (b) data from Kruger National Park, South Africa (from du Toit 1990).

including suni) in Prins *et al.* (2006). However, these are simply more sophisticated detection and visualization techniques, and are still subject to all the caveats outlined in the introduction to this section on interspecific competition.

Kleptoparasitism: African wild dogs and spotted hyaenas

As we have seen, many field observations of 'competition' have relied heavily upon observing changing numbers over time. One species arrives in an area, or a resident species increases its numbers, while another species declines in abundance. An interesting example with carnivores is provided by the increase in abundance of the spotted hyaena (*Crocuta crocuta*) in the Serengeti–Mara ecosystem, and the subsequent decline of the African wild dog (*Lycaon pictus*). As we saw in Chapter 4, spotted hyaena are the most abundant large predator in the Serengeti–Mara system. Between 1969 and 1976 the spotted hyaena population increased by 50 per cent, probably as a response to the herbivore increase, following the removal of the rinderpest virus. Some estimates for the subsequent population of hyaena in the Serengeti were shown in Table 4.7. Over this period the wild dog has shown a decline in numbers, with all remaining dog packs finally disappearing from the Serengeti system in the early 1990s (Fig. 5.28). During the early 1970s part of this decline in wild dog numbers was certainly due to hunting (wild dogs were regarded as vicious vermin to be exterminated). However, the reason for the later slow decline is less certain, but interference competition with spotted hyaena may have been important.

Both these predators hunt in packs and run down their prey (mainly Thomson's gazelle, wildebeest and zebra) (Table 5.3). Spotted hyaena frequently follow dog packs when they go hunting from a den (during the breeding season) and steal their kill (Lawick and Lawick-Goodall 1970; Lawick 1973). Dogs can defend a kill, but about four dogs are required to keep off one hyaena. Although food 'on the hoof' is probably not a

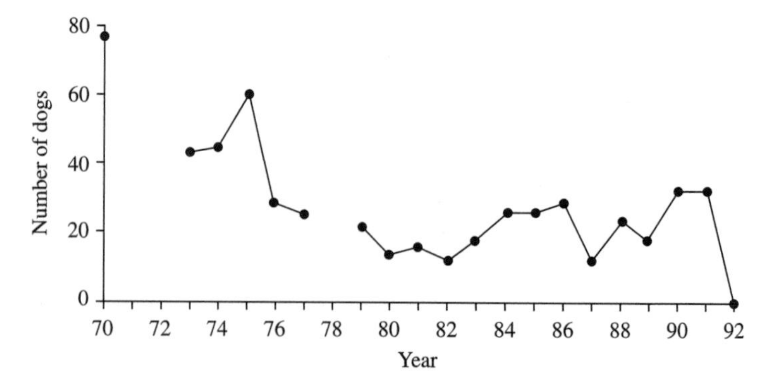

Fig. 5.28 Number of wild dogs in the Serengeti ecosystem.

limiting resource for large carnivores in the Serengeti, food in the form of 'kills' maybe. Particularly during the breeding season, dog packs are hunting not only for themselves, but also for the dominant bitch and her pups back at the den. Losing kills, to competitors such as hyaenas, may have a serious effect upon pup survival. Because the wild dogs ability to defend kills depends on the ratio of dogs to hyaenas, this effect may have become gradually more serious over the last thirty years. Additional evidence supporting this explanation comes from the correlation ($r = -0.92, P = 0.01$) between hyaena density and dog density over a series of ecosystems in eastern and southern Africa (Fig. 5.29a)(Creel and Creel 1996). Aggressive interactions at wild dog kills confirm that the frequency, and impact, of interference competition increases where hyaena density is high and where visibility is good)(Creel and Creel 1996). Wild dogs also appear to suffer competition (and predation) from lions, and there is a similar negative correlation ($r = -0.91, P = 0.03$) between wild dog density and lion density in a number of savannah systems (Fig. 5.29b).

Kleptoparasitism between spotted hyaenas and lions also occurs. In the absence of adult male lions, hyaenas can drive female and subadult lions off their kills, provided they outnumber the lions by a factor of four. Cooper (1991) found that in Chobe National Park, Botswana, where there was a shortage of adult male lions, the groups of female and subadult lions lost almost 20 per cent of their food to hyaenas. Losses were most frequent for those lions living in small groups. In contrast, Trinkel and Kastberger (2005) found that in Etosha National Park, Namibia, spotted hyaenas were unable to prevent kleptoparasitism by lions, and failed to acquire kills from lions. This appeared to be due to the low ratio of hyaenas to lions, and the presence of male lions. Another predator that may suffer interference from hyaenas and lions are cheetah. Looking at distributional data, Durant

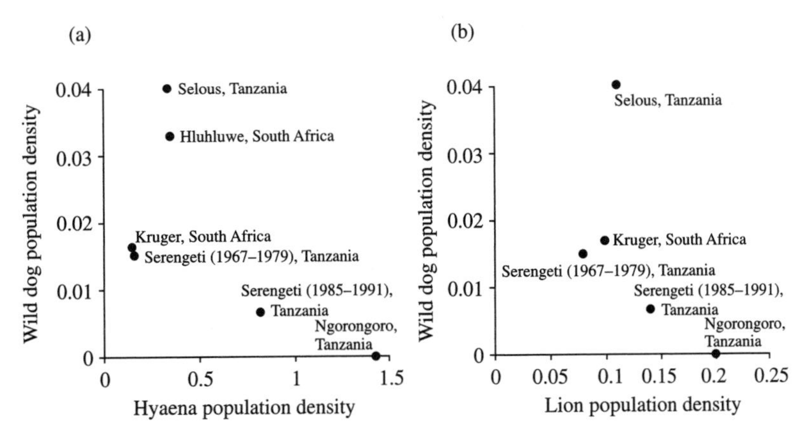

Fig. 5.29 Correlation between population density of wild dogs and (a) spotted hyaena and (b) lion, over a number of African ecosystems (from Creel and Creel 1996).

(1998) suggested that, in the Serengeti National Park, cheetah avoid areas were hyaena and lions are common, even though these areas have lots of prey. This behaviour is strongest when cheetah are most likely to suffer loss of prey to these two larger predators—that is when hunting.

Mutualistic interactions (+ +)

In savannah systems, mutualisms are frequently mentioned but much more rarely studied. In truth this is a general comment on most ecological systems (Vandermeer 1984). Mutualistic interactions in African savannahs include nodulated legumes (e.g. between root bacteria and *Acacia* species), ruminant digestion (between cellulose digesting bacteria and ruminants), pollination (e.g. between honey bees and *Acacia* species, fig wasps and *Ficus* species), seed dispersal (e.g. between indehiscent *Acacia* and many herbivores), protection (between ant species and *Acacia* species), guiding behaviour (between the honey guide, *Indicator indicator* and the honey badger), and scavengers (e.g. between vultures and large carnivores).

Typical of a mutualism lacking detailed scientific study, but none the less described repeatedly in savannah texts, is that of the greater honey guide (*Indicator indicator*) and the ratel, or honey badger (*Mellivora capensis*) (Chapter 3). Nice accounts can be found in Kingdon (1977) and Estes (1991). The greater honey guide has been observed to solicit help from honey badgers, humans and baboons. When it sees a potential follower it approaches within 5 to 15 m, chirring (a sound similar to shaking a box of matches rapidly) and fanning its tail to expose the white outer feathers. The guiding bird may lead its follower anything from a few metres to 2 kilometres, and the journey may take up to half an hour. Once a bee-hive has been located the bird sits quietly in a nearby tree until its fellow honey hunter has eaten from the hive and retired. It then approaches the broken hive and feeds on beeswax, grubs and eggs. The honey badger is reputed to have a most unusual method of dealing with the bees before it breaks open the hive with its powerful claws (Kingdon 1977). According to African honey hunters the honey badger 'backs up' to the hive, and uses its anal glands to fumigate the bees. The bees either flee or become inactive. This mutualism is facultative rather than obligate since ratels are known to find bee-hives without the aid of *Indicator*, and non-guiding birds are found with beeswax in their crops.

An important example of mutualism is that between *Acacia* and herbivores. *Acacia* pods are of two types. (1) Dehiscent, in which the pod splits and the seeds are dispersed by wind. (2) Indehiscent, in which the pods remain on the tree until removed by browsers, or as in *A. tortilis* the pods drop to the floor but do not split. The seed pods of indehiscent *Acacia* are eaten by a wide variety of wildlife including elephant, white rhinoceros, giraffe, eland, kudu, impala, steenbok, several rodents and ostrich. The seeds from indehiscent pods have a very hard coat which enables some of them to travel through the gut of herbivores and pass out unharmed in the faeces. However, many are destroyed by chewing and digestion. The herbivores benefit by eating the nutritious pods, but the *Acacia* species benefit by having their seeds dispersed and fertilized. For the *Acacia* this association is more or less obligate. Because their seeds have a very tough outer coat they do not germinate unless they pass through the gut of a herbivore. Miller (1995) found that undigested seeds of *A. tortilis* had a germination rate of 7 per cent, while seeds retrieved from the stomach of kudu had a rate of 48 per cent. *Acacia tortilis* seed from kudu dung had a germination rate of 60 per cent.

Feeding facilitation and the grazing succession

The possibility of feeding facilitation, through increased access to resources, was first suggested by Vesey-Fitzgerald (1960). His observations were made in tall floodplain grasslands in the Rukwa Valley, Tanzania, where trampling and feeding by elephants exposed medium-height grasses to buffalo. Buffalo in turn generated shorter grass that was eaten by topi. Further evidence to support a grazing succession was provided by Gwynne and Bell (1968) and Bell (1970, 1971) who examined the movement of buffalo, zebra, wildebeest, topi and Thomson's gazelle down the catenas (Chapter 2) of the western Serengeti. In the wet season these species concentrate on the higher ground where short grasses provide the best grazing. As the dry season approaches they move down the catena, in decreasing body size, to feed upon the taller, but poorer quality, grasses on the lower ground. However, perhaps the most well-known example of a grazing succession is that provided by the Serengeti seasonal migrations already mentioned (Bell 1971). First to migrate are zebra, followed by wildebeest and finally Thomson's gazelle. Again the movement is in decreasing order of body size. The successive nature of this movement, at least in the Kirawa area of the western Serengeti ecosystem is shown in Figure 5.30, using data from Bell (1969). Peak numbers of zebra moved through a 300 m transect in April, peak numbers of wildebeest in July, and peak numbers of Thomson's gazelle in September. The original idea was that zebra (a hind-gut fermenter) lead the succession, opening up the herb layer by trampling and increasing the relative frequency of grass leaf by consuming stems. They therefore increase the suitability of the vegetation structure for wildebeest (a large ruminant). These in turn reduce the quantity of grass leaf, facilitating the use of herbs by Thomson's gazelle (a small ruminant). One way to see this idea is as a multispecies extension of the 'grazing lawn' idea mentioned earlier in this chapter, and indeed the possibility of self-facilitation has been raised (Owen-Smith 1988). In effect, herbivores manipulating their plant food to their own, or in this case other species', advantage. It's a good story, but is it true?

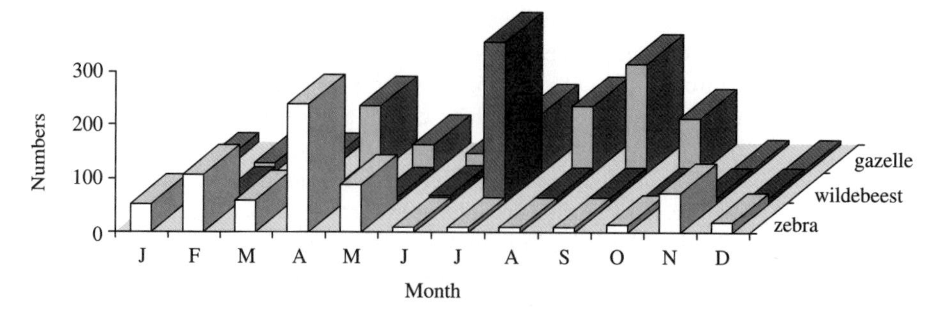

Fig. 5.30 Average daily numbers of zebra, wildebeest and Thomson's gazelle recorded in a 300 m transect at Kirawira, in the western Serengeti. Note that the numbers of wildebeest are divided by 10 to accomadate them on the same graph (data from Bell 1969).

Of course we know that such facilitation could only be facultative, since not all the wildebeest follow the zebra (Sinclair and Norton-Griffiths 1982; Sinclair 1985, see below) and most of the migrating Thomson's gazelle do not follow the zebra and wildebeest into the northern Serengeti and Masai Mara (Fig. 5.22). In addition, only the two following species could gain an advantage. Zebra cannot gain any mutualistic advantage since they are at the front, 'leading' the succession. Zebra may however gain an advantage in another way—by avoiding competition (Sinclair 1985). In June 1980, a systematic aerial survey of the complete Serengeti zebra population (using a grid of 1 km^2) recorded 5,555 zebra of which 71 per cent were within 1 km of the wildebeest herds. The great majority of these zebra had a well defined position on the leading edge of the herds (Fig. 5.31) confirming that they do indeed precede the wildebeest in the migration. Of course, not all the zebra were found with the wildebeest herds during this wet season, 29 per cent were scattered over a large area many kilometres away. In the Masai Mara, during the dry season, this proportion increased to 55 per cent in July, and 50 per cent in August, suggesting a change from significant association with wildebeest in the wet season to significant avoidance in the dry season. Sinclair maintains that this change in zebra behaviour is brought about by a continuous conflict between two other interactions—predation and interspecific competition. Wildebeest are the preferred prey of all the large carnivore species in the Serengeti ecosystem (Schaller 1972; Kruuk 1972). Therefore zebra (and other ungulate species) could reduce the risk of predation on themselves by staying close to wildebeest. However, zebra may also suffer by being too close to wildebeest because of disturbance and interspecific competition. Sinclair suggests therefore that when food is not limiting (in the wet season) the zebra move close to the wildebeest for protection, but non-the-less stay ahead of the main herd to reach the available food first. During the dry season when grass is limiting (Chapter 4) pressure from interspecific competition may outweigh that from predation, and the zebra move further away from the wildebeest herds. Sinclair is

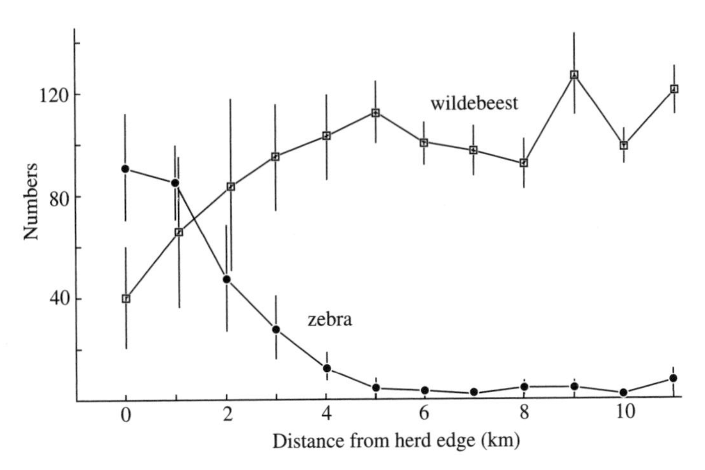

Fig. 5.31 Location of zebra (●) along transects through the large wildebeest herds (□), on the Serengeti Plains in the wet season. Numbers are averages of 8 transects. All wildebeest and zebra within a 200 m radius of the vehicle were counted (from Sinclair 1985).

suggesting that what we are seeing in the Serengeti migrations (at least between zebra and wildebeest) is a response to predation and interspecific competition, rather than facilitation.

McNaughton (1976) did find evidence that Thomson's gazelle were attracted to areas on the Serengeti plains where previously grazing wildebeest had improved the quality of the grass sward. What is more, the Thomson's gazelle maintained this preference for up to six months after the passage of the wildebeest. Unfortunately, this local facilitation does not appear to have translated into a population response. There has been no detectable increase in Thomson's gazelle following the eruption of the wildebeest population in the 1960s (Fig. 5.23). In fact, quite the opposite. An excellent review on this topic is provided by Arsenault and Owen-Smith (2002).

Acacia ants

In sub-Saharan Africa 1,686 species of ants have been recorded (Bolton 1995), but at least 3x this number probably remain to be described. Many plants provide food for ants in the form of extra floral nectaries. These sites of sugary secretions are well away from any flower, often in the leaf petioles or blades. The ants are attracted to the plant and their presence results in some protection from herbivores for the host plant. For example, in Mexican species, Janzen (1966) showed that 2.7 per cent of shoots with ants had herbivorous insects, while 38.5 per cent of shoots without ants had insect herbivores. This type of plant-insect interaction is particularly well developed in savannahs, with several species of *Acacia* tree. In about 10 per cent of spiny acacia species, the base of their spines becomes greatly expanded, forming a pseudogall (Fig. 5.32a). These are not true galls because they are not induced by the insect. The ants chew a hole in the new, soft, green pseudogall and hollow out the inside as a site for a mini-colony. As the gall grows older it darkens, and becomes harder and more protective. The colony of ants established on an *Acacia* tree aggressively remove any insect herbivore, frequently consuming them, and even attack mammalian herbivores, such a giraffe. One of the problems that the tree has with this relationship however, is that it needs insects to visit in order to pollinate its flowers. Intriguingly, work on *Acacia zanzibarica*, in Mkomazi Game Reserve in northern Tanzania, suggests that the ants are mainly active at dawn and dusk, while most pollinating bees and butterflies are most active during the middle part of the day (Willmer, Stone, and Mafunde 1998). Because of this diurnal rhythm, bees and butterflies almost never meet an ant.

A particularly detailed study of these ant-*Acacia* mutualistic systems has been carried out in the Laikipia District of central Kenya (Young,

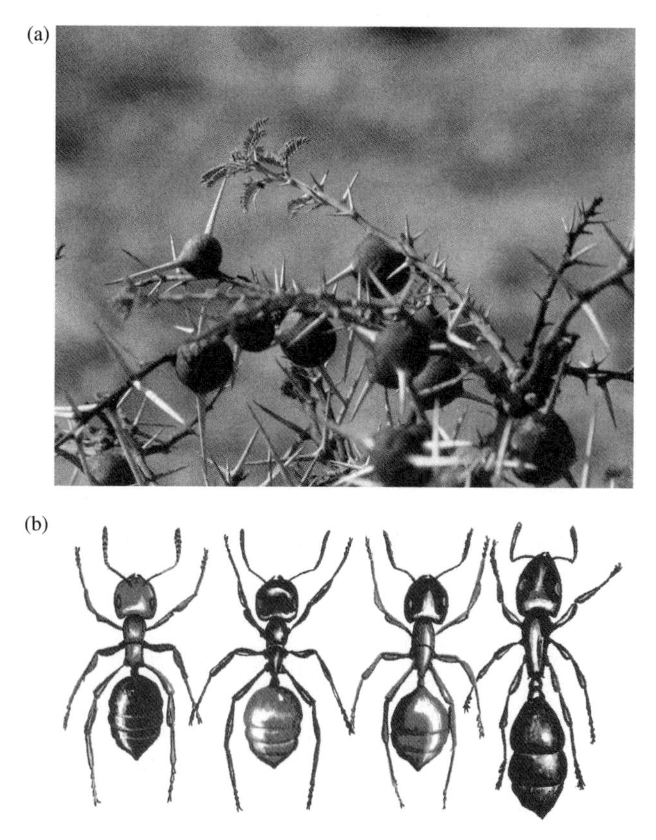

Fig. 5.32 Ants and Acacia (a) whistling thorn, *Acacia drepanolobium* (b) the four *Acacia drepanolobium* ants (head colour/thorax colour/abdomen colour, R = red and B = black), left to right: *Crematogaster mimosae* (RRB), *C. nigriceps* (BBR), *C. sjostedti* (BRR) and *Tetraponera penzigi* (BBB) ((photograph a by Jo Shorrocks, drawing b by Dino Martins).

Stubblesfield, and Isbell 1997; Stanton, Palmer, and Young 2002). In this semi-arid bush savannah *Acacia drepanolobium*, the whistling thorn, often makes up 95 per cent of the trees (about 2000 trees/ha). Since virtually all whistling thorns over 0.5m tall are occupied by ants it has been estimated that acacia-ants may account for up to 25 per cent of the animal biomass in this ecosystem. In most plants there is a pair of scale-like appendages at the base of the leaf-stalk, or petiole. In whistling thorn these stipules are developed into a formidable pair of thorns, up to 7cm long (Chapter 2). Between 5 per cent and 40 per cent of these paired thorns share a hollow inflated base that is 1.5–3.5 cm in diameter (Fig. 5.32a). Ants produce winged, reproductive forms that disperse to found new colonies. These newly mated, winged females, explore the exterior surfaces of whistling thorn looking for new, unoccupied galls. Having found a suitable swollen thorn, the female removes her wings, initiates a conversion of her flight muscles into food and chews an entry hole into the swollen thorn. She then plugs up the hole from the inside and remains sealed within the swollen thorn until her first workers are produced. Acting together, ants and sharp thorns form a quite effective defence against browsing by small to medium-sized herbivores. Madden and Young (1992) reported a negative correlation between the number of swarming ants and the duration of feeding bouts for giraffe calves ($r = -0.91$, $P < 0.05$). Adult giraffe feeding seemed less sensitive with a positive, but non-significant, correlation with ant swarming ($r = 0.41$, $P = 0.10$). Figure 5.33 shows the results of a nice manipulative field experiment in which four different types of *Acacia drepanolobium* branch were offered to goats (Stapley 1998). Branch type 0 had thorns and ants removed, type 1 had ants but no thorns, type 2 had thorns but no ants, and type 3 had both ants and thorns (the natural condition). Both ants and thorns appear to have an effect upon the mean amount of vegetation eaten, and the mean number of bites taken, by the goat. In the Laikipia District, four species of ants are the main occupants of these acacia pseudogalls (Fig. 5.32b). *Tetraponera penzigi*, *Crematogaster nigriceps*, and *C. mimosae* depend entirely on the interiors of swollen thorns for nesting space, and specialize on using *A. drepanolobium*. *Crematogaster sjostedti* also occurs on a less common swollen-thorn acacia called *A. seyal* variety *fistula* (chapter 2). This last acacia-ant nests mainly within dead acacia tissue, but its workers may also occupy swollen thorns. With the three *Crematogaster* species, a single colony of ants usually spans several trees, while with *T. penzigi* a colony occupies a single whistling thorn. The four ant species are very intolerant of each other and fight over the possession of trees. Observations suggest that a hierarchy exists with the order *C. sjostedti* > *C. mimosae* > *C. nigriceps* > *Tretraponera penzigi*.

Scavenging vultures: mammalian carnivores facilitating birds?

Although there are about 97 'bird of prey' species in Africa (Zimmerman *et al.* 1996), one of the most obvious and recognizable groups in savannahs

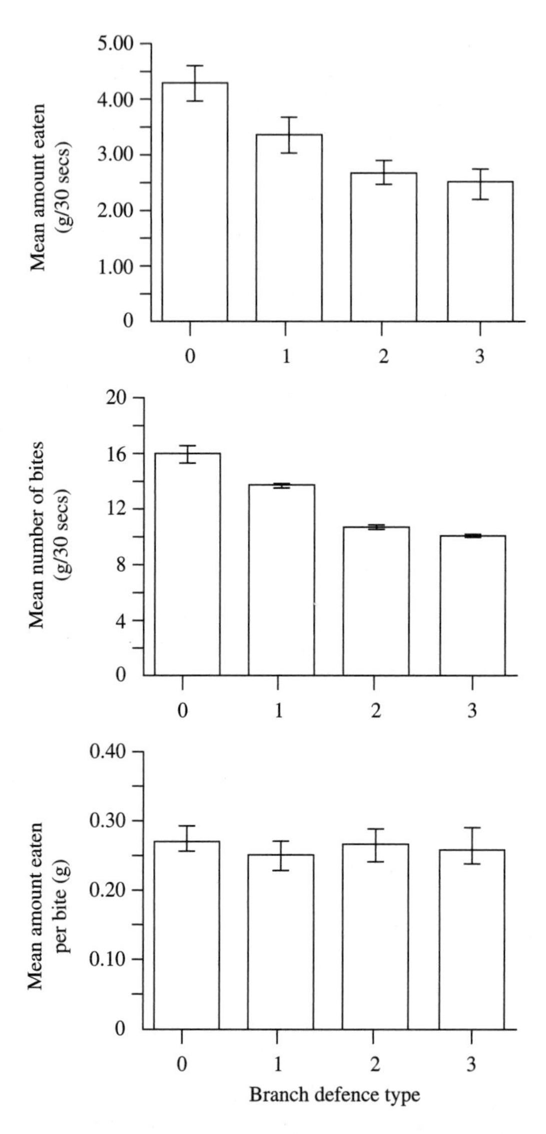

Fig. 5.33 The effect of ants and thorns on browsing in *Acacia drepanolobium*. 0 = no ants or thorns, 1 = ants but no thorns, 2 = thorns but no ants, 3 = both ants and thorns (from Stapley 1998 *Oecologica*, 115, pp. 401–405. With kind permission of Springer Science and Business Media.)

are the vultures, with ten African species, eight of which are sub-Saharan. They are large, open country birds, adapted to carrion-feeding and to extended periods soaring on long, usually broad, wings. Except for the lammergeier feathering is reduced, or absent, on the face, neck and legs. In many areas, most of these species coexist and are frequently found together at a carcass, where they compete for food (Kruuk 1972). Their different body size and beak structure allows them to partition the carcasses they locate. The small, black and white, Egyptian vulture (*Neophron percnopterus*) (body weight 1.8 kg), inhabits the arid savannahs of the Sahel, down the east of Africa to northern Tanzania. It also occurs in northern Namibia and Angola. It feeds on carrion and meat scraps of any kind, and bird's eggs. It cannot compete at carcasses with the larger vultures and waits on the outskirts to snatch scraps. It nests on cliffs. The other small species, the hooded vulture (*Necrosyrtes monachus*) (body weight 1.9 kg) is very widespread, occurring in most of the sub-Saharan savannahs. It also feeds on all kinds of carrion, and is a scavenger in towns. Several pairs nest together, in trees. Both these small species have long, thin beaks that are relatively weak. In addition to scraps of meat from carcasses, they take termites, beetles, snakes, and lizards. They usually fly at low altitudes and are non-territorial. The three, large, griffon vultures (*Gyps* species) are all widespread and specialize on the soft flesh and intestines of large dead mammals. Their beaks are long, with a sharp cutting edge. The tongue is barbed for gripping soft tissue. They are non-territorial, fly at high altitudes, rely on soaring, and fly long distances each day. They frequently collect in large numbers at suitable carcasses. The African white-backed griffin (*Gyps africanus*) (body weight 5.3 kg) is the commonest large vulture, found in all sub-Saharan savannahs from the Sahel to southern Africa. Like the other two species it is gregarious at roosts, and at carrion, but more solitary when breeding. The other two *Gyps* species have a more restricted range, but are still common. They are both colonial cliff breeders. Ruppel's griffon vulture (*Gyps rupellii*) (body weight 7.4 kg) is found south of the Sahara from Senegal to Sudan, Kenya and Tanzania and roosts at night on cliff faces. This species has the highest credible altitude record for any bird, one having been killed by a jet aircraft at 11,300 m. The Cape griffin vulture (*Gyps coprotheres*) is restricted to southern Africa. The last three species, below, have large beaks, enabling them to tear the meat from large carcasses. Although mainly eating flesh, they can also eat sinews, bones and skin, parts normally left by the griffin vultures. They all appear to be territorial. The fleshy pink head, and neck, with large fleshy lappets make the lappet-faced vulture (*Torgus tracheliotus*) (body weight 6.2 kg) one of the easiest to recognize. It nests singly in trees. The widespread white-headed vulture (*Trigonoceps occipitalis*) eats mostly carrion, but is also suspected of killing hares, gazelle calves and raids flamingo colonies. It nests singly in trees such as *Acacia*. The lammergeier, or bearded vulture (*Gypaetus barbatus meridionalis*), is a graceful, long-winged, soaring bird that will consume almost every scrap of meat, skin or bones at carrion. Its tongue is adapted for removing marrow from long bones, and its most remarkable feeding habit, is dropping bones on rocks to split them.

There are basically three ways in which a carnivore, feeding on large ungulates can obtain food. It can hunt, steal kills of other hunters (kleptoparasitism), or it can scavenge from the remains of carcasses of animals killed by other predators, or that have died from other causes. In the Serengeti ecosystem Houston (1979), estimated that approximately 40,094,000 kg of ungulates die each year, mainly (70 per cent) from starvation and disease. He also estimated that the five major predators (lion, spotted hyaena, cheetah, leopard and wild dog) consumed about 14,362,750 kg and that kleptoparasitism was of minor importance (but see earlier). This implies that, each year, there is approximately 26 million kg of dead animals, not eaten by mammalian carnivores, available for avian scavengers. In the Serengeti, almost all these are vultures, the only other common scavenger species being the marabou stork (*Leptoptilos crumeniferus*) (body weight 5.5 kg) which, however, is unable to tear flesh from a carcass and relies on other species to do this. It also has alternative food, taking fish, frogs, and other small vertebrates. The figure of 26 million kg of carrion is a huge resource. Even if some 15 per cent of this is consumed by fly larvae and bacteria, it leaves enough food to easily explain the huge numbers of vultures seen in the Serengeti, and presumably all other savannah ecosystems. Houston (1979) estimated that it would require an average of 25,000 griffin vultures to consume this carrion, a figure consistent with estimated vulture numbers in the Serengeti. This relationship between vultures and large carnivores may be truly mutualistic. Not only do vultures gain from the kills of carnivores but there is anecdotal evidence that lions frequently 'home in' on areas where they have seen vultures descending.

6 The savannah community and its conservation

This chapter deals with a number of 'community level' topics, for which there is savannah data. We have already looked at individuals (Chapters 2 and 3), populations (Chapter 4), and simple interactions between species (Chapter 5). However, in much the same way that a population is more than a sum of its constituent individuals, a community is more than a sum of its constituent populations. It is the sum plus the interaction between these populations. We will therefore revisit competition, herbivory, predation, parasitism, mutualism, and scavenging. However, in previous chapters the focus was on interactions between species that occupied the same trophic level (interspecific competition), or between members of adjacent trophic (feeding) levels (herbivore-plant, carnivore-herbivore, parasite-host). These were 'direct' effects, as opposed to the cascading, indirect, unexpected, effects that become apparent when communities of interacting species are examined. Sih *et al.* (1985) looked at 100 experimental studies of predation and found that approximately one-third showed 'unexpected' effects. In addition, when we examine communities, new' emergent, properties and topics arise such as food web dynamics, species diversity, and the 'rules' for assembling communities. Ecology at the community level is difficult because the database involved is enormous and can become unmanagable. We will discover that ecologists have attempted to circumvent this problem by studying energy flow through 'trophic groups', by using summaries of species diversity such as 'rank-abundance models', and by concentrating their studies on only a subgroup of the community—like the herbivore community.

Of course, communities have no scale except for the one that we (humans) impose upon them. The 'local savannah community' looks quite different to a rinderpest virus, a rumen bacteria, an *Acacia* ant, a migrating wildebeest, or an elephant. We humans tend to see savannahs as large areas of grassland with *Acacia* trees, and large herds of mammalian herbivores and carnivores. It is mostly at this scale that I will view savannahs in this chapter. An exception is the first topic, energy flow, because here microbial decomposers are most important. Surprisingly, most of the energy passing through a savannah community passes through the detritivore system, not the herbivore system.

One final point. Many ecologists use the term community only for the biotic components of a system, while using the term ecosystem for the biotic plus abiotic components. However, I agree with Begon *et al.* (1996) that in practice the two terms are interchangeable, since it is impossible to ignore abiotic aspects when considering species interactions, e.g. rain, grass, and herbivores.

Energy flow and food webs

In an attempt to find simple ways to describe and study complex communities, many early ecologists adopted a dynamic trophic approach. Linderman (1942) attempted to quantify the idea of food chains and food webs by considering the transfer of energy between trophic levels. This theme continues to the present day with ecosystem productivity forming a major catalyst for the International Biological Programme (IBP) and the Global Change and Terrestrial Ecosystems (GCTE) component of the Geosphere—Biosphere Programme (IGBP)(Steffan *et al.* 1992).

At the bottom of the food chain are the primary producers: green plants. These autotrophs alone are responsible for 'primary production', by capturing the incident solar radiation from the sun. This primary productivity of a community is the rate at which biomass is produce per unit area, by plants. It is conventionally expressed either as energy (e.g. J m^{-2} day^{-1}) or dry organic matter (e.g. kg ha^{-1} $year^{-1}$). The total fixation of energy, by photosynthesis, is called gross primary production (GPP) and GPP minus the energy lost through respiration is known as net primary production (NPP). So $NPP = GPP - R$. Savannahs have a range of NPP from 20 to 2000 g m^{-2} per

Table 6.1 Annual net primary production and biomass estimates for a selection of the world's biomes.

Biome	NPP (g m^{-2})	World NPP (10^9 t)	Biomass (kg m^{-2})
Tropical forest	2,200	37.4	45
Temperate deciduous forest	1,200	8.4	30
Savannah	900	13.5	4
Boreal forest	800	9.6	20
Cultivated land	650	9.1	1
Temperate grassland	600	5.4	1.6
Desert and semi desert	90	1.6	0.7

Data from Whittaker 1975. The high forest biomass is a product of accumulated dead biomass (wood).

year, with a mean of about 900 g m^{-2} per year. Table 6.1 shows NPP for a range of biomes (along with their total world NPP), illustrating that savannahs have an intermediate productivity. Notice however, that below ground productivity (roots etc.) is almost always underestimated. Also notice that the productivity to biomass ratio is low in forests because a large part of the biomass is dead. Several factors limit primary production. First of all, terrestrial communities use solar radiation inefficiently. Between 0 and 5 J of solar energy strikes each 1 m^2 of the earth's surface every minute. If all this radiation was used by green plants there would be a prodigious production of plant material. This is not the case because firstly only about 44 per cent of the incident short-wave radiation occurs at wavelengths suitable for photosynthesis. Secondly, even this suitable radiation is not used efficiently. Photosynthetic efficiency is defined as the percentage of incoming photosynthetically active radiation (PAR) incorporated into above-ground NPP. It varies from 0.01 per cent to 3 per cent (Cooper 1975). Figure 6.1 shows a flow diagram depicting the path of solar radiation at Nylsvley, a woodland (*Burkea*/*Combretum*) savannah in southern Africa (Scholes and Walker 1993). Notice that only a very small amount of the solar radiation is available for photosynthesis. Because of cloud cover, Nylsvley receives, on average, only 75 per cent of its potential annual total of 4371 sunshine hours. The total annual solar radiation at canopy level is about 7316 MJ m^{-2}, which is 61 per cent of the radiation received above the atmosphere.

In addition to this inefficient use of solar radiation, other factors limit NPP. Foremost among these is water. In arid regions, which include many African savannahs, there is an approximately linear increase in NPP with increase in precipitation. Figures 2.1 and 2.2 are an obvious example from African savannahs. A further limiting factor is potential evaporation (PET).

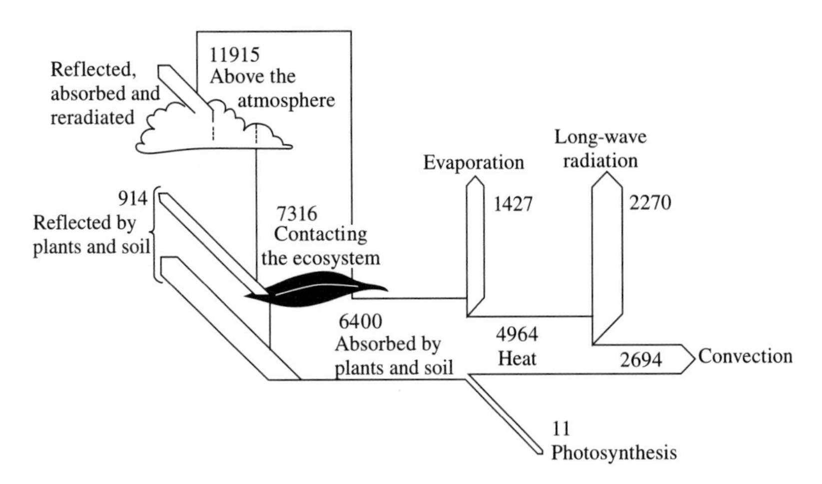

Fig. 6.1 Flow diagram showing the fate of annual solar radiation in a woodland savannah in southern Africa. Numbers are MJ m^{-2} (after Scholes and Walker 1993).

This is an index of the theoretical maximum rate at which water might evaporate into the atmosphere (mm per year) given the prevailing radiation, average vapour pressure deficit in the air, wind speed and temperature. The latter factors all affect transpiration of water from the leaf. Therefore (PET−precipitation) provides a crude index of 'drought' or how far the water availability for plant growth falls below what might be transpired by actively growing vegetation.

Its relationship with above ground NPP is shown in Figure 6.2 for a variety of community types in temperate North America, with African savannahs added for comparison. Water shortage affects plant productivity, leading to less dense vegetation. This in turn exposes more bare ground, leading to a wastage of incoming solar radiation. In fact, this wastage of solar energy striking bare ground is one of the major causes of low productivity per area in arid areas (table 6.1). This becomes clear when productivity per unit of leaf biomass, rather than productivity per unit of area, is calculated. For deciduous forest this is 2.22 g g^{-1} $year^{-1}$, compared to an equal productivity of 2.33 g g^{-1} $year^{-1}$ for deserts. Grassland have a productivity per unit of leaf equal to 1.21 g g^{-1} $year^{-1}$.

Energy flow

Primary production is only the start of the flow of energy through the community. After the producers (green plants) come consumers and

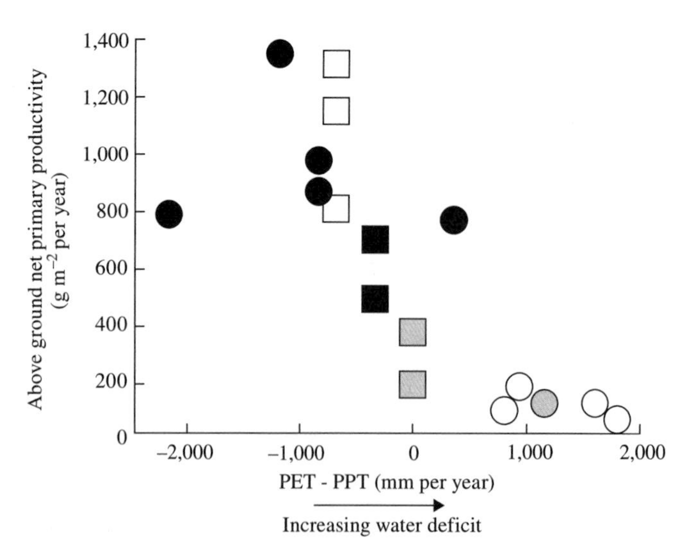

Fig. 6.2 Relationship between above ground productivity and 'PET-precipitation for a variety of community types. Circles = North America, black = woodlands, white = desert and grey = grassland. Squares = African savannahs, black = two *Burkea* sites in South Africa, white = three *Loudetia* sites in Lampto, West Africa, and grey = two sites in Kruger, South Africa (data from Webb *et al.* 1983, Hartley and Walker 1982, and du Toit *et al.* 2003).

decomposers. The former are further divided into primary consumers (e.g. grasshoppers, baboons, wildebeests), and secondary consumers (baboons, spotted hyaenas, lions). In fact these food chains can originate in either live plants (green grass, wildebeest, lion) or dead plants and animals (wood, termite, bat-eared fox; carcass, hyaena; carcass, vulture, serval). A simplified view of this further complication is shown in Figure 6.3 for a generalized African savannah system, along with the energy lost through respiration and the part played by microrganisms in this trophic network. Notice that most of the primary production does not go through the herbivore system but through the scavenger and decomposer system (right-hand side of Fig. 6.3). This is because the faeces and dead bodies are lost to the grazer system (and enter the decomposer system), while the faeces and dead bodies of the decomposer system are simply returned to the dead organic matter box at the base (Fig. 6.3). Fig. 6.4 shows some actual data for the Nylsvley ecosystem in southern Africa (Scholes and Walker 1993). Figure 6.4a again shows that most above ground primary production is consumed by decomposers and fire, with only a very small component finding its way into herbivores. Figure 6.4b shows the fate of this consumed energy for an invertebrate and a vertebrate herbivore. Notice that most of the energy taken in (I), is channelled into excretion (E) or respiration (R). Only a small proportion goes into production (P). In the case of the grasshopper, a significant proportion

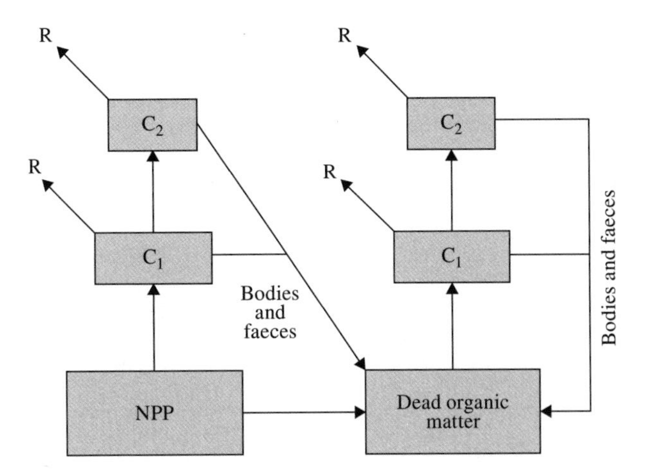

Fig. 6.3 A generalized model of savannah trophic structure. NPP = net primary production, R = respiration, C_1 = consumer level one, C_2 = consumer level two. On the NPP side of the diagram, C_1 could be grasshopper, baboon or wildebeest, and C_2 could be baboon again (they are omnivorous), spotted hyaena or lion. On the dead organic matter side of the diagram C_1 could be termite, spotted hyaena again (because they make kills and scavenge), vultures, or microrganisms, and C_2 could be ardwolf, bat-eared fox (both take termites), serval (which is known to take vultures from time to time), or microrganisms again.

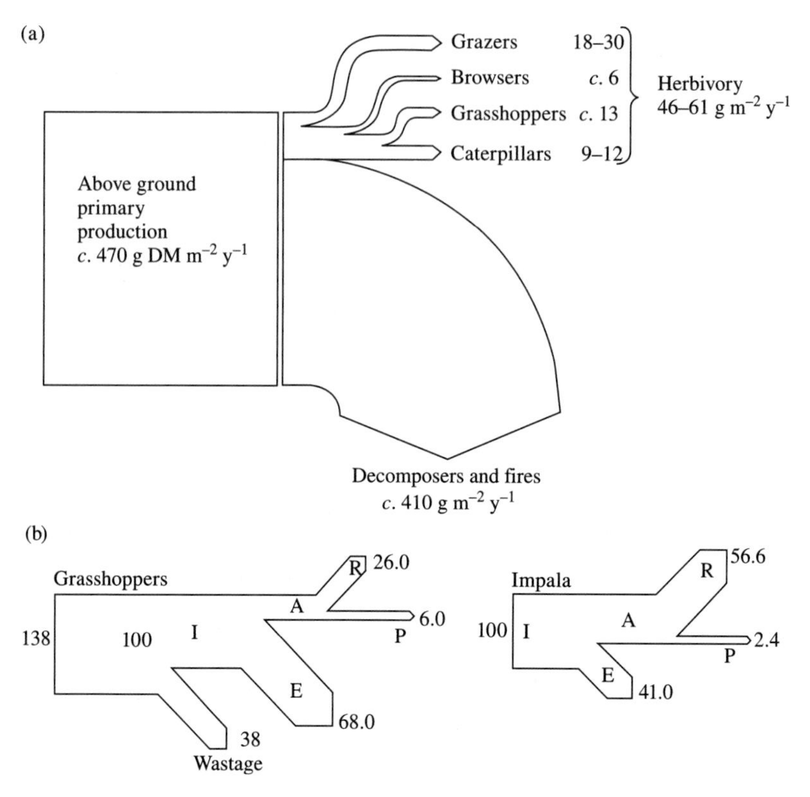

Fig. 6.4 Secondary consumption in Nylsvley, a wooded savannah in southern African. (a) pathways showing the disappearance of plant material. (b) partitioning of energy consumption by two herbivores into intake (I), excretion (E), production (P) and respiration (R). The fraction not excreted is assimalated (A) by the digestive system (after Scholes and Walker 1993).

of the material removed from the plant is not ingested, but wasted, because it falls to the ground during the feeding process.

As organic matter is consumed, energy is transferred to the next trophic level. However this transfer is inefficient, leading to the familiar pyramid of numbers within a community. The biomass of herbivores is much less than that of the vegetation that supports it, and the biomass of carnivores even less than that of the herbivores. Indeed, this is one of the reasons frequently put forward to explain the length of food chains. There simply are not enough lions to support another predator level above them (if you don't count disease). There are in fact three transfer efficiencies between primary production and the secondary consumers above. These are (1) consumption efficiency, (2) assimilation efficiency, and (3) production efficiency. Consumption efficiency is the percentage of the total available productivity consumed (ingested) by the next trophic level. For example, the

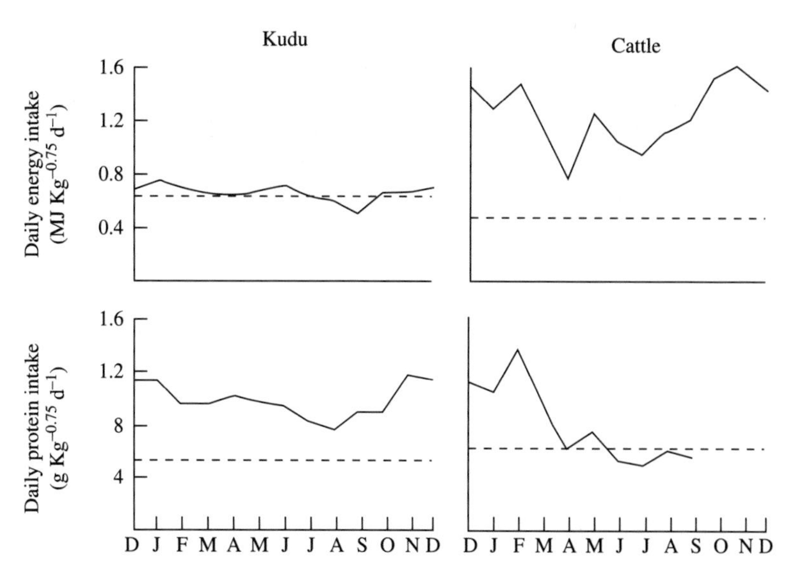

Fig. 6.5 Seasonal intake of protein and energy in a browser (kudu) and a grazer (cattle) (from Scholes and Walker 1993).

percentage of NPP finding its way into the guts of herbivores. In general this is low, between about 0.5 and 20 per cent. Assimilation efficiency is the percentage of food taken into the gut of consumers that is assimilated across the gut wall. In general this is about 20 to 50 per cent for herbivores and 80 per cent for carnivores. Production efficiency is the percentage of assimilated energy that is incorporated into new biomass. For invertebrates this is about 30 to 40 per cent. For ectotherm vertebrates (reptiles and amphibians) it is about 10 per cent and for endotherm vertebrates (birds and mammals) about 1 to 2 per cent. The much lower rate for endotherms is because they expend energy in regulating their body temperature. This suggests that the average efficiency of a mammalian herbivore, like a wildebeest, is approximately $0.1 \times 0.35 \times 0.015 = 0.05 = 5$ per cent of grass production. By the time this 5 per cent of NPP has become incorporated into new lion biomass it will only be about 0.006 per cent of NPP. No wonder the pyramid of numbers decreases so rapidly. Most terrestrial communities described so far typically have food chains of length three or four (Begon *et al.* 1996), and this is also true of African savannahs. Almost all the mammalian food chains are three in length (e.g. grass, wildebeest, lion or grass, termite, aardwolf). Many avian food chains are of length four (grass, grasshopper, insect-eating bird, raptor).

Of course energy flow graphs like those of Figure 6.4 are annual averages. Incoming solar energy varies throughout the year (even in the tropics) because of cloud cover, and NPP varies because of factors such as seasonal water availability. Not only production varies, but consumption also.

Owen-Smith (1982) suggested that grazing large mammals face a period of protein shortage during the dry season (see also Chapter 4) while browsers have a period of energy shortage at this time. Figure 6.5 shows that kudu at Nylsvley (Scholes and Walker 1993) are able to satisfy their protein requirements at all times of the year, but that they are unable to meet their energy requirements for a two month period in late winter. This is because the forage consumed by browsers has a high protein content relative, to grass, even in the dry season. However, because many browse species are deciduous, forage is scarce at this time.

Food webs

Since the early 1970s there has been an increasing interest among community ecologists in the dynamics of food webs (May 1972, 1973; Pimm 1991). This interest has included several topics, including the relationship between complexity and stability, the importance of 'connectance' (the fraction of all possible pairs of species that interact directly), the length of food chains and the possibility that food webs are compartmentalized. Most of these topics involve the idea that food webs (community structure) are in some way constrained to be like they are. We have already referred to the constraint of energy transfer efficiency, between trophic levels, that leads to the pyramid of numbers and constrains the length of food chains. These discussions depend critically on the quality of data that are available across a range of communities, and to date this quality of information is not available for savannah communities. However, it is well worth mentioning the field experiments of McNaughton (1977, 1985). One of the food web topics that has provided considerable scope for debate is that of 'complexity and stability'. The conventional wisdom was that more complex communities were

Table 6.2 The influence of grazing, by buffalo, on biomass (% of control) and species diversity in two areas differing in plant diversity.

Species diversity	Not grazed	Grazed	Statistical sig.
Biomass			
Species-poor plot		75.9 (69.3)	
Species-rich plot		66.9 (11.3)	
Statistical sig.		Not sig. ($P < 0.005$)	
Diversity			
Species-poor plot	1.069	1.357	Not significant
Species-rich plot	1.783	1.302	$P < 0.005$
Statistical sig.	$P < 0.005$	Not significant	

From McNaughton 1977. Species diversity is measured using the Shannon index of diversity $(H) = -\Sigma p_i \ln p_i$, where p_i is the frequency of the ith species, and ln is the natural logarithm. This index takes into account the number of species, and their frequency. The latter is deemed important because a community with, say, 10 equally abundant species is considered more diverse than one with 10 species, one of which comprised 91% and the other nine 1% each. $H = 0$ when only one species is present and $H_{max} = \ln S$, where S = number of species.

more stable, and MacArthur (1955) and Elton (1958) brought together a variety of observations to support this idea. Unfortunately, neither 'complexity' (= more species, more interactions between species, greater average interaction) or 'stability' were rigorously defined. What is more, the original observations of MacArthur and Elton, although consistent with the idea that more complex communities are more stable, could also be seen in terms of other reasonable, alternative, explanations (Begon *et al.* 1996). This issue was made more contentious when in 1972 Robert May introduced some mathematical descriptions (models) of food webs in which more complex webs were actually less stable.

In the Serengeti National Park, McNaughton (1977) chose two grassland areas that differed in the diversity of their plant communities. One was species-poor and one was species rich. Within each area, African buffalo were allowed to graze in some plots and were excluded from others. The results are shown in Table 6.2. The more diverse plant community was subject to a greater diversity modification by the grazing buffalo. However, this greater diversity impact was translated into a reduced effect on a functional property—green biomass. Although the amounts eaten were similar, growth of uneaten species in the more diverse community compensated for the greater consumption. Compensation by the community for grazing off take was 83 per cent in the species rich community but only 9 per cent in the species poor community. McNaughton believes that the buffalo study provides confirmation of the idea that plant community diversity stabilizes functional properties of the community to environmental perturbations. In fact the adjustments in species abundances in the more diverse community (resulting in greater diversity (H) changes) mediate functional stability. In other words, more diverse communities have a greater array of species that are able to react and compensate for what other species are doing. In 1985, McNaughton reported a more extensive study of the effects of grazing by zebra, buffalo, wildebeest, and Thomson's gazelle on grassland in the Serengeti. He considered what he called biomass resistance, as a measure of community stability. This was defined as the proportion of the above ground vegetation not eaten during either a single passage of the herd (when wildebeest was the main grazer) in the dry season, or a 12 day period in the wet season (all species). The results are shown in Figure 6.6. Notice that in the dry season (Fig. 6.6a) wildebeest had a much greater impact on the vegetation. In both experiments grazing had a much greater impact (less vegetation uneaten) on areas of low diversity, suggesting that community stability is greater with higher diversity. The complexity and stability debate is a critical one because if more complex communities (more species, more species diversity) are more stable (persist longer, resist disturbance better) it is important to try and maintain this complexity if we wish to see savannah ecosystems survive.

Another food web issue, already alluded to, involves cascading, indirect, interactions and their effects. Until we have more detailed information on

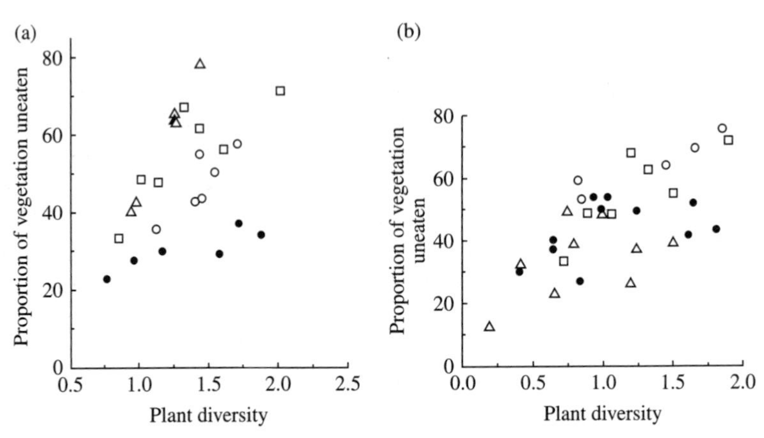

Fig. 6.6 Resistance of vegetation to grazing in the Serengeti National Park, Tanzania. (a) dry season data, ($P < 0.05$), (b) wet season data, ($P < 0.001$). Zebra (○), buffalo (□), wildebeest (●) and Thomson's gazelle (△) (from McNaughton 1985).

savannah food webs, with interaction strengths between all species, and descriptive models to simulate the cascading interactions, we cannot be precise about these indirect interactions. Of course, at a level much lower than a complete food web, the model of Dublin *et al.* (1990), referred to in chapter 5, was an attempt to investigate these more complex multispecies interactions. However, to see something of the potential issues involved and the likely consequences of indirect interactions, we can consider Figure 6.7.

This is a simplified food web for the Serengeti–Mara ecosystem taken from several sources, but mainly from Sinclair and Norton-Griffiths (1979) and Sinclair and Arcese (1995). No strengths of interaction are indicated, only their positive or negative effects, and many species are absent. However, as an example of an indirect interaction consider what is implied if rinderpest is reduced or eliminated from the system. With the removal of rinderpest, wildebeest numbers will increase, green grass will decrease and consequently there will be less dry grass to burn. This will result in fewer fires (or less extensive/intense fires), and therefore greater survival of young trees. More young trees will mean more mature trees, which should have a positive effect on the giraffe population. The implication of the trophic networks in Figure 6.7 is that many years after the removal of rinderpest there may (other things being equal) be an increase in numbers of giraffe. In fact, such an increase has been reported by Grimsdall (1979). In one area of the Serengeti estimates of giraffe numbers increased from 2,780 ± 634 (1971) to 4,970 ± 916 (1976). Pellow (1977) also had evidence of increasing giraffe numbers in the central area of the Serengeti.

In fact a start has been made to look at some of these interactions within the Serengeti–Mara ecosystem. In the first Serengeti book (Sinclair and Norton Griffiths 1979), Hilborn and Sinclair (1979) constructed a model of

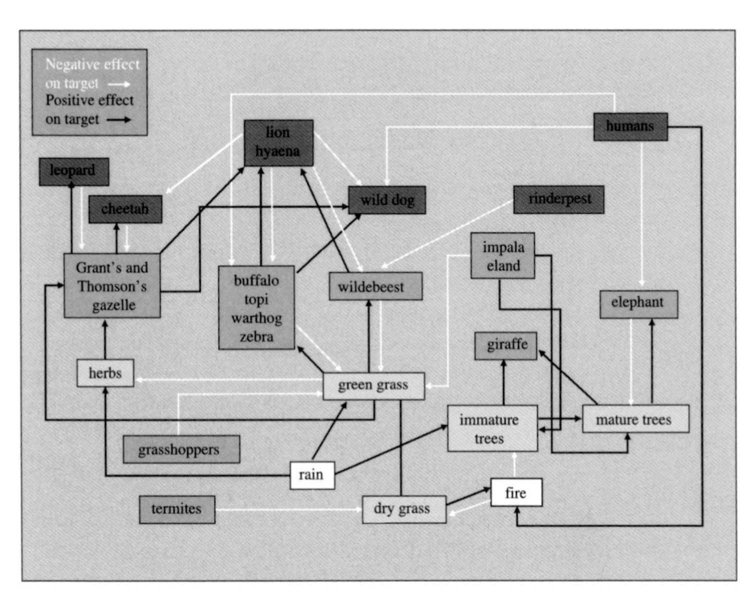

Fig. 6.7 Simplified trophic web of the Serengeti–Mara ecosystem. ▭ physical factor, ▭ plants, ▭ herbivores, ▭ carnivores.

the Serengeti ecosystem. This model focused on the interaction of rainfall, grass, wildebeest, and the predators, lion and spotted hyaena. The rainfall, grass, wildebeest component was a less sophisticated fore runner of their later model, already reported in Chapter 4 (Mduma, Hilborn, and Sinclair 1998). To this basic single species model they added predation. To simulate the interaction between predator and prey they used Holling's disc equation (Chapter 4, equation 3, Figure 4.16) modified into its multiprey form (Charnov 1974). Data on total quantity of prey consumed in a year (kg) came from Schaller (1972) and prey preference came from work in Nairobi National Park (Foster and Kearney 1967; Foster and McLaughlin 1968) for wildebeest, zebra, impala and hartebeest. For predator cub survival they used a linear relationship with food availability. Some of their results are shown in Figure 6.8. The two 'rate of increase' curves are for different amounts of annual rainfall and show quite clearly the effect of rain, via grass, on wildebeest population growth (see also Chapter 4). In 1995 Hildborn *et al.* commented that the '. . . model did predict quite accurately how the wildebeest would respond to the rainfall regime that has occurred in the 14 years since the model was constructed'. Where the curves cross the $N_{t+1}/N_t = 0$ line at high wildebeest density we see the effects of food limitation (Chapter 4). The curves crossing the zero line at low density is probably a result of predation. If this is a correct description of real events it suggests that predators (lions and hyaenas) may well have a crucial effect on wildebeest population size, at very low density.

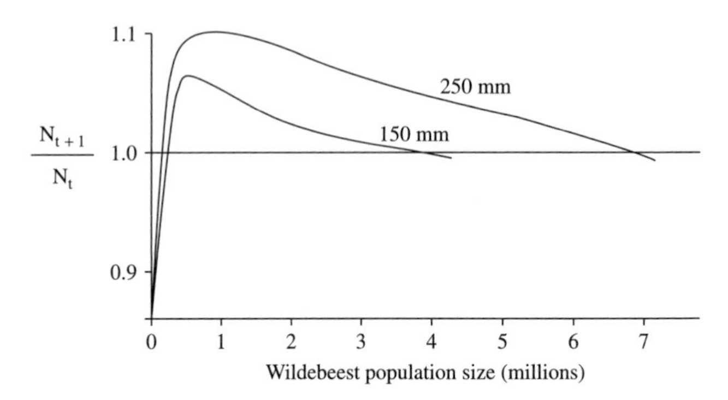

Fig. 6.8 Rate of change of wildebeest population plotted against wildebeest population size for two dry-season rainfall regimes (from Hilborn and Sinclair 1979).

At an ecological workshop held at the Serengeti Wildlife Research Centre in December 1991, Ray Hilborn and 25 other researchers spent four days trying to construct a second model of the Serengeti ecosystem (Hilborn *et al.* 1995). This model was built primarily for research coordination and policy evaluation. While some of the issues it addressed were clearly biological in nature (e.g. herbivore and carnivore population dynamics, species loss, vegetation change), some were more management orientated (e.g. visitor capacity, hunting and poaching, economics and cash flow). The spatial scale of the model divided the ecosystem into ten areas (four of which include the Serengeti N.P., one the Masai Mara and five the areas surrounding the Serengeti. The temporal scale divided the year into an eight month wet season and a four month dry season, although some parts of the model did not use this intra-annual scale (e.g. human population growth). The model was subdivided into five sub models: vegetation, ungulates, predators, inside park, and outside park. Simulations started in 1960 and continued until 2020.

The **vegetation sub model** used dry and wet season rainfall to predict dry season old grass and new dry season green grass. These were reduced by the percentage of each of the ten areas that was burnt, under cultivation or woodland. In the **ungulate sub model** six 'species' were described: wildebeest, zebra, Thomson's gazelle, elephant, buffalo, and 'brown animals'. The latter collectively refers to topi, impala and hartebeest. For each 'species', the number of animals born was simply made proportional to the population size i.e. $N_t \times$ birth rate. The population size each year (N_{t+1}) was made equal to that in the previous year (N_t) $\times$ dry season survival + births − hunter kills − disease − predator kills. The key dynamic factor is dry season survival which was assumed to be related to the amount of dry season grass per individual. The **predator sub model** included both calculation of the kill rate of prey items and the population dynamics of the predators. Kill rate was once again modelled by a disc equation in which the variables are

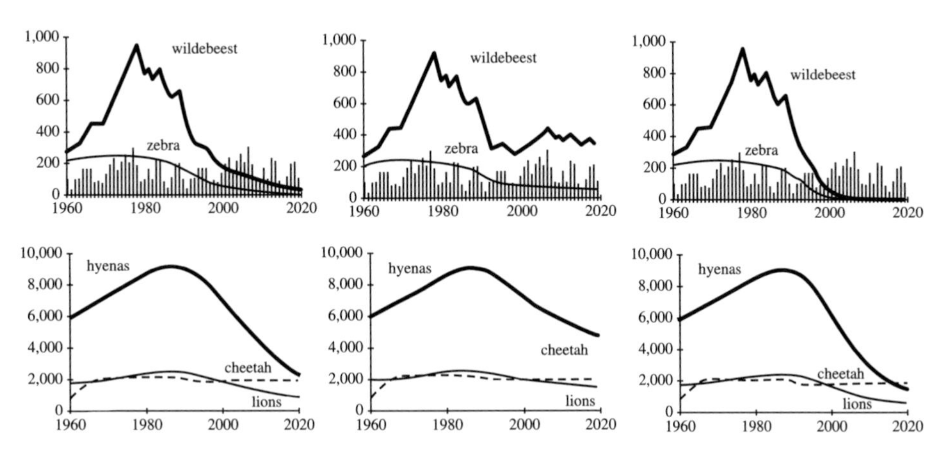

Fig. 6.9 Output Serengeti model. (a) current (1991) values for input variables, (b) increased anti poaching budget, (c) greatly reduced anti poaching budget. Histograms are rainfall, both actual and predicted (from Hilborn *et al.* 1995).

'predator attack success' and 'prey handling time' for each predator (lion, hyaena, cheetah, leopard, and wild dog). Population dynamics were simply $N_{t+1} = N_t \times (1 + \text{birth} - \text{mortality} - \text{number poached})$, with both birth and mortality related to predator density. The **inside park sub model** had two major components, 'tourism quality/growth' and 'park revenue'. Tourism quality goes up as more animals are seen and down as more tourists are present. Animals seen was not simply a product of numbers. For example, predator numbers were weighted with regard to their 'tourist value', with lions considered one-half that of cheetahs and leopards, and hyaenas as zero! Tourist growth rate was related to tourism quality and the general growth rate of the tourism industry. Revenue was related to the number of tourists × the park fee ($15.00) + the fee from overnight stays in hotels ($5.00), with 50–75 per cent of the revenue allocated to the park's operating revenue and the rest going to the Tanzanian government. The anti poaching budget (APB) of the park was a percentage of the park's operating revenue. The **outside park sub model** had three major components: poacher effort and kill, human population growth (HPG) and changes in land use. Poaching was assumed to increase with HPG and decrease with APB. The number of kills was further influenced by species vulnerability. HPG was assumed to have a constant rate and land in wilderness was assumed to decrease in proportion to HPG. Figure 6.9 shows some of the biological output from the model. The dramatic effect that more or less anti poaching has on the system is clearly seen, emphasizing the important role of tourism in maintaining the park budget (see later in this chapter).

Of course, one problem with community models such as this is that much detail is omitted. Factors put into the model are regarded by the modellers as 'important'. For example, grass production was assumed to be largely determined by rainfall and only dry season mortality of ungulates is included. If these prior decisions are incorrect the model will be misleading. Nonetheless, there must be more attempts to describe these savannah communities by modelling.

Not all species within a community have equal roles to play in its dynamics. This would be suspected just from an examination of the frequency distribution of abundances, even without a knowledge of the species interactions. In most communities, a few species are very abundant, while many species are much less common and even rare (Magurran 2004). Savannah communities are no exception (Fig. 6.10). However, some species have an impact on community dynamics far in excess of their numerical abundance and Paine (1966, 1980) called these 'keystone' species. Sinclair and Byrom (2006) state that wildebeest in the Serengeti ecosystem, have such a keystone role, although it is not clear that this is in excess of their numerical abundance. However, they certainly do play a pivotal role in the dynamics of the large mammal community, as Figure 6.7 implies. In addition, recent work has detected the impact of wildebeest on less obvious components of the ecosystem. By their impact on both the physical structure and species composition of the grasses and herbs of the Serengeti plains, grazed areas can support a density of butterflies two orders of magnitude greater than ungrazed areas (Sinclair and Byrom 2006). With grasshoppers the effect is negative, probably because both are feeding on grasses. In ungrazed areas, 49 species of grasshopper have been recorded, while in grazed areas it can be as low as only one species. Wildebeest may even affect the bird community. Their grazing, through its effect upon grass composition and height, may affect both feeding and nest sites. Of the eight commonest bird species feeding in the grass layer of grazed sites, seven showed reductions of 50 to 80 per cent in ungrazed sites (Sinclair and Byrom 2006). Figure 6.7 may look complex, but all the potential unexpected, cascading, effects are certainly not present.

A similar keystone role has been attributed to elephants (e.g. Laws 1970, Waithaka 1996). They have the ability to modify the habitat greatly, particularly in their destruction of trees, shrubs, and saplings. We saw in Chapter 5 (Fig. 5.8) that, in association with fire, they have the ability to convert woodland savannah into grassland savannah. In addition, studies carried out in the Aberdares, in Kenya (Waithaka 1996), have shown that elephants create and expand gaps in forests and in the process open up a more productive and diversified ground layer that is subsequently exploited by a wide range of other animals. This elephant effect on savannah and forest vegetation was particularly evident in many protected areas in the 1980s (Chapter 5) when elephant numbers in East Africa averaged five times those outside parks. Elephants are also a key species in maintaining and promoting the tourism industry. In a survey conducted in Kenya in the late 1980s, tourists rated the elephant the most important species. These kinds of community issues will be examined later in this chapter, when the impact of tourism and agriculture are examined.

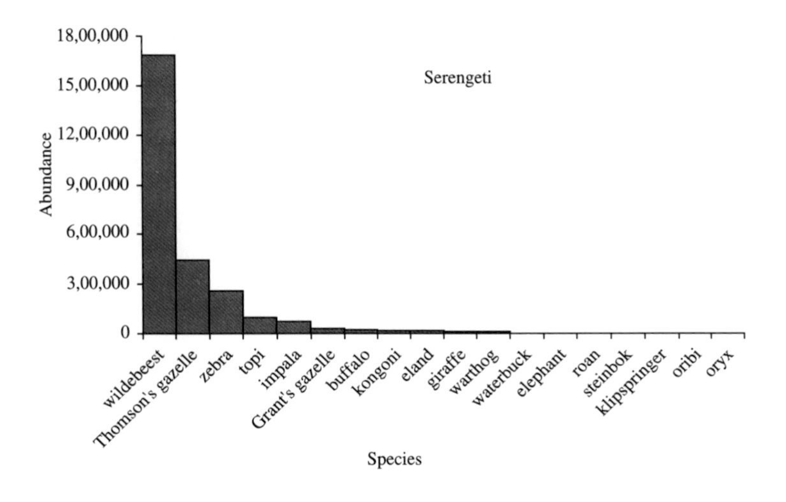

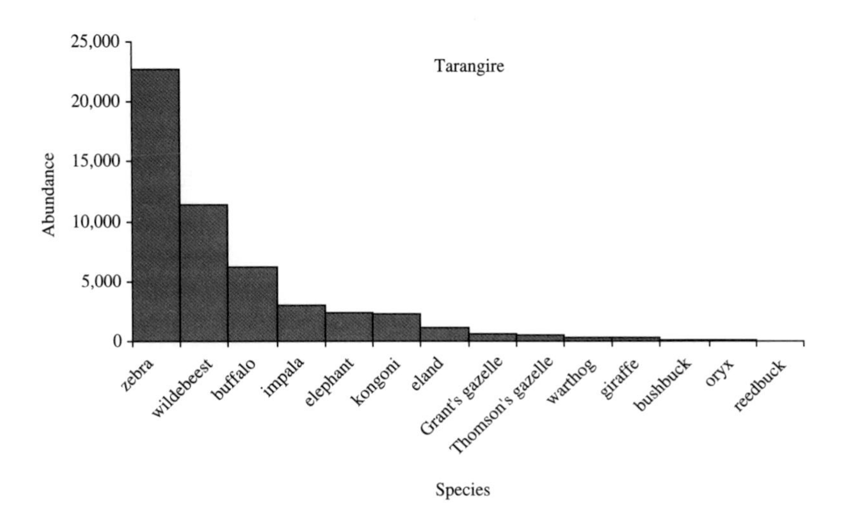

Fig. 6.10 Rank abundance of ungulates in the Serengeti ecosystem in 1991 (top) and the Tarangire ecosystem in 1990 (bottom).

Assembly rules

In Chapter 5, when I talked about interspecific competition, I commented that many ecologists searched for 'community patterns' in order to uncover the action of competition. Niche differences, discussed in Chapter 5, were one such pattern. Now I will examine some other patterns.

The idea that communities have assembly rules (they are not just random assemblages), and therefore show patterns in the type of species they contain, has a long history in ecology. If interspecific competition is important in structuring communities, niches should be more spaced out along a resource dimension than expected by chance. Niche differences, particularly those associated with diet, are frequently manifested as morphological differences. Therefore this over dispersion of niches should reveal itself as an over dispersion of morphological measurements. Such morphology might include bill size in birds and body size in savannah ungulates. Hutchinson (1959) collected a number of examples, from both vertebrates and invertebrates, of sequences of potential competitors in which individuals from adjacent species had body weight ratios of approximately two. The implication is that the larger species should be about twice as big in order to 'escape' interspecific competition.

As we saw in Chapter 5 (Figure 5.25) there is a relationship between energy turnover and body size (energy turnover $= K + W^{0.75}$). Given a series of animals, such as ruminants, with comparable digestive systems, the smaller animals require a diet with a higher proportion of protein and soluble carbohydrate at the expense of fibre. There is in fact a negative relationship between body mass and the percentage of non-stem in the diet of grazers (Owen-Smith and Cumming 1993) (Figure 6.11). Larger species bulk feed on whole plants, while smaller species selectively graze on the more nutritious leaves and shoots. Because grasses tend to employ structural defence mechanisms against herbivores (Chapter 2), while dicotyledons (herbs and trees) tend to employ unpalatable secondary metabolites, this influence of diet on body size extends to browsers compared to grazers. To process grass you require not only suitably resistant teeth but also a large rumen (Chapter 3). The rumen of grazers therefore tend to be larger than those of browsers (Hofmann 1968, 1973) and this translates into the

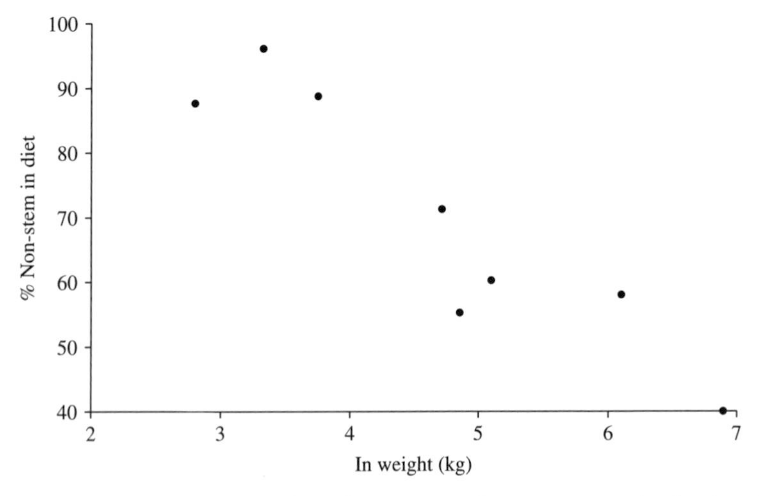

Fig. 6.11 Relationship between body weight and percentage of non-stem grass in the diet of grazing ungulates (data from Owen-Smith and Cumming 1993).

frequently observed body size difference between browsers and grazers (Fig. 6.12). Given this close relationship between diet and size in savannah ungulates it would seem that they are a likely candidate for Hutchinson's x2 assembly rule.

Prins and Olff (1998) examined 95 African grazers with a body weight heavier than 2 kg. Most of these were mammals, but they included two species of geese (Egyptian and spur-winged) and the leopard tortoise (*Geochelone pardalis*). Figure 6.13 shows their results for their total African data and for a Serengeti subgroup. If sequential species obey Hutchinson's x2 rule and the ranked data are plotted against log body weight, the line should be linear and the slope should be $0.693 = e^{0.693} = 2$. Prins and Olff (1998) also calculated the expected line, of rank vs. ln body weight, if each body weight had the same probability of occurring (the null model if species do not interact). They argue that if every body mass, between the minimum (W_{min}) and the maximum (W_{max}) had an equal probability of occurring then a non-linear line results. Weight (W) would have a uniform distribution and the expected difference between consecutive species (the weight ratio, WR) would be $((W_{max} - W_{min})/SR)$ where SR is the number of species in the assemblage. The expected value of the body mass of the *i*th species (W_i) with rank R_i would be

$$W_i = W_{min} + (R - 1)/((W_{max} - W_{min})/(SR - 1)),$$

or taking natural logarithms $\ln(W_i) = \ln(W_{min} + (R-1)/((W_{max}-W_{min})/(SR-1)))$. This $\ln(W_i)$ will be a non-linear function of R_i. This 'neutral' line, indicating no competitive spacing, is also shown on Figure 6 13. Notice that the weight ratio, in combination with the range of body weights $(W_{max}-W_{min})$ will therefore determine how many species are found. That is, species richness $SR = (W_{max}-W_{min})/WR$. If the smallest and largest

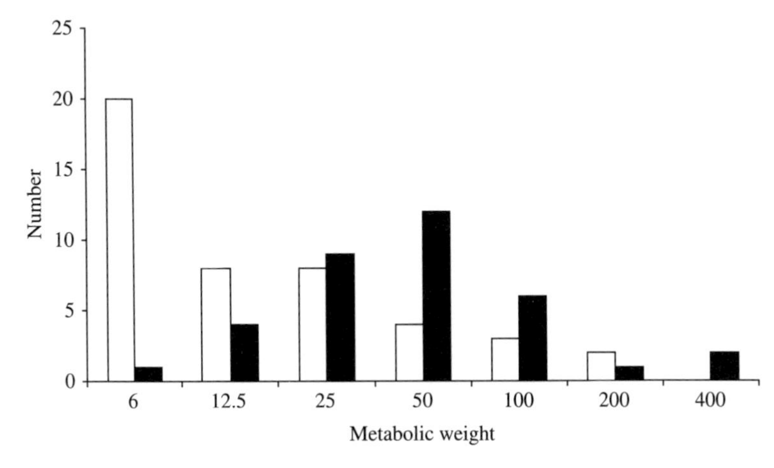

Fig. 6.12 Frequency distribution of body sizes for African grazing (dark bars) and browsing (white bars) ungulates. Metabolic weight is $kg^{0.75}$ (data from Owen-Smith and Cumming 1993).

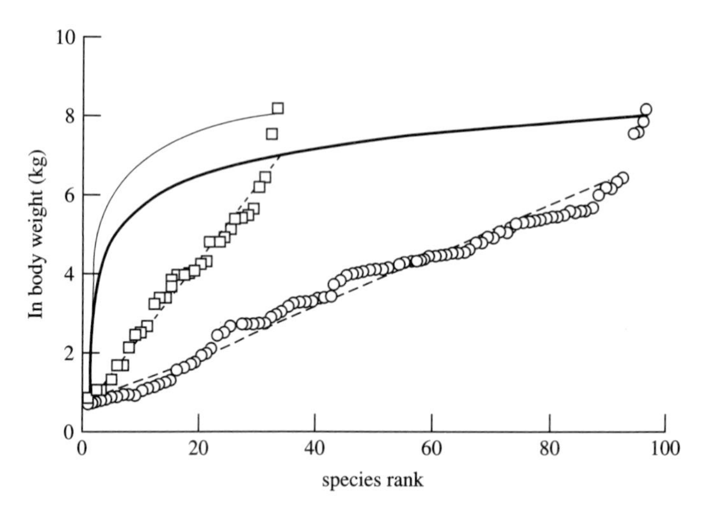

Fig. 6.13 Ranked ln body mass of grazers (W_i) plotted against rank number (R_i) for the whole of Africa (95 species) (circles and thick line) and for the Serengeti ecosystem (33 species) squares and thin line) (from Prins and Olff 1998).

species are rather constant (as they are in most African savannahs) this means that the weight ratio and species richness will tend to be inversely related.

The results in Figure 6.13 suggest that for the whole of Africa data, the weight ratio is only 1.066 and not the two suggested by Hutchinson (1959). However, the line is linear suggesting a regular dispersion of species along this feeding axis. Of course we might expect that species taken from the whole continent of Africa might be more closely packed than expected. Competition would not 'push' species apart that lived in separate savannah ecosystems, in different parts of African. In this connection the Serengeti data in Figure 6.13 is interesting. Again the ranked—body weight line is linear suggesting an even placement of species, but the weight ratio is now 1.21, suggesting that in a smaller area species might have to be more ecologically separated. Prins and Olff (1998) have pointed out that the total African data in Figure 6.13 is also interesting because there appears to be a gap between buffalo (632 kg) and the last four largest species (white rhinoceros 1875 kg, hippopotamus 1900 kg, forest elephant 2575 kg and savannah elephant 3550 kg)(Chapter 3), often referred to as mega-herbivores (Owen-Smith 1988). Does this gap in the community pattern of grazing herbivores suggest that there are 'missing' species that could fit into this grazing resource axis. From Figure 6.13 the number of 'missing' species can be estimated to be between 6 and 10 (Prins and Olff 1998). It is intriguing to speculate that the missing very large herbivores could be several species that have gone extinct. These include the proboscid *Deinotherium bozasi* (in the early Pleistocene) and four 'elephants' (*Elephas recki*, *E. iolensis*, *E. zulu*

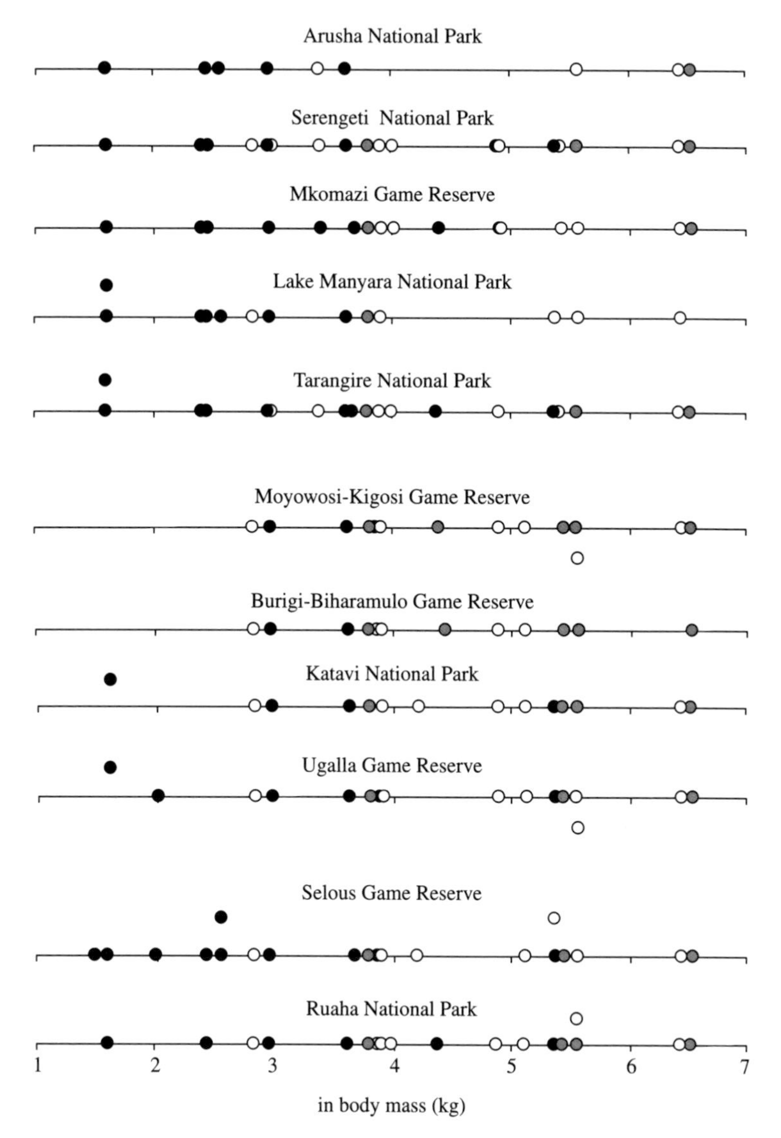

Fig. 6.14 Body weight of Tanzanian bovids plotted along a in(weight) axis, with diet indicated as grazer (white), browser (black) or mixed feeder (grey). Each point represents one species. Where two species have identical weights, one is indicated off the line.

and *Loxodonta atlantica*) which went extinct between a million and 400,000 years ago. Other mega-grazers that went extinct during the Pleistocene were *Hippopotamus gorgops*, a giant hartebeest (*Megalotrogus priscus*) and the giant buffalo (*Pelorovis antiquus*)(Klein 1988, Owen-Smith 1988). The predicted number of missing species (6 to 10) therefore tallies quite well with the number of recently extinct species (8)(Prins and Olff 1998).

Table 6.3 Weight ratios for 11 Tanzanian protected areas with savannah habitats.

Protected area	Weight ratio	Area (ha)
Selous GR	1.30	5,233,000
Serengeti NP	1.27	3,425,100
Ruaha NP	1.29	2,595,000
Ugalla GR	1.36	850,000
Tarangire NP	1.26	680,000
Katavi NP	1.36	525,300
Burigi-Biharamulo GR	1.32	350,000
Mkomazi GR	1.39	250,000
Moyowosi-Kigosi GR	1.32	130,000
Arusha NP	1.80	13,700
Lake Manyara NP	1.45	10,833

I will end this examination of savannah assembly rules by presenting some unpublished data that Fiona Wragg and I have collated from Tanzanian protected areas (Wragg 2002). This data includes all the Tanzanian grazing and browsing bovids, and they are displayed, in Figure 6.14, along a weight axis for those reserves with savannah habitats. Arusha NP to Tarangire NP are northern Tanzanian reserves, with 'classic' grass and shrub savannah. Moyowosi-Kigosi GR to Ugalla GR are western reserves with woodland (miombo) savannah. Selous GR and Ruaha NP are central Tanzanian reserves, on the 'border' between miombo and grass/shrub savannah (see Fig. 1.9). Notice that as already stated (Fig. 6.12), browsers are usually smaller than grazers. The weight ratios are shown in table 6.3. All are again less than Hutchinson's x2 rule but show significant linear regressions of ln weight vs. rank, implying that the body weights of these bovids are more regularly spaced than you would expect by chance. There do appear to be community patterns, suggesting assembly rules, for both grazing and browsing savannah herbivores.

Island biogeography

Larger islands contain more species. Although the specific relationship between number of species (S) and island area (A) can vary, it has traditionally been viewed as a power function:

$$S = cA^{z}$$

where c is a constant giving the number of species when A has a value of 1 (often specific to a taxon), and z encapsulates the relationship between species number and area. By taking logarithms of both sides of this 'species-area' equation we get:

$$\log S = \log c + z \cdot \log A.$$

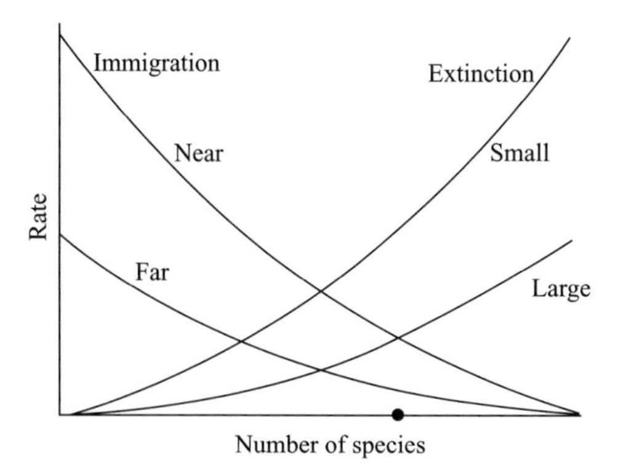

Fig. 6.15 MacArthur and Wilson' (1967) equilibrium theory of island biogeography showing the rate of immigration and extinction plotted against the number of resident species on the island. The predicted equilibrium between immigration and extinction, for a large island and a near source of colonists, is shown by the ● symbol.

That is, on a double log plot ($\log S$ vs. $\log A$) we see a straight line relationship between S and A, with z the slope of the line. But why is this interesting island relationship relevant to African savannahs?

The species–area relationship has been found to apply to a wide variety of taxonomic groups (e.g. plants, insects, fish, birds and mammals) and to a wide variety of types of island, including 'virtual' islands (e.g. lakes, mountain tops, and protected areas). Sadly, intact African savannahs are increasingly only found in protected areas. These National Parks, Game Reserves and private reserves are 'islands' of relatively intact savannah fauna and flora in a 'sea' of degraded habitat. What does island biogeography tell us about the numbers of species in these savannah islands? However, before we consider African savannah islands specifically it will be useful to examine briefly the reasons that have been put forward for the island species–area relationship. There have been two main themes, not necessarily unconnected.

The original explanation , put forward by MacArthur and Wilson (1967), is known as the 'dynamic theory of island biogeography'. It proposed that species number (on an island) is a dynamic balance between two forces, immigration and extinction, and that these are in turn modified by island size and isolation. Figure 6.15 summarizes the situation proposed by the dynamic theory. Two example immigration curves and two example extinction curves are shown. When an island is empty of species any new arrival will contribute to the rate of immigration of new species. As species number on the island increases the number of new species arriving will fall and in theory will reach zero when all the potential colonists have arrived. Islands that are near the source of the colonists (the mainland for real islands) and large islands (large targets) will have higher rates of immigration. Extinction rate will rise when there are more species on the island.

This is thought to occur because with more species competitive exclusion becomes more likely and the population sizes will, on average, be lower. This latter will make them more susceptible to chance extinction. A similar reasoning would suggest that smaller islands would have higher rates of extinction. For example, the rate of ungulate extinction in Tanzanian reserves correlates significantly with reserve area ($r = -0.93$, $P < 0.02$) (Newmark 1996). Where the extinction curve and the immigration curve cross gives the dynamic equilibrium species number for a particular island. The theory therefore explains why smaller islands have fewer species and suggests that more isolated islands will also have fewer species.

The second major theme put forward to explain the species–area relationship is that of habitat diversity (Lack 1969, 1976). Essentially this states that larger islands will have more habitats and an area with more habitats will have more species. Of course this explanation could be incorporated within the dynamic theory because habitat diversity, linked to island size, could be one of the reasons for the different extinction curves in Figure 6.15.

Protected savannah areas

One of the earliest attempts to apply island biogeography ideas to savannah reserves was that of Miller and Harris (1977). They examined thirteen East African savannah parks and looked at the relationship between area and species number. Their results are shown in Figure 6.16a. The fitted relationship is $S = 35A^{0.02}$. The regression is not significant ($F = 0.17, P = 0.69$). Soule, Wilcox and Holtby (1979) published another set of data for 20 East African areas (Figure 6.16b) in which the species–area relationship was $S = 18A^{0.13}$, and this time the regression was significant ($F = 24.67$, $P = 0.0001$). Soule and colleagues make the point that without 'active' intervention, newly isolated communities such as National Parks and other protected areas tend to show species loss. This is known as relaxation. In effect, the isolated area travels 'down' the species area curve, from a higher point (larger area and more species), to a lower point (smaller area, fewer species). As protected areas become smaller and more isolated the rate of species loss increases because populations are smaller and therefore more prone to extinction (wild dogs in Chapter 5) and recolonization is less likely because of intervening, unsuitable, habitat. Of course the traditional theory of island biogeography suggests that the species going extinct will be lost at random, whereas in reality certain species are more prone to extinction. These might include, for example, species with larger territorial requirements, species with special habitat requirements and perhaps species more susceptible to competition. Smaller 'islands' will probably have fewer habitats and this will almost certainly be a major factor in the higher extinction rate found in smaller areas. Soule and colleagues estimate that simply due to benign neglect many of the smaller African savannah reserves would lose up to 10 to 20 per cent of their large mammal species in the fifty years

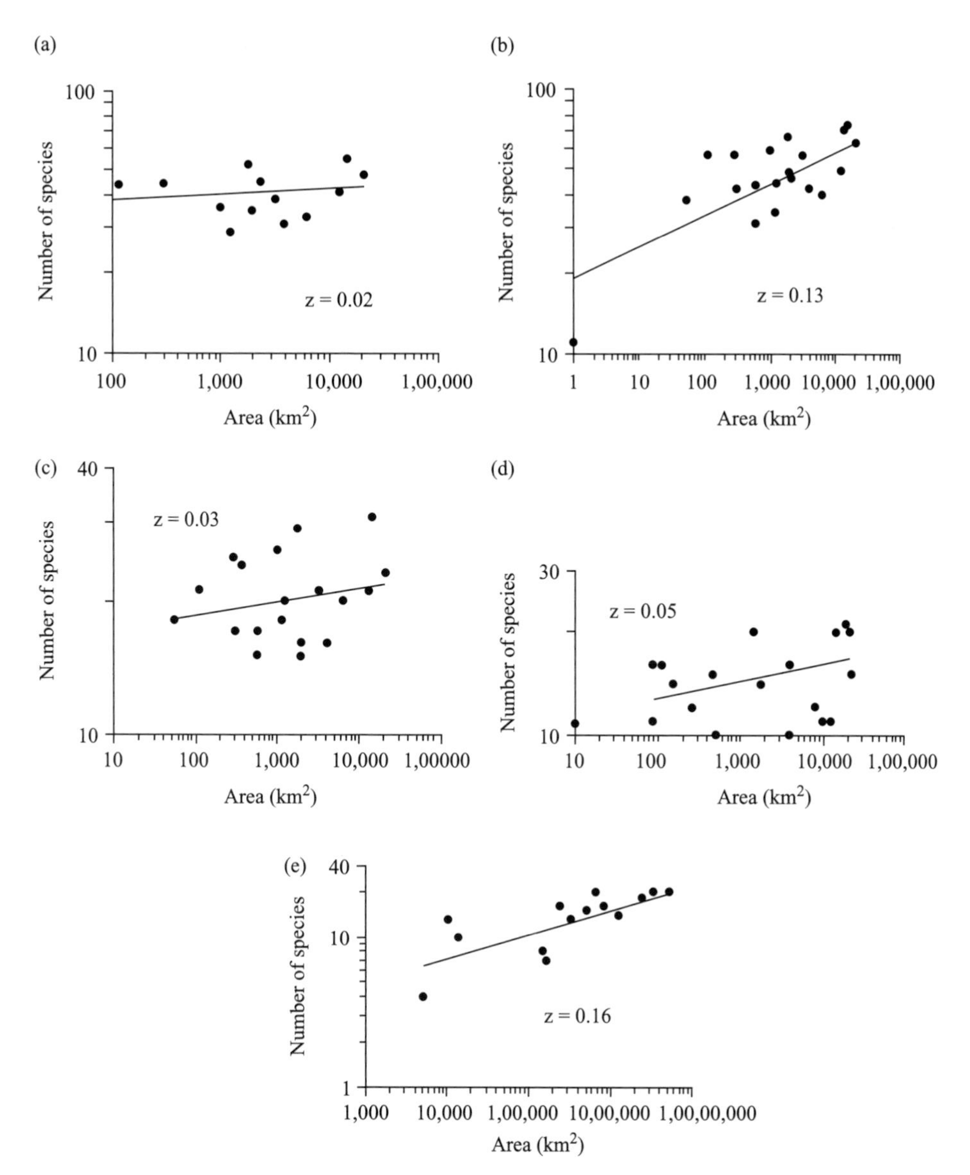

Fig. 6.16 Species-area graphs from five African savannah studies. Each graph shows the value of z for the data collected. (a) 'large mammals' in 13 East African parks (Miller and Harris 1977), (b) 'large mammals excluding insectivores, rodents, bats, and lagomorphs' in 20 East African areas (Soule, Wilcox and Holtby 1979), (c) 'ungulate species' in 19 East African parks and reserves (Western and Ssemakula 1981), (d) 'large herbivores' in 20 African reserves (East 1983), (e) 'bovids' in 14 Tanzanian ecological areas (Wragg 2002). The text in ' ' indicates the species used, in the author's own words.

following isolation. If this is true it suggests that protected areas will either have to be enormous or that they will require active intervention to maintain an intact fauna.

In response to the Soule predictions, Weston and Ssemakula (1981) published their own dataset of the ungulate species in nineteen eastern African parks and reserves (Fig. 6.16c). The species–area relationship was $S = 16A^{0.03}$ and the regression was not significant ($F = 0.93$, $P = 0.35$), suggesting that smaller parks do not have fewer species. These authors are also more pessimistic about the survival of large mammal species in reserve islands. They believe that the extinction rates used by Soule *et al.* were too high, having been derived from studies on 'true' oceanic islands. Savannah reserves (at least in East Africa) continue to experience immigration because, Weston and Ssemakula claim, '*extensive lifestock ranching is not necessarily inimical to wildlife*' They conclude that the '*significance of island biogeography to the design of nature reserves is limited, at least in the savannahs*'. Finally a geographically more extensive dataset was published by East (1983). He looked at 'large herbivores' (Fig. 6.16d). The species–area relationship was $S = 11A^{0.04}$ and the regression is not significant ($F = 1.81$, $P = 0.19$).

Of course one of the problems here is that different studies seem to show both significant or non-significant species–area relationships. It's not easy to see why. There are differences in the data, even the size of the same protected areas are frequently different. In some cases the species chosen are difficult to discover because terms like 'large mammals' and 'large herbivores' are used. I will end this section on species–area curves by again presenting some unpublished data that Fiona Wragg and I have collated from Tanzanian protected areas (Wragg 2002). This dataset looked specifically at bovids because a specific species–area curve probably only exists for ecologically similar species. It also looks at fourteen contiguous ecological areas, rather than the 24 gazetted parks for which we have data. So for example, the 'Serengeti island' = Serengeti NP + Ngorongoro CA + Maswa GR + Masai Mara NR + Ikorongo-Grumeti GCA and 'Selous island' = Selous GR + Mikumi NP. This is quite important since two adjacent reserves clearly constitute one 'island'. The data was compiled from a very extensive literature search (over 200 sources), and confirmed by using the IUCN African Antelope Database (East 1998). The species–area relationship is shown in Figure 6.16e. The species–area relationship is $S = 2A^{0.16}$ and the regression is very significant ($F = 16.15$, $P = 0.002$). Although not shown in Figure 16 we also had data for the larger set of all ungulates. Here the species–area relationship is $S = 1.5A^{0.19}$, and the regression is again very significant ($F = 14.73$, $P = 0.002$). Both these results suggest that there is a species–area relationship for both bovids and ungulates, at least in Tanzania, suggesting that the kind of effect suggested by Soule *et al.* (1979) might occur for these savannah species. It should be noted that none of our protected areas had the full compliment of Tanzanian bovid species (35). Extrapolating Figure 6.16e until $S = 35$ suggests that an area of more than

200 million km^2 would be need to preserve this number of species. In other words, an area larger than Tanzania. This suggests strongly that protected areas cannot be run on the principle of 'benign neglect'. That is, they cannot simply be established and left. Even in the largest protected areas such as the Serengeti complex and Selous complex in Tanzania, or the Kruger ecosystem in southern Africa, species will be absent or eventually lost. Intervention and management is required.

Conserving savannah ecosystems

Savannah wildlife has a value—ecological, scientific, financial, and aesthetic. It also has a cost—loss of human life, loss of property, crops, livestock, and income. These conflicting issues are briefly examined here, at the end of this book. For a more extensive account the reader should look at Prins, Grootenhuis and Dolan (2000) *Wildlife Conservation by Sustainable Use*. This conflict between humans and wildlife is often one between the specific interests of people living close to wildlife, with the less localized interests of wildlife conservation in Europe and North America. To really appreciate these issues people in the UK, for example, should imagine how they would feel about wildlife issues if bears and packs of wolves were reintroduced to Wales, North Yorkshire and Scotland. They would be an exciting tourist attraction but how would we cope with them killing sheep or the occasional person. This is the real issue with conservation on the ground, and if wildlife is to survive, this type of human conflict issue has to be resolved. We cannot simply enclose savannahs by fences, and hope everything will be OK.

Harcourt, Parks, and Woodroffe (2001) in their examination of the species–area relationship for African reserves discovered that small reserves tend to be sited in regions of high human population density (Fig. 6.17a). Given the problems experienced by small reserves, mentioned in the last section, does this mean that small reserves suffer a double jeopardy? Do they lose species because of their small size (smaller populations, fewer habitats) and because of human interference (hunting, poaching, surrounding, degraded, land making them more isolated)? To examine the latter effect data is available for eight protected areas, in six African countries (Hwange in Zimbabwe, Kruger in South African, Moremi in Botswana, Etosha in Namibia, Selous and Serengeti in Tanzania and Masai Mara and Nairobi in Kenya), for three carnivore species (wild dog, lion, and spotted hyaena). Human density correlates significantly with the percentage mortality caused by humans (Fig. 6.17b). This suggests that humans have a serious negative affect on wildlife (see also Fig. 6.7) and in fact the major conservation issues that we will now consider are human induced. There will, of course, always be local conservation issues that are important but cannot be examined in a small book of this kind. We will now look at

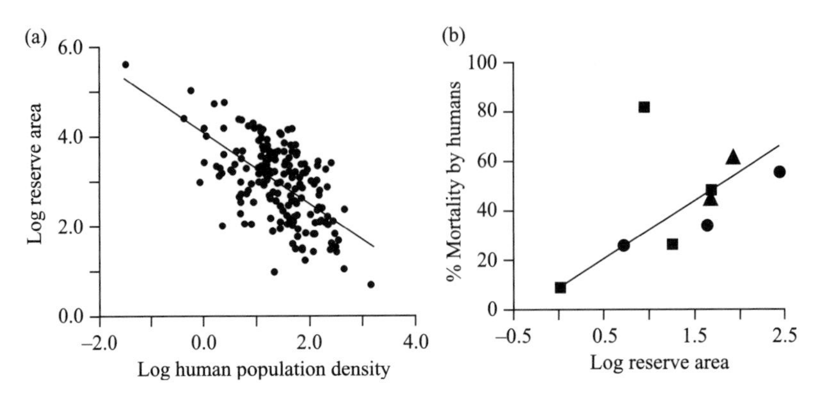

Fig. 6.17 (a) The relationship between log area (km^2) of IUCN category I and II African reserves and log local (within 50 km radius) human density (km^2). The regression line is log area = 4 − 0.75log density ($F = 78.4$, $P < 0.0001$). (b) Percentage mortality caused by humans on three carnivores, wild dog (●), lion (■) and hyaena (▲). The regression line is with the statistical outlier Hwange, Zimbabwe, omitted (from Harcourt, Parks, and Woodroffe 2001).

hunting and poaching, habitat destruction and wildlife–human conflicts, and wildlife tourism. The last three are closely linked since ranching, for example, often creates conflicts and habitat destruction, while an alternative to such ranching in more arid areas may well be wildlife tourism.

Hunting and poaching

As an example of the effect that hunting and poaching can have on wildlife I will outline a series of events that took place in the Serengeti National Park in the last century, and that also illustrate how tourism may interact with wildlife protection. Some reference has already been made to this in Chapter 5. In April 1977 the international border between Tanzania and Kenya was closed, and it remained closed until about 1986, when it was reopened partially for tourism. However, the main tourist route between the Masai Mara Reserve, in Kenya, and the Serengeti National Park, in Tanzania, remained closed. This border closure immediately affected the number of tourists coming into the Serengeti National Park. In 1976 it was 70,000 and in 1977 it was only 10,000 (Fig. 6.18a). Tourist numbers remained at this low level throughout the 1980s and only started to increase again in the 1990s. One consequence for the Serengeti N.P. was a drop in income from visitors, leading to a drop in the operating budget of the Park. This dropped continuously from 1982 to 1985 and remained low until 1987 (Fig. 6.18b). As a result, the anti-poaching effort (in terms of patrol days) dropped by the mid-1980s to 60 per cent of that prior to the border closure (Sinclair 1995).

At about this time there had also been a considerable increase in the human population between the Park and Lake Victoria to the west, often

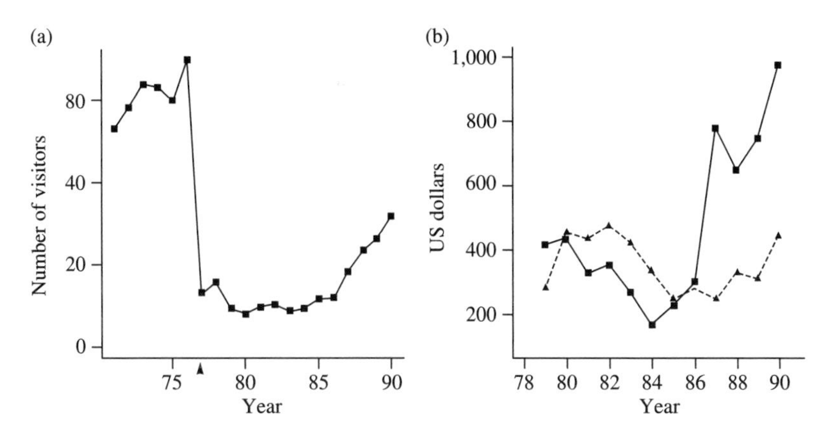

Fig. 6.18 (a) annual foreign visitor numbers. (b) income (solid line) and operating budget (dashed line) in US dollar equivalents for the Serengeti National Park, Tanzania. The arrow indicates the border closure (from Sinclair 1995).

an increase approaching 15 per cent per year. This human population increase, coupled with the lack of anti-poaching patrols, resulted in an invasion of northern and western Serengeti by poachers. The first species to be affected was the black rhinoceros which was hunted for its horn. It lost 52 per cent of its Serengeti population in the first year of border closure, 1977 (Fig. 6.19a). By 1980 this rhino was effectively extinct in the Serengeti and this is still the situation today. The occasional individual is seen in the southern part of the Park, having 'commuted' from the Ngorongoro Crater. About thirty black rhinos are still found in the Masai Mara. Elephant and buffalo were the next two species to show an effect of the border closure. The elephant, like the black rhino which was hunted for its horn, was subjected to 'trophy' hunting for tusks. Buffalo were hunted for their 'bush meat'. Several hundred elephants moved to the Masai Mara where poaching, although present, was less severe. This difference in the Serengeti and Mara elephant populations eventually had a distinct effect upon the vegetation of the two regions (Chapter 5). Both rhino and elephant have been placed on the CITES list of protected species. This had a very specific positive effect on the elephant poaching in the Serengeti (Fig. 6.19a) since there was an abrupt reduction in poaching in 1989 when the CITES ban on the ivory trade was imposed. It also appeared to have a good effect on elephant poaching generally, particularly throughout eastern and southern Africa. In Kenya, for example, elephant populations increased by 30 per cent between 1989 and 1997. The price of ivory dropped and the international trade in ivory collapsed.

The Convention on International Trade in Endangered Species (CITES) forms a core piece of international legislation protecting wildlife, particularly the African Elephant. It is a United Nations administered Treaty which

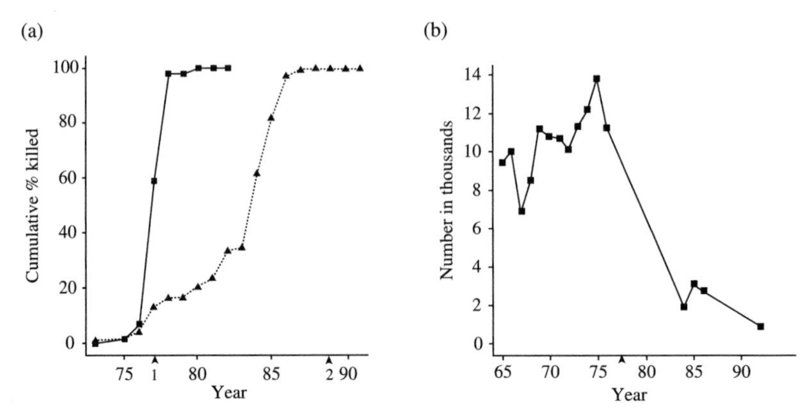

Fig. 6.19 (a) Cumulative % of total number killed by poachers and found by Serengeti National Park staff for black rhinoceros (squares) and elephant (triangles). Arrow 1 indicates border closure, arrow 2 indicates ivory trade ban. (b) Census of buffalo numbers in northern Serengeti. No censuses were conducted during 1977–83. Arrow indicates border closure. (from Sinclair 1995).

was set up in 1975 to control the international trade in wild flora and fauna in order to protect against over exploitation through commercial trade. It first entered into force on 1 July 1975, and now over 160 nations have signed the treaty. However, CITES relies on specific management authorities from signatory countries to regulate and control the trade. There are three levels of regulation, depending on which of three appendices a threatened species is listed.

Appendix I includes species that are threatened with extinction and that are, or may be, affected by international trade. These species are prohibited from being traded internationally for commercial purposes (e.g. pet trade, food industry, jewellery or ornament making, skins or traditional medicines). Some trade is allowed for non-commercial purposes (e.g. to go to educational facilities, or for scientific purposes). Savannah examples of Appendix I species (as of 14 June 2006) are cheetah, leopard, black and white rhinoceros, Grevy's zebra, and African elephant (except the populations of Botswana, Namibia, South Africa, and Zimbabwe, which are included in Appendix II).

Appendix II includes species that, although not necessarily threatened with extinction, may become so unless trade is regulated. Species may also be classified as Appendix II if their parts or products cannot be easily distinguished from those of other Appendix I or II species. Trade is allowed in Appendix II species, but is strictly regulated and only when it has been found that it will not be detrimental to the survival of the species. Savannah examples of Appendix II species are the African elephant (except for the African countries listed above), and lion.

Appendix III includes species that any country has identified as being exploited and/or threatened within their country, and needs the help of

Table 6.4 Numbers of species and subspecies in the CITES Appendices.

	Appendix I	Appendix II	Appendix III
Mammals	228 spp. + 21 sspp. + 13 popns	369 spp. + 34 sspp. + 14 popns	57 spp. + 11 sspp.
Birds	146 spp. + 19 sspp. + 2 popns	1,401 spp. + 8 sspp. + 1 popn	149 spp.
Reptiles	67 spp. +3 sspp. +4 popns	508 spp. + 3 sspp. + 4 popns	25 spp.
Amphibians	16 spp.	90 spp.	–
Fish	9 spp.	68 spp.	–
Invertebrates	63 spp. + 5 sspp.	2,030 spp. + 1 ssp.	16 spp.
Plants	298 spp. + 4 sspp.	28,074 spp. + 3 sspp. + 6 popns	45 spp. + 1 ssp. + 2 popns
Totals	827 spp. + 52 sspp. + 19 popns	32,540 spp. + 49 sspp. + 25 popns	291 spp. + 12 sspp. + 2 popns

other countries to regulate international trade in it. Savannah examples of Appendix III species are African civets (Botswana), and aardwolf (Botswana).

Roughly 5,000 species of animals and 28,000 species of plants are protected by CITES against over-exploitation through international trade. These include some whole groups, such as primates, cetaceans (whales, dolphins, and porpoises), sea turtles, parrots, corals, cacti, and orchids. But in some cases only a subspecies or geographically separate population of a species (for example the population of just one country) is listed. Notice that the CITES lists include species because of trade, not just because they are endangered. Table 6.4 shows the approximate numbers of species that are included in the CITES Appendices, as of present (14 June 2006).

Although legislation, such as the CITES agreements, have helped to combat poaching, particularly for trophies, they have certainly not prevented it. Even when trophy hunting is banned (e.g. black and white rhino) it has not prevented their severe reduction. There appears to be only two ways to reduce indiscriminate hunting: (1) set up protected areas, or (2) use the wildlife resources and consequently ensure its protection as an investment. This latter approach has been tried in Zimbabwe.

In Zimbabwe, the Community Areas Management Project for Indigenous Resources (CAMPFIRE) was launched in 1987 to give rural communities control over harvesting their natural resources. The project was set up with the philosophy that by allowing the selective harvesting of wildlife, rural people would convert less land to agriculture and police against poaching themselves. The project has been a success but sport hunting has been the main 'use', with 90 per cent of generated income coming from this source, and 64 per cent of this coming from hunting elephants. However, evidence still suggests that elephant populations in Zimbabwe have continued to increase at about 5 per cent, or 3,000 animals per year. The idea of using wildlife as a sustainable resource, that never-the-less maintains the savannah ecosystem intact is one of the conservation strategies that believes in the saying 'use it or lose it' . Another is tourism and we will approach that topic by way of the human-wildlife conflicts imposed via agriculture and ranching.

Habitat destruction and wildlife-human conflict

Wildlife-human conflicts occur frequently in rural areas. African farmers sometimes lose considerable amounts of their crops to wild animals (Simons and Chirambo 1991). In Nyami Nyami District, in Zimbabwe, crop damage accounted for 96 per cent of all complaints about animal problems (Hoare 1995). Crop loss in Laikipia District, Kenya, was estimated at between 10 and 24 per cent of the total maize crop (Thoules 1994) and Ngure (1995) records that Kenyans around Tsavo National Park, in 1991, lost crops worth an average of US$76 per farm. Farmers close to Kasungu National Park, Malawi, lost 10 per cent of their crops to wildlife (Deodatus and Lipiya 1991). However, there is an important and intriguing misconception, among farmers, about the wildlife responsible for this damage. Table 6.5 shows a comparison of ranked raid frequency on farms and numbers of complaints made at Kasungu National Park, Malawi. The two ranking are significantly different.

Sixty per cent of the complaints made by farmers, in this six-month period, related to elephants, while in reality they ranked third after baboons and bush pig in raid frequency. In Nyami Nyami District in Zimbabwe, elephant accounted for 78–80 per cent of all complaints, with buffalo second at 15–18 per cent (Hoare 1995). Large animals are frequently associated in the minds of farmers with 'wildlife', local National Parks and restrictions on economic activities. They see themselves in conflict with these wildlife species. Smaller species, such as rodents, are taken for granted and simply regarded as 'agricultural pests' rather like fungi, nematodes, and insects. Although exact figures are difficult to obtain it is known, for example, that

Table 6.5 Comparison of raid frequency (average monthly number of raids) on farms, and number of complaints made at Kasungu National Park, Malawi, January to June 1990.

Species	Mean raid frequency	Raid ranking	Number of complaints	Complaint ranking
bushpig	9.63	1	5	2
baboon	5.01	2	0	5
elephant	3.01	3	15	1
vervet monkey	1.71	4	2	4
hippopotamus	0.67	5	3	3
buffalo	0.33	6	0	5

Data from Deodatus and Sefu 1992.

near seven Tanzanian protected areas the main pest species were primates (51.9 per cent), bush pig (13.3 per cent), and rodents (10.6 per cent) (Newmark *et al.* 1994). In savannah areas where crop growing could be profitable the conservation solution has to be large reserves along with a well supervized system of compensation for farmers. Revenue from tourism and systems like CAMPFIRE will also help to change the attitude of local farmers. But the local population must be involved in this conservation process.

In those savannah areas where traditionally cattle rather than crops are reared, the conservation issues are slightly different. Here there is the possibility of using the same area for both wildlife and livestock, a situation not really possible with agricultural crops. However there are still issues of conflict. There is a widespread belief that grazing wildlife such as zebra and wildebeest compete with cattle for grass (e.g. Pratt and Gwynne 1977). What is the evidence for this? Prins (2000) reviewed the available literature and came to the conclusion that despite the considerable diet overlap between the two groups (an essential prerequisite for competition) there was little evidence for food limitation of livestock populations. He suggested that competition is largely asymmetrical and diffuse, with cattle having a competitive effect on wildlife, but wildlife having little or no effect on cattle. However, controlled replicate field experiments were lacking. Since then, Young, Palmer, and Gadd (2005) have published the results of a long-term enclosure experiment in Laikipia, Kenya. The experiments ran from 1995 to 2002 and consisted of a number of experimental enclosures using a series of semi-permeable barriers to differentially exclude cattle (*Bos indicus*), mega-herbivores (elephants and reticulated giraffe), and all 'wildlife' (herbivores > 15 kg = common zebra, Grevy's zebra, buffalo, eland, Grant's gazelle, hartebeest, oryx and steenbok). Each plot measured 200 m × 200 m (4 ha), and the following experimental treatments were obtained:

1. all large mammals excluded (O)
2. only cattle allowed (C)
3. only wildlife allowed (W)
4. wildlife and cattle allowed (WC)
5. wildlife and mega-herbivores allowed (MW)
6. all large herbivores allowed (MWC).

Dung counts (as a measure of species use) in the enclosures showed that the barriers were effective in excluding the target species, and zebra dung showed a number of significant patterns (Fig. 6.20). There was a 44 per cent increase in the presence of zebra dung (550 dung piles/ha in MWC and WC vs. 805 in MW and W) ($F = 23.85$, $P = 0.0028$), although zebra dung density was essentially the same in plots with and without mega-herbivores (680 in MW and MWC vs. 670 in W and WC) ($F = 0.013$, $P = 0.91$). However there was an interesting, and significant, cattle x mega-herbivore interaction. In plots without mega-herbivores there was 79 per cent more zebra dung when cattle were excluded (485 in WC vs. 865 in W), while in

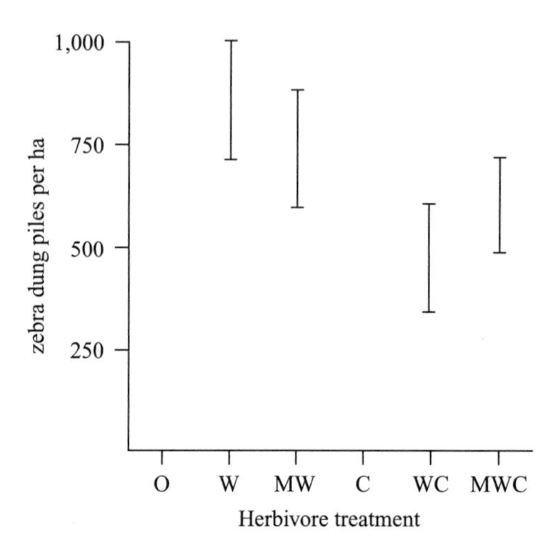

Fig. 6.20 The density of zebra dung in 2000/20001 in plots from which cattle and mega-herbivores had been excluded since September 1995. Error bars (±1 SE) (from Young *et al.* 2005).

plots with mega-herbivores, but cattle excluded, there was only a 22 per cent increase (610 in MWC vs. 745 in MW) ($F = 11.39$, $P = 0.015$). Young *et al.* (2005) also carried out vegetation surveys on their experimental plots and interestingly the results paralleled those for the zebra dung. Zebra dung density was strongly correlated with grass cover (suggesting they can track resources), which was negatively associated with cattle presence. In conjunction with published reports of strong dietary overlap (Casebeer and Koss 1970, Hoppe *et al.* 1977, Voeten and Prins 1999) this suggests that zebra compete with cattle for food. However, the implication is that this competition is highly asymmetrical and that wildlife (zebra) suffer more from the competition with cattle than *visa versa*. Compared to the total exclosure plots, cattle alone reduced grass cover by 33 per cent, and there was no further reduction in grass cover when wildlife were added. In contrast, wildlife alone (mostly zebra) reduced grass cover only 14 per cent and wildlife plus mega-herbivores reduced cover 21 per cent. It appears that cattle can fully compensate for the absence of zebra and other wildlife. Here compensation means the lack of a decrease in a shared resource (grass) when one competitor, but not the other, is excluded. In the absence of cattle, zebra and other wildlife do not compensate fully, in terms of grass cover. This controlled field experiment therefore confirms the conclusions of Prins (2000), previously mentioned.

This conclusion might appear to indicate that livestock ranches can allow wildlife to flourish on their property without fear of it affecting their cattle. Unfortunately not all property owners believe that this is true. For example, Heath (2000) who examined the estimated costs of ranching in Laikipia.

Table 6.6 Estimated profit/loss from different sized cattle herds on a 20,000 ha ranch in Laikipia, Kenya. All money in US$. Income is based on each cattle having an average weight of 400 kg and an average market price of 1.05US$ per kg. Direct costs include herders and veterinary inputs such as dips, salt and drugs. Overheads include management, ancillary staff, vehicle and equipment maintenance, fuel, office, bank, insurance and land rates.

Cattle	1,500	2,000	2,500	3,000
Income	126,000	168,000	210,000	252,000
Expenditure				
Direct costs	26,170	34,909	43,615	52,338
Overheads	116,541	129,457	142,403	156,643
Cash profit/loss	−16,711	3,634	23,982	43,019

After Heath 2000.

On a typical 20,000 ha ranch in the Laikipia District of Kenya he suggests that the 1,475 wildlife represent the metabolic equivalent of 1,095 cattle and that even after taking into consideration the varying degrees of overlap in diet this still amounts to some 830 cattle. Despite the claims of people like Prins (2000), Heath maintains that water shortage may often be the limiting factor in these semi-arid savannahs that are used for cattle rearing. In fact ranching in these areas is not a very secure occupation anyway. Heath (2000) gives his estimates of the anticipated profit/loss for different sized cattle herds on a 20,000 ha ranch in Laikipia (Table 6.6). Size appears to be crucial. Small ranches are very near the loss margin, and in Laikipia they often remove the wildlife from their ranches. In contrast large ranches tolerate wildlife, and even use tourism to supplement their income. They have formed the Laikipia Wildlife Forum. The Forum was established in 1992 by private and communal landowners with common interests in managing, conserving and profiting from wildlife resources. The organization was created in response to an initiative by the Kenya Wildlife Service, designed to engage landowners and land users in the conservation and management of wildlife in non-protected areas. In addition it also protects essential environmental resources such as river flow, as well as improving the livelihood and security of local people. Membership of the Forum currently consists of 36 large-scale ranches, 47 community groups, 50 tour operators, 54 individuals and eight interest groups. Since an estimated 60–70 per cent of wildlife in Kenya is found outside protected areas, this may well be the way forward for many of these semi-arid savannahs, in both eastern and southern Africa, that are outside National Parks. A good example of one such ranch within the Forum is Lewa Downs, in the Meru District of Kenya.

Lewa Wildlife Conservancy (LWC) is situated in Laikipia District, in the northern foothills of Mount Kenya, and includes an estimated 2,200 km^2. The main vegetation type is *Acacia* savannah, grassland and some indigenous forest. Lewa was once a cattle ranch; it then became a heavily guarded black rhino sanctuary, and it is now the headquarters for a non-profit wildlife conservancy. The Craig/Douglas family first came to Lewa Downs in 1922, and managed it as a cattle ranch for over fifty years. In the 1980s the emphasis at LWC changed and wildlife conservation became the prime objective. Cattle are still farmed but in smaller numbers. In addition to being a private conservation area for all wildlife, Lewa Downs supports a variety of programmes. These include an endangered species programme (black rhino, elephant, Grevy's zebra, cheetah, lion, leopard, hyaena, and wild dogs), a research programme (including Grevy's zebra, carnivores and general monitoring of wildlife) and a community development programme (health, education and farm development). Income comes from farming, tourism, and donations. It has an annual budget of 2.2 million dollars and provides employment for some 450–500 people. A large part of the budget is allocated to the community projects (24 per cent in 2006), and the endangered species programme, particularly the black rhino (24 per cent in 2006) and Grevy's zebra (12 per cent in 2006) programmes.

Wildlife tourism

The desire of overseas tourists to experience African wildlife, in its natural habitat, has resulted in Africa having one of the strongest tourist growth markets, with most destinations showing consistently above average increases in arrivals and receipts. Between 2000 and 2005, international tourist arrivals to Africa increased from 28 million to nearly 40 million—an average growth of 5.6 per cent a year, compared to a world-wide 3.1 per cent a year. In the same period Africa's International tourism receipts doubled from US$ 10.5 billion to US$ 21.3 billion. Total travel and tourist revenues for the whole of Africa were expected to generate $73.6 bn of GDP in 2005, equivalent to 8.8 per cent of the regional economy. The tourist sector also provided 3,877,200 jobs directly and a total of 10,647,000 jobs indirectly in 2005, or 6.8 per cent of all employment in sub-Saharan Africa. According to UNWTO figures for 2005, the number of visitors to sub-Saharan Africa from outside the region increased 13 per cent, to 23.1 m. The most rapid growth was recorded in Kenya, where the number of visitors increased 26 per cent on 2004, and Mozambique, where the number of visitors increased a massive 37 per cent. Many of the African countries that support large areas of savannah habitat are high on the tourist list. Table 6.7 compares six countries for which figures are available for 2003.

Most of this tourism is due to wildlife, and interest in African wildlife centres on savannah ecosystems with their charismatic large herbivores and carnivores. In Botswana it is estimated that wildlife viewing accounted for about half of the total overseas' tourist expenditure and generated up to $3 million income for the government in 1990. About half of this was from entry fees and about a fifth from tax revenues (Modise 1990). In Kenya, in 1995, about 70 per cent of the tourist income ($500 million) could be

Table 6.7 Tourism figures for 6 African countries with extensive savannah ecosystems, for 2003.

Country	Tourist arrivals (000s)	Tourism receipts (US$ M)	Tourism as % exports
Botswana	975	309	n/a
Kenya	927	611	17.1
South Africa	6,505	5,232	11.5
Tanzania	552	441	28.1
Zambia	578	149	11.2
Zimbabwe	2,068	44	n/a

n/a = not available
Source: Jonathan Mitchell and Caroline Ashley, Overseas Development Institute, 2006.

attributed to wildlife tourism, and government revenues from this were in excess of $20 million (Republic of Kenya 1996). Key wildlife species are especially highly valued. For example, in Kenya the net global returns to the 1.5 million wildebeest of the Masai Mara National Reserve have been estimated at between $125 and $150 per animal per year (Norton-Griffiths 1995). In Amboseli National Park the viewing value of a lion has been estimated to be in excess of $0.5 million (Thresher 1981) and the flamingos of Lake Nakuru National Park have an annual tourist value of between $3 million and $5 million (Earnshaw and Emerton 2000).

If wildlife is worth so much, via tourism, isn't this a way to ensure that wildlife is protected? Isn't it an investment worth protecting for its financial return, in addition to its aesthetic and scientific value? The answer is yes. But the financial returns must filter down to the local farmers, ranchers, hunters, and poachers who would otherwise see a better profit in the demise of wildlife. Use the money from tourism, and perhaps hunting schemes such as CAMPFIRE, to benefit local communities. The problem with conserving savannah ecosystems (and other ecosystems) is that we do not simply have to convince those of us that already value their scientific worth, and their exciting beauty. We also have to convince those people who see a financial gain in destroying them. Convince this latter group that they should be protected and African savannahs will be safe for posterity. Well, there is global climate change!

References

Abrams P. (1983). The theory of limiting similarity. *Annual Review of Ecology and Systematics*, **14**, 359–76.

Anderson J. M. and Coe M. J. (1974). Decomposition of elephant dung in an arid, tropical environment. *Oecologia*, **14**, 111–25.

Arsenault R. and Owen-Smith N. (2002). Facilitation versus competition in grazing herbivore assemblages. *Oikos*, **97**, 313–18.

Aubréville A. (1963). Classification des formes biologiques des plantes vasculaires en milieu tropical. *Adonsonia*, **3**, 221–6.

Bailey T. N. (1993). *The African leopard. Ecology and behaviour of a solitary felid.* Columbia University Press, New York.

Begon M., Harper J. L. and Townsend C. R. (1996). *Ecology. Individuals, populations and communities.* 3rd edn. Blackwell Science, Oxford.

Bell R. H. V. (1969). The use of the herb layer by grazing ungulates in the Serengeti National Park, Tanzania. PhD thesis, University of Manchester.

Bell R. H. V. (1970). The use of the herb layer by grazing ungulates in the Serengeti. In A. Watson, ed. *Animal populations in relation to their food resources*, pp. 111–23, Blackwell Scientific Publications, Oxford.

Bell R. H. V. (1971). A grazing ecosystem in the Serengeti. *Scientific American*, **224(1)**, 86–93.

Bell R. H. V. (1982) The effect of nutrient availability on community structure in African ecosystems. In B. J. Huntley and B. H. Walker, eds. *Ecology of Tropical Savannas*, Springer-Verlag, New York.

Belsky A. J. (1990). Tree/grass ratios in East African savannas: a comparison of existing models. *Journal of Biogeography*, **17**, 483–9.

Belsky A. J. (1992). Effects of grazing, competition, disturbance and fire on species composition and diversity in grassland communities. *Journal of Vegetation Science*, **3**, 187–200.

Ben-Shahar R. (1990). Resource availability and habitat preferences of three African ungulates. *Biological Conservation*.

Ben-Shahar R. and Skinner J. D. (1988). Habitat preferences of African ungulates derived by uni- and multivariate analyses. *Ecology*, **69**, 1479–86.

Bertram B. C. R. (1979). Serengeti predators and their social systems. In A. R. E. Sinclair and M. Norton-Griffiths, eds. *Serengeti. Dynamics of an ecosystem*, pp. 221–48, University of Chicago Press, Chicago and London.

Birkett A. (2002). The impact of giraffe, rhino and elephant on the habitat of a black rhino sanctuary in Kenya. *African Journal of Ecology*, **40**, 276–82.

Birkett A. and Stevens-Wood B. (2005). Effect of low rainfall and browsing by large herbivores on an enclosed savannah habitat in Kenya. *African Journal of Ecology*, **43**, 123–30.

Bothma J. du P. and Walker C. (1999). *Larger carnivores of the African savannas.* Springer-Verlag, Berlin Heidelberg New York.

Braack L. and Kryger P. (2003). Insects and savanna heterogeneity. In J. T. DuToit, K. H. Rogers and H. C. Biggs, eds. *The Kruger experience. Ecology and management of savanna heterogeneity*, pp. 263–75, Island Press, Washington.

Braithwaite R. W. (1991). Australia's unique biota: implications for ecological processes. In P. A. Werner, ed. *Savanna ecology and management*, pp. 3–10, Blackwell Scientific Publications.

Brashares J. S., Garland T. and Arcese P. (2000). Phylogeny analysis of coadaptation in behaviour, diet, and body size in the African antelope. *Behavioural Ecology*, **4**, 452–63.

Braun H. M. H. (1973). Primary production in the Serengeti: purpose, methods and some results of research. *Annual of the University of Abidjan*, **6**, 171–88.

Briand Peterson J. C. (1972). An identification system for zebra (*Equus burchelli*, Gray). *East African Wildlife Journal*, **10**, 59–63.

Broten M. D. and Said M. (1995). Population trends of ungulates in and around Kenya's Mara Reserve. In A. R. E. Sinclair and P. Arcese, eds. *Serengeti II: dynamics, management and conservation of an ecosystem*, pp. 169–93, Chicago University Press, Chicago.

Buckland S. T., Anderson D. R., Burnham K. P., Laake J. L., Borchers D. L., and Thomas L. (2001). *Introduction to distance sampling. Estimating abundance of biological populations.* Oxford University Press.

Cailleux A. (1953). *Biogéographie mondiale.* PUF, Paris.

Cambefort Y. (1984). Community structure and role in dung burial of forest and savanna Scarabaeidae dung beetles in Ivory Coast. 27 International Congress of Entomology, Hamburg, Abstract Volume, p. 339.

Caro T. M. (1994). *Cheetahs of the Serengeti plains.* University of Chicago Press, Chicago and London.

Casebeer R. L. and Koss G. G. (1970). Food habits of wildebeest, zebra, hartebeest and cattle in Kenya Masailand. *East African Wildlife Journal*, **8**, 25–36.

Caswell H., Reed R., Stephenson S. N. and Werner P. A. (1973). Photosynthetic pathways and selective herbivory: a hypothesis. *American Naturalist*, **107**, 465–80.

Caughley G. (1974). Bias in aerial survey. *Journal of Wildlife Management*, **38**, 921–33.

Caughley G. (1976). The elephant problem—an alternative hypothesis. *East African Wildlife Journal*, **14**, 265–84.

Chalmers M. I. (1961). Protein synthesis in the rumen. In D. Lewis, ed. *Digestive Physiology and Nutrition in the Ruminant*, pp. 205–22, Butterworth & Co., London.

Coates Palgrave K. (1983). *Trees of Southern Africa.* Struik Publishers, Cape Town.

Coe M. J., Cumming D. H. Phillipson J. (1976). Biomass and production of large African herbivores in relation to rainfall and primary production. *Oecologia*, **22**, 341–54.

Cole M. M. (1986). *The savannas. Biogeography and geobotany.* Academic Press, London.

Connell J. H. (1980). Diversity and the coevolution of competitors, or the ghost of competition past. *Oikos*, **35**, 131–38.

Connell J. H. (1983). On the prevalence and relative importance of interspecific competition: evidence from field experiments. *American Naturalist*, **122**, 661–96.

Cooper J. P. (ed.)(1975). *Photosynthesis and productivity in different environments.* Cambridge University Press, Cambridge.

Cooper S. M. (1991). Optimal hunting group size: the need for lions to defend their kills against loss to spotted hyaenas. *African Journal of Ecology*, **29**, 130–6.

Crawley, M. J. (1983). *Herbivory. The dynamics of animal-plant interactions.* Blackwell Scientific Publications, Oxford.

Creel S. and Creel N. M. (1996). Limitation of African wild dogs by competition with larger carnivores. *Conservation Biology*, **10**, 526–38.

Creel S. and Creel N. M. (2002). *The African wild dog. Behaviour, ecology and conservation.* Princeton University Press, Princeton and Oxford.

Croze H. (1974a). The Seronera bull problem. Part 1. The elephants. *East African Wildlife Journal*, **12**, 1–28.

Croze H. (1974b). The Seronera bull problem. Part 2. The trees. *East African Wildlife Journal*, **12**, 29–48.

Dangerfield J. M. (1990). The distribution and abundance of *Cubitermes sankurensis* (Wassmann)(Isoptera; Termitidae) within a Miombo woodland site in Zimbabwe. *African Journal of Ecology*, **28**, 15–20.

Dankwa-Wiredu B. and Euler D. L. (2002). Bushbuck (*Tragelaphus scriptus* Pallas) habitat in Mole National Park, northern Ghana. *African Journal of Ecology*, **40**, 35–41.

Dean W. R. J. (2000). *The Birds of Angola: An annotated checklist.* BOU Checklist No. 18. British Ornithologists' Union, Herts, UK.

Demment M. W. and Van Soest P. J. (1985). A nutritional explanation for body-size patterns of ruminant and nonruminant herbivores. *American Naturalist*, **125**, 641–72.

Deodatus E. D. and Lipiya A. K. (1991). Wildlife pest impact around Kasungu N. P. FP: MLW/87/010 Field Document No. **10**, FAO, Malawi.

Deodatus E. D. and Sefu L. (1992). National survey of wildlife pests. FO: MLW/87/010 Field Document No. **24**. FAO, Malawi.

Deshmukh I. K. (1984). A common relationship between precipitation and grassland peak biomass for East and southern Africa. *African Journal of Ecology*, **22**, 181–6.

Dharani N. (2006). *Field Guide to Acacias of East Africa.* Struik Publishers, South Africa.

Diamond J. M. (1975). Assembly of species communities. In M. L. Cody and J. M. Diamond, eds. *Ecology and evolution of communities*, Belknap Press, Cambridge, MA.

Diamond J. M. (1986). Overview: Laboratory experiments, field experiments, and natural experiments. In J. Diamond and E. J. Case, eds. *Community Ecology*, pp. 3–22.

Dorst J. and Dandelot P. (1972). *Larger mammals of Africa.* Harper Collins, London.

Drent R. H. and van der Wal R. (1999). Cyclic grazing in vertebrates and the manipulation of the food resource. In H. Olff, V. K. Brown, and R. H. Drent, eds. *Herbivores: between plants and predators*, pp. 271–99, Blackwell Science, Oxford.

Dublin H. T. (1995). Vegetation dynamics in the Serengeti-Mara ecosystem: the role of elephants, fire and other factors. In A. R.E. Sinclair and P. Arcese, eds. *Serengeti II: dynamics, management and conservation of an ecosystem*, pp. 71–90, Chicago University Press, Chicago.

Dublin H. T., Sinclair A. R. E. and McGlade J. (1990). Elephants and fire as causes of multiple stable states in the Serengeti-Mara woodlands. *Journal of Animal Ecology*, **59**, 1147–64.

Dublin H. T., Sinclair A. R. E., Boutin S., Anderson E., Jago M., and Arcese P. (1990). Does competition regulate ungulate populations? Further evidence from Serengeti, Tanzania. *Oecologia*, **82**, 283–8.

Duncan P. (1975). Topi and their food supply. PhD thesis, University of Nairobi.

Durant S. M. (1998). Competition refuges and coexistence: an example from Serengeti carnivores. *Journal of Animal Ecology*, **67**, 370–86.

Du Toit J. T. (1990). Feeding-height stratification among African browsing ruminants. *African Journal of Ecology*, **28**, 55–61.

Du Toit J. T. (2003). Large herbivores and savanna heterogeneity. In J. T. DuToit, K. H. Rogers and H. C. Biggs, eds. *The Kruger experience. Ecology and management of savanna heterogeneity*, pp. 292–309, Island Press, Washington.

Du Toit J. T., Provenza F. D. and Nastis A. (1991). Conditioned taste aversion: how sick must a ruminant get before it learns about toxicity in food? *Applied Animal Behaviour Science*, **30**, 35–46.

East R. (1983). Application of species-area curves to African savannah reserves. *African Journal of Ecology*, **21**, 123–8.

East R. (1984). Rainfall, soil nutrient status and biomass of large African savannah mammals. *African Journal of Ecology*, **22**, 245–70.

East R. (1998). *African antelope database 1998*. IUCN/SSC Antelope Specialist Group. IUCN, Gland, Switzerland and Cambridge, UK. 434pp.

Elliott J. P., McTaggart-Cowan I. and Holling C. S. (1977). Prey capture by the African lion. *Canadian Journal of Zoology*, **55**, 1811–28.

Eloff F. C. (1984). Food ecology of the Kalahari lion Panthera leo vernayi. *Koedoe*, **1984** (Suppl.), 249–58.

Elton C. S. (1958). *The ecology of invasion by animals and plants*. Methuen, London.

Elton C. S. and Nicholson A. J. (1942). The ten-year cycle in numbers of the lynx in Canada. *Journal of Animal Ecology*, **11**, 215–44.

Eltringham S. K. (1974). Changes in the large mammal community of Mweya Peninsula, Rwensori National Park, Uganda, following removal of hippopotamus. *Journal of Applied Ecology*, **11**, 855–66.

Estes R. D. (1991). *The behaviour guide to African mammals*. University of California Press, Berkley.

Ferrar A. A. and Walker B. H. (1974). An analysis of herbivore/habitat relationships in Kyle National Park, Rhodesia. *Journal of the South African Management Association*, **4**, 137–47.

Field C. R. (1968). A comparative study of the food habits of some wild ungulates in the Queen Elizabeth Park, Uganda: preliminary report. In MA Crawford ed. *Comparative nutrition of wild large mammals. Symposium of the Zoological Society (London)*, **21**, 135–51.

Field C. R. (1972). The food habits of wild ungulates in Uganda by analysis of stomach contents. *East African Wildlife Journal*, **10**, 17–42.

Fischer F. and Linsenmair K. E. (2001). Spatial and temporal habitat use of kob antelopes (*Kobus kob kob* Erxleben 1777) in the Comoé National Park, Ivory Coast as revealed by radio tracking. *African Journal of Ecology*, **39**, 249–56.

FitzGibbon C. D. (1989). A cost to individuals with reduced vigilance in groups of Thomson's gazelle hunted by cheetahs. *Animal Behaviour*, **37**, 508–10.

Ford J. (1971). *The role of trypanosomiases in Africa ecology*. Clarendon Press, Oxford.

Ford J. and Clifford H. R. (1968). Changes in the distribution of cattle and of bovine trypanosomiasis associated with the spread of tsetse flies (*Glossina*) in south-west Uganda. *Journal of Applied Ecology*, **5**, 301–37.

Foster J. B. and Kearney D. (1967). Nairobi National Park game census, 1966. *East African Wildlife Journal*, **5**, 112–20.

Foster J. B. and McLaughlin R. (1968). Nairobi National Park game census, 1967. *East African Wildlife Journal*, **6**, 152–4.

Freeland W. J. (1990). Large herbivorous mammals: exotic species in northern Australia. *Journal of Biogeography*, **17**, 445–9.

Freeland W. J. (1991). Large herbivorous mammals: exotic species in northern Australia. In P. A. Werner, ed. *Savanna ecology and management*, pp. 101–5, Blackwell Scientific Publications.

Fryxell J. M. (1995). Aggregation and migration by grazing ungulates in relation to resources and predators. In A. R. E. Sinclair and P. Arcese, eds. *Serengeti II: dynamics, management and conservation of an ecosystem*, pp. 257–73, Chicago University Press, Chicago.

Gadgil M. and Meher Homji V. M. (1985) Land use and productive potential of Indian savanna. In J. C. Tothill and J. J. Mott, eds. *Ecology and management of world savannas*, pp. 107–13, Australian Academy of Sciences, Canberra.

Gandar M. V. (1982a). The dynamics and trophic ecology of grasshoppers (Acridoidea) in a South African savanna. *Oecologia*, **54**, 370–8.

Gandar M. V. (1982b). Trophic ecology and plant/herbivore energetics. In B. J. Huntly and B. H. Walker, eds. *Ecology of tropical savannas*, pp. 514–43, Springer, Berlin.

Gaston K. J. (1996). Species richness: measure and measurement. In K. J. Gaston, ed. *Biodiversity. A biology of numbers and difference*, 77–113, Blackwell Science.

Geertsema A. (1981). The servals of Gorigor. *Wildlife News*, **16**, 4–8.

Georgiadis N., Hack M. and Turpin K. (2003). The influence of rainfall on zebra population dynamics: implications for management. *Journal of Applied Ecology*, **40**, 125–36.

Goodall J. (1988). *In the shadow of man.* Houghton Mifflin Company. Boston.

Gradstein F. M., Ogg J. G., and Smith A. G. (2004). *A Geologic Time Scale 2004.* 589 pp. Cambridge University Press, Cambridge.

Grimsdell J. J. R. (1979). Appendix A. Changes in populations of resident ungulates. In A. R. E. Sinclair and M. Norton-Griffiths, eds. *Serengeti: dynamics of an ecosystem*, pp. 353–9, Chicago University Press, Chicago.

Grzimek B. and Grzimek M. (1960a). *Serengeti shall not die.* Hamish Hamilton, London.

Grzimek M. and Grzimek B. (1960b). Census of plains animals in the Serengeti National Park, Tanganyika. *Journal of Wildlife Management*, **24**, 27–37.

Gwynne M. D. and Bell R. H. V. (1968). Selection of grazing components by grazong ungulates in the Serengeti National Park. *Nature (London)*, **220**, 390–3.

Harcourt A. H., Parks S. A. and Woodroffe R. (2001). Human density as an influence on species-area relationships: double jeopardy for small African reserves? *Biodiversity and Conservation*, **10**, 1001–26.

Haxeltine A. and Prentice I. C. (1996). An equilibrium terrestrial biosphere model based on ecophysiological constraints, resource availability and competition among plant functional types. *Global Biogeoche Cyc*, **10**, 693–709.

Helsa B. I., Tieszen H. L. and Boutton T. W. (1985). Seasonal water relations of savanna shrubs and grasses in Kenya, East Africa. *Journal of Arid Environments*, **8**, 15–31.

Hilborn R. and Sinclair A. R. E. (1979). A simulation of the wildebeest population, other ungulates, and their predators, pp. 287–309, In A. R. E. Sinclair and M. Norton-Griffiths, eds. *Serengeti: dynamics of an ecosystem*, pp. 1–30, Chicago University Press, Chicago.

Hilborn R., Georgiadis N., Lazarus J., Fryxell J. M., Broten M. D., Mbano B. N. N., Murray M. G., Sinclair A. R.E., Durant S. M., Mwasaga B., Maige M. K. S., Arcese P., Albon S., Hofer H., Kapela M., Dobson A., East M., Nkya H., Dublin H. T., Packer C., Campbell K. L. I., Gascoyne S. C., Creed S. R., Hertz P., Creel N. M. and Caro T. M. (1995). A model to evaluate alternative management policies for the Serengeti-Mara ecosystem. In A. R. E. Sinclair and P. Arcese, eds. *Serengeti II: dynamics, management and conservation of an ecosystem*, pp. 617–37, Chicago University Press, Chicago.

Hines D. A. and Eckman K. (1993). Indigenous multipurpose trees of Tanzania: uses and economic benefits for people. Food and Agriculture Organization (FAO) of the United Nations, FAO Corporate Document Repository.

Hirst S. M. (1969). Populations in a Transvaal lowland nature reserve. *Zool. Afr.*, **4**, 199–230.

Hitchins P. M. (1968). Some preliminary findings on the population structure and status of the black rhinoceros *Diceros bicornis* in Hluluwe Game Reserve, Zululand. *Lammergeyer*, **9**, 26–8.

Hitchins P. M. (1969). Influence of vegetation types on sizes of home ranges of black rhino in Hluluwe Game Reserve. *Lammergeyer*, **3**, 81–5.

Hoare R. (1995). Options for the control of elephants in conflict with people. *Pachyderm*, **19**, 54–63.

Hofmann R. R. (1968) Comparisons of the rumen and omasum structure in East African game ruminants in relation to their feeding habits. *Symposium of the Zoological Society of London*, **21**, 179–94.

Hofmann R. R. (1973). *The ruminant stomach*. East African Literature Bureau, Nairobi.

Hoffmann A. and Klingel H. (2001). Spatial and temporal patterns in *Lemniscomys striatus* (Linnaeus 1758) as revealed by radio-tracking. *African Journal of Ecology*, **39**, 351–6.

Holling C. S. (1959). The components of predation as revealed by a study of small-mammal predation of the European pine sawfly. *Canadian Entomologist*, **91**, 293–320.

Holt J. A. (1987). Carbon mineralisation in semi-arid northeastern Australia: the role of termites. *Journal of Tropical Ecology*, **3**, 255–63.

Holt R. D. (1987). Prey communities in patchy environments. *Oikos*, **50**, 276–90.

Hopkins B. (1962). Vegetation of the Olokemeji Forest Reserve, Nigeria. I. General features of the reserve and the research sites. *Journal of Ecology*, **50**, 559–98.

Hoppe P. P., Qvortrup S. A. and Woodford M. H. (1977). Rumen fermentation and food selection in East African zebu cattle, wildebeest, Coke's hartebeest and topi. *Journal of Zoology*, **181**, 1–9.

Houston D. C. (1979). The adaptations of scavengers. In A. R. E. Sinclair and M. Norton-Griffiths, eds. *Serengeti. Dynamics of an ecosystem*, pp. 263–86, University of Chicago Press, Chicago and London.

Huntley B. J. (1978). Ecosystem conservation in southern Africa. In M. J. A. Werger, ed. *Biogeography and ecology of southern Africa*, W. Junk, The Hague.

Hutchinson G. E. (1959). Homage to Santa Rosalia, or why are there so many kinds of animals? *American Naturalist*, **93**, 145–59.

Jackson C. H. N. (1933). On the true density of tsetse flies. *Journal of Animal Ecology*, **2**, 204–9.

Jansen D. H. (1966). Coevolution of mutualism between ants and Acacia in Central America. *Evolution*, **20**, 249–75.

Jarman P. J. (1971). Diets of large mammals in the woodlands around Lake Kariba, Rhodesia. *Oecologia (Berlin)*, **8**, 157–78.

Jarman P. J. (1972). Seasonal distribution of large mammal populations in the unflooded middle Zambezi valley. *Journal of Applied Ecology*, **9**, 283–99.

Jarman P. J. (1974). The social organisation of antelope in relation to their ecology. *Behaviour*, **48**, 215–67.

Kat P. W., Alexander K. A., Smith J. S. and Munson L. (1995). Rabies and African wild dogs in Kenya. *Proceedings of the Royal Society of London B*, **262**, 229–33.

Kingdon J. (1977). *East African mammals.* IIIa (carnivores). University of Chicago Press.

Kingdon J. (1982). *East African mammals.* IIIc (*bovids*). University of Chicago Press.

Kingdon J. (1997). *The Kingdon field guide to African mammals.* Academic Press.

Kingdon J. (2004). *The Kingdon pocket guide to African mammals.* A&C Black, London.

Klein R. G. (1988). The archaeological significance of animal bones from Acheulean sites in southern Africa. *African Archeological Review*, **6**, 3–25.

Knoop, W. T. and Walker B. H. (1985). Interactions of woody and herbaceous vegetation in southern African savanna. *Journal of Ecology*, **73**, 235–53.

Krause J. and Ruxton G. D. (2002). *Living in groups.* Oxford University Press, Oxford, UK.

Krebs C. J., Boutin S., Boonstra R., Sinclair A. R. E., Smith J. N. M., Dale M. R. T., Martin K. and Turkington R. (1995). Impact of food and predation on the snowshoe hare cycle. *Science*, **269**, 1112–15.

Kruuk H. (1972). *The spotted hyaena.* University of Chicago Press, Chicago and London.

Lack D. (1969). The numbers of bird species on islands. *Bird Study*, **16**, 193–209.

Lack D. (1976). *Island birds.* Blackwell Scientific Publications, Oxford.

Lamprey H. F. (1963). Ecological separation of the large mammal species in the Tarangire Game Reserve, Tanganyika. *East African Wildlife Journal*, **1**, 63–92.

Lawick H. van (1973). *Solo. The story of an African wild dog.* Collins, London.

Lawick H. van and Lawick-Goodall J. van (1970). *Innocent killers.* Collins, London.

Laws R. M. (1966). Age criteria for the African elephant *Loxodonta africana. East African Wildlife Journal*, **4**, 1–37.

Laws R. M. (1970). Elephants as agents of habitat and landscape change in East Africa. *Oikos*, **21**, 1–15.

Lebrum J. (1960). Sur la richesse de la flore de divers territoires africains. *Acad. R. Sci. Outre-Mer, Bull. Séances*, **6**, 669–90.

Le Roux X., Bariac T. and Mariotti A. (1995). Spatial partitioning of the soil water-source between grass and shrub components in a West African humid savanna. *Oecologia*, **104**, 147–55.

Leuthold W. (1966). Variation in territorial behaviour of Ugandan kob, *Adenota kob thomasi* (Neumann 1896). *Behaviour*, **27**, 214–57.

Leuthold W. and Sale J. B. (1973). Movements and patterns of habitat utilization in Tsavo National Park, Kenya. *East African Wildlife Journal*, **11**, 369–84.

Lindeque M. and Lindeque P. M. (1991). Satellite tracking of elephants in north-western Namibia. *African Journal of Ecology*, **29**, 196–206.

Linderman R. L. (1942). The trophic-dynamic aspect of ecology. *Ecology*, **23**, 399–418.

MacArthur R. H. (1955). Fluctuations of animal populations and a measure of community stability. *Ecology*, **36**, 533–6.

MacArthur R. H. and Levins R. (1967). The limiting similarity, convergence and divergence of coexisting species. *American Naturalist*, **101**, 377–85.

MacArthur R. H. and Wilson E. O. (1967). *The theory of island biogeography*. Princeton University Press, Princeton, NJ.

MacDonald D. (2001). *The New encyclopedia of mammals.* OUP.

McKeon G. M., Day K. A., Howden S. M., Mott J. J., Orr D. M., Scattini W. J. and Weston E. J. (1991) Northern Australian savannas: management and pastoral production. In P. A. Werner, ed. *Savanna ecology and management*, pp. 11–28, Blackwell Scientific Publications.

McNaughton S. J. (1976). Serengeti migratory wildebeest: facilitation of energy flow by grazing. *Science*, **191**, 92–4.

McNaughton S. J. (1977). Diversity and stability of ecological communities: a comment on the role of empiricism in ecology. *American Naturalist*, **111**, 515–25.

McNaughton S. J. (1979a). Grazing as an optimization process: grass-ungulate relationships in the Serengeti. *American Naturalist*, **113**, 691–703.

McNaughton S. J. (1979b). Grassland-herbivore dynamics. In A. R. E. Sinclair and M. Norton-Griffiths, eds. *Serengeti. Dynamics of an ecosystem*, pp. 46–81, University of Chicago Press, Chicago and London.

McNaughton S. J. (1985). Ecology of a grazing ecosystem: the Serengeti. *Ecological Monographs*, **55**, 259–94.

McNaughton S. J. and Georgiadis N. J. (1986). Ecology of African grazing and browsing mammals. *Annual Revue of Ecology and Systematics*, **17**, 39–65.

Madden D. and Young T. P. (1992). Symbiotic ants as an alternative defense against giraffe herbivorory in spinescent *Acacia drepanolobium*. *Oecologia*, **91**, 235–8.

Magurran A. E. (2004). *Measuring biological diversity.* Blackwell Publishing, Oxford.

May R. M. (1972). Will a large complex system be stable? *Nature*, **238**, 413–14.

May R. M. (1973). On relationships among various types of population models. *American Naturalist*, **107**, 46–57.

Mduma S. A. R. (1996). Serengeti wildebeest population dynamics: regulation, limitation and implications for harvesting. PhD thesis, University of British Columbia.

Mduma S. A. R., Sinclair A. R. E. and Hilborn R. (1999). Food regulates the Serengeti wildebeest: a 40-year record. *Journal of Animal Ecology*, **68**, 1101–22.

Mduma S. A. R., Hilborn R. and Sinclair A. R. E. (1998). Limits to exploitation of Serengeti wildebeest and implications for its management. In D. M. Newbery, H. H. T. Prins and N. Brown, eds. *Dynamics of Tropical Communities.* 37th Symposium of the British Ecological Society, pp. 243–65, Blackwell Science Ltd, Oxford.

Menaut, J. C. (1971). *Etude de quelques peuplements ligneux d'une savane guineenne de Cote d'Ivoire.* Thesis, University of Paris, Paris, 141pp.

Menaut, J. C. (1983) The vegetation of African savannas. In F. Bourliere, ed. *Tropical savannas*, pp. 109–50, Elsevier, Amsterdam.

Meyer V. W., Crewe R. M., Braack H. T., Groeneveld H. T. and van der Linde M. J. (2000). Intracolonial demography of the mound-building termite *Macrotermes natalensis* (Haviland) (Isoptera, Termitidae) in the northern Kruger National Park, South Africa. *Insectes Sociaux*, **47**, 390–7.

Miller M. F. (1995). *Acacia* seed survival, seed germination and seedling growth following pod consumption by large herbivores and seed chewing by rodents. *African Journal of Ecology*, **33**, 194–210.

Miller R. I. and Harris (1977). Isolation and extirpation in wildlife reserves. *Biological Conservation*, **12**, 311–15.

Mills M. G. L. (1984). Prey selection and feeding habits of the large carnivores in the southern Kalahari. *Koedoe*, **1984** (Suppl.), 281–94.

Mills M. G. L. (1990. *Kalahari hyaenas: comparative behavioural ecology of two species.* Unwin Hyman, London.

Mills M. G. L. and Biggs H. C. (1993). Prey apportionment and related ecological relationships between large carnivores and prey in the Kruger National Park. *Symposium of the Zoological Society of London*, **65**, 253–68.

Mills M. G. L. and Funston P. J. (2003). Large carnivores and savanna heterogeneity. In J. T. DuToit, K. H. Rogers and H. C. Biggs, eds. *The Kruger experience. Ecology and management of savanna heterogeneity*, pp. 370–88, Island Press, Washington.

Mills M. G. L. and Hofer H. (compliers) (1998). *Hyaenas. Status survey and conservation action plan.* IUCN/SSC Hyaena Specialist Group. IUCN, Gland, Switzerland and Cambridge, UK. 154pp.

Mills M. G. L. and Shenk T. M. (1992). Predator-prey relationships: the impact of lion predation on wildebeest and zebra populations. *Journal of Animal Ecology*, **61**, 693–702.

Mills M. G. L., Biggs H. C. and Whyte I. J. (1995). The relationship between rainfall, lion predation and population trends in African herbivores. *Wildlife Research*, **22**, 75–88.

Misra R. (1983). Indian savannahs. In F. Bourlière, ed. *Tropical savannas*, volume 13, pp. 151–66. Elsevier, Amsterdam.

Mitchell B. L., Skenton J. B. and Uye J. C. M. (1965). Predation on large mammals in the Kafue National Park, Zambia. *Zool. Afr.*, **1**, 297–318.

Mordelet P., Menaut J. C. and Mariotti A. (1997). Tree and grass rooting patterns in an African humid savanna. *Journal of Vegetation Science*, **8**, 65–70.

Morris S., Humphreys D. and Reynolds D. (2006). Myth, marula and elephant: an assessment of voluntary ethanol intoxication of the African elephant (*Loxodonta africana*) following feeding on the fruit of the marula tree (*Sclerocarya birrea*). *Physiological and Biochemical Zoology*, **78**.

Mott J. J., Williams J., Andrew M. H. and Gillison A. N. (1985) Australian savanna ecosystems. In J. C. Tothill and J. J. Mott, eds. *Ecology and management of the world's savannas*, (ed.), pp. 56–82, Australian Academy of Sciences, Canberra.

Murray M. G. (1995). Specific nutrient requirements and migration of wildebeest. In A. R. E. Sinclair and P Arcese, eds. *Serengeti II: dynamics, management and conservation of an ecosystem*, pp. 231–56, Chicago University Press, Chicago.

Newmark W. D., Manyanza D. N., Gamassa D. G. M. and Sariko H. I. (1994). The conflict between wildlife and local people living adjacent to protected areas in Tanzania: human density as a predictor. *Conservation Biology*, **8**, 249–55.

Ngure N. (1995). People-elephant conflict management in Tsavo, Kenya. *Pachyderm*, **19**, 20–5.

Nix H. A. (1981). The environment of Terra Australis. In A. Keast, ed. *Ecological biogeography of Australia*, pp. 103–33, Junk, The Hague.

Noad T. and Birnie A. (1989). *Trees of Kenya.* Noad and Birnie, Nairobi.

Norton-Griffiths M. (1973). Counting migratory wildebeest using two-stage sampling. *East African Wildlife Journal*, **11**, 135–49.

Norton-Griffiths M. (1978). *Counting Animals.* 2nd edn. African Wildlife Leadership Foundation, Nairobi. Available from the African Wildlife Foundation, PO Box 48177, Nairobi, Kenya.

Norton-Griffiths M. (1979). The influence of grazing, browsing, and fire on the vegetation dynamics of the Serengeti. In A. R. E. Sinclair and M. Norton-Griffiths, eds. *Serengeti. Dynamics of an ecosystem*, pp. 310–52, University of Chicago Press, Chicago and London.

Nowell K. and Jackson P. (eds)(1996). *Wild cats*. IUCN Status Survey and Conservation Action Plan, IUCN, Gland, Switzerland and Cambridge, UK. 382pp.

O'Connor T. G., Haines, L. M. and Snyman H. A. (2001). Influence of precipitation and species composition on phytomass of a semi-arid African grassland. *Journal of Ecology*, **89**, 850–60.

Ogutu J. O. and Owen-Smith N. (2005). Oscillations in large mammal populations: are they related to predation or rainfall? *African Journal of Ecology*, **43**, 332–9.

Ojasti J. (1983). Ungulates and large rodents of South America. In F. Bourlière, ed. *Tropical savannas*, volume 13, pp. 427–39. Elsevier, Amsterdam.

Ottichilo W. K., De Leeuw J., Skidmore A. K., Prins H. H. T. and Said M. Y. (2000). Population trends of large non-migratory wild herbivores and livestock in the Masai Mara ecosystem, Kenya, between 1977 and 1997. *African Journal of Ecology*, **38**, 202–16.

Owen-Smith N. (1988). *Megaherbivores. The influence of very large body size on ecology.* Cambridge University Press, Cambridge, UK.

Owen-Smith N. (1990). Demography of a large herbivore, the greater kudu, in relation to rainfall. *Journal of Animal Ecology*, **59**, 893–913.

Owen-Smith N. and D. H. M. Cumming (1993). Comparative foraging strategies of grazing ungulates in African savannah grasslands. *Proceedings of the XVII International Grassland Congress.*

Owen-Smith N. and Mason D. R. (2005). Comparative changes in adult vs. juvenile survival affecting population trends of African ungulates. *Journal of Animal Ecology*, **74**, 762–73.

Owen-Smith N. and Ogutu J. O. (2003). Rainfall influences on ungulate population dynamics in the Kruger National Park. In J. T. DuToit, K. H. Rogers and H. C. Biggs, eds. *The Kruger experience. Ecology and management of savanna heterogeneity*, pp. 310–31, Island Press, Washington.

Owen-Smith N., Mason D. R. and Ogutu J. O. (2005). Correlates of survival rates for 10 African ungulate populations: density, rainfall and predation. *Journal of Animal Ecology*, **74**, 774–88.

Packer C. (1990). Serengeti lion survey. Report to TANAPA, SWRI, MWEKA, and the Game Depertment. Arusha, Tanzania: Tanzanian National Parks.

Paine R. T. (1966). Food web complexity and species diversity. *American Naturalist*, **100**, 65–75.

Paine R. T. (1980). Food webs: linkage, interaction strength and community infrastructure. *Journal of Animal Ecology*, **49**, 667–85.

Pearsall W. H. (1957). Report on an ecological survey of the Serengeti National Park, Tanganyika. *Oryx*, **4**, 71–136.

Pellow R. (1977). *Serengeti Research Institute, Annual Report*, 1975–76, pp. 49–67, Arusha, Tanzania National Parks.

Pellow R. (1981). *The giraffe (Giraffa camelopardalis tippelskirchi Matschie) and its Acacia food resource in the Serengeti National Park*. PhD thesis, University of London.

Pennycuick L. (1975). Movements of migratory wildebeest population in the Serengeti area between 1960 and 1973. *East African Wildlife Journal*, **13**, 65–87.

Penzhorn B. L. (1971). A summary of the re-introduction of ungulates into South African National Parks (to 31 December 1970). *Koedoe*, **14**, 145–59.

Pienar U. de V. (1961). A second outbreak of anthrax amongst game animals in the Kruger National Park. *Koedoe*, **4**, 4–17.

Pienar U. de V. (1963). The large mammals of the Kruger National Park—their distribution and present-day status. *Koedoe*, **6**, 1–263.

Pienar U. de V. (1964). The small mammals of the Kruger National Park—a systematic list and zoogeography. *Koedoe*, **7**, 1–25.

Pienar U. de V. (1967). Epidemiology of anthrax of wild animals and the control of anthrax epizootics in the Kruger National Park, South Africa. *Fed. Proc.*, **26**, 1496–502.

Pienar U. de V. (1969). Predator–prey relationships amongst the larger mammals of the Kruger National Park. *Koedoe*, **12**, 108–76.

Pimm S. L. (1991). *The balance of nature?* University of Chicago Press, Chicago and London.

Prins H. H. T. (1996). *Ecology and Behaviour of the African Buffalo. Social inequality and decision making.* Chapman and Hall, London.

Prins H. H. T. (2000). Competition between wildlife and livestock in Africa. In H. H. T. prins, J. G. Grootenhuis and T. T. Dolan, eds. *Wildlife Conservation by Sustainable Use.* Kluwer Academic Publications, Wageningen, Netherlands.

Prins H. H. T. and Olff H. (1998). Species richness of African grazer assemblages: towards a functional explanation. In D. M. Newberry, H. H. T. Prins and N. Brown, eds. *Dynamics of tropical communities*, Blackwell Scientific, Oxford.

Prins H. H. T., de Boer W. F., van Oeveren H., Correia A., Mafuca J. and Olff H. (2006). Co-existence and niche segregation of three small bovid species in southern Mozambique. *African Journal of Ecology*, **44**, 186–98.

Raunkier, C. (1934). *The Life Forms of Plants and Statistical Plant Geography.* Clarendon Press, Oxford, 632pp.

Robertson H. G. (1998). Ants (Hymenoptera: Formicidae) of Mkomazi. In M. Coe, N. McWilliam, G. Stone and M. Packer, eds. *Mkomazi. The ecology, biodiversity and conservation of a Tanzanian savanna*, pp. 337–60, Royal Geographic Society (with the Institute of British Geographers), London.

Roca A. L., Georgiadis N., Pecon-Slattery J. and O'Brien S. J. (2001) Genetic evidence for two species of elephant in Africa. *Science*, **293**, 1473–77.

Royama T. (1996). A fundamental problem in key factor analysis. *Ecology*, **77**, 87–93.

Sankaran M., Hanan N. P., Scholes R. J., Ratnam J., Augustine D. J., Cade B. S., Gignoux J., Higgins S. I., Le Roux X., Ludwig F., Ardo J., Banyikwa F., Bronn A., Bucini G., Caylor K. K., Coughenour M. B., Diouf A., Ekaya W., Feral C. J., February E. C., Frost P. G. H., Hiernaux P., Hraber H., Metzger K. L., Prins H. H. T., Ringrose S., Sea W., Tews J., Worden J. and Zambatis N. (2005). Determinants of woody cover in African savannas. *Nature*, **438**, 846–9.

Sarmiento G. (1998). Biodiversity and water relations in tropical savannas. In O. T. Solbrig, E. Medina and J. F. Silva, eds. *Biodiversity and savanna ecosystem processes. A global perspective*, pp. 61–75, Springer-Verlag, Berlin.

Sarmiento, G. and Monasterio, M. (1983) Life forms and phenology. In F. Bourliere, ed. *Tropical savannas*, pp. 109–50, Elsevier, Amsterdam.

Schaller G. B. (1972). *The Serengeti lion.* University of Chicago Press, Chicago and London.

Schoener T. W. (1983). Field experiments on interspecific competition. *American Naturalist*, **122**, 240–85.

Scrimshaw N. S., Taylor C. E. and Gordon J. E. (1968). Interactions of nutrition and infection. *WHO publication* **57**, Geneva.

Seber G. A. F. (1982). *The Estimation of animal abundance and related parameters.* Macmillan, New York.

Seghieri J. (1995). The rooting patterns of woody and herbaceous plants in a savanna: are they complimentary or in competition? *African Journal of Ecology*, **33**, 358–65.

Shorrocks B. (1978). *The genesis of diversity.* Hodder and Stoughton, London.

Shorrocks B. and Cokayne A. (2005). Vigilance and group size in impala (*Aepyceros melampus* Lichtenstein): a study in Nairobi National Park, Kenya. *African Journal of Ecology*, **43**, 91–6.

Shorrocks B. and Croft D. (2006). Necks and networks. Mpala Research Newsletter, **3**, 3.

Sillero-Zubiri C., King A. A. and Macdonald D. W. (1996). Rabies and mortality in Ethiopian wolves (*Canis simensis*). *Journal of Wildlife Diseases*, **32**, 80–6.

Simon N. (1962). *Between the sunlight and the thunder: the wildlife of Kenya.* Collins, London.

Simons H. W. and Chirambo P. C. (1991). Wildlife pest impact around Liwonde National Park, March-April (1990). FO: MLW/87/010 Field Document No. **11**, FAO, Malawi.

Sinclair A. R. E. (1974). The natural regulation of buffalo populations in East Africa. III. Population trends and mortality. *East African Wildlife Journal*, **12**, 185–200.

Sinclair A. R. E. (1975). The resource limitation of trophic levels in tropical grassland ecosystems. *Journal of Animal Ecology*, **44**, 497–520.

Sinclair, A. R. E. (1977). *The African buffalo. A study of resource limitation of populations.* University of Chicago Press, Chicago and London.

Sinclair, A. R. E. (1979). Dynamics of the Serengeti ecosystem: process and pattern. In A. R. E. Sinclair and M. Norton-Griffiths, eds. *Serengeti: dynamics of an ecosystem*, pp. 1–30, Chicago University Press, Chicago.

Sinclair A. R. E. (1985). Does interspecific competition or predation shape the African ungulate community? *Journal of Animal Ecology*, **54**, 899–918.

Sinclair A. R. E. (1995). Serengeti past and present. In A. R.E. Sinclair and P. Arcese, eds. *Serengeti II: dynamics, management and conservation of an ecosystem*, pp. 3–30, Chicago University Press, Chicago.

Sinclair A. R. E. (1995). Population limitation of resident herbivores. In A. R. E. Sinclair and P. Arcese, eds. *Serengeti II: dynamics, management and conservation of an ecosystem*, pp. 194–219, Chicago University Press, Chicago.

Sinclair A. R. E. (2000). Adaptation, niche partitioning and coexistence in the African Bovidae: clues to the past. In E. S. Vrba and G. B. Schaller, eds, *Antelopes, deer and relatives: fossil record, behavioural ecology, systematics and conservation.* Yale University Press, New Haven.

Sinclair A. R. E. and Arcese P. (eds)(1995). *Serengeti II: dynamics, management and conservation of an ecosystem.* Chicago University Press, Chicago.

Sinclair A. R. E. and Byrom A. E. (2006). Understanding ecosystem dynamics for conservation of biota. *Journal of Animal Ecology*, **75**, 64–79.

Sinclair A. R. E. and Norton-Griffiths M. (eds.)(1979). *Serengeti: dynamics of an ecosystem.* Chicago University Press, Chicago.

Sinclair A. R. E. and Norton-Griffiths M. (1982). Does competition or facilitation regulate migrant ungulate populations in the Serengeti? A test of hypotheses. *Oecologis*, **53**, 364–9.

Smith P. P. (1998). A reconnaissance survey of the vegetation of the North Luangwa National Park, Zambia. *Bothalia*, **28**, 197–211.

Smuts G. L. (1978). Interrelations between predators, prey and their environment. *Bio-Science*, **28**, 316–20.

Solbrig O. T. (1996). The diversity of savanna ecosystems. In O. T. Solbrig, E. Medina and J. F. Silva, eds. *Biodiversity and savanna ecosystem processes. A global perspective*, pp. 1–27, Springer-Verlag, Berlin.

Soulé M. E., Wilcox B. A. and Holtby C. (1979). Benign neglect: a model of faunal collapse in the game reserves of East Africa. *Biological Conservation*, **15**, 259–72.

Spinage C. A. (1973). A review of ivory exploitation and elephant population trends in Africa. *East African Wildlife Journal*, **11**, 281–9.

Stanton M. L., Palmer T. M. and Young T. P. (2002). Competition-colonization trade-offs in a guild of African acacia-ants. *Ecological Monographs*, **72**, 347–63.

Stapley L. (1998). The interaction of thorns and symbiotic ants as an effective defence mechanism of swollen-thorn acacias. *Oecologia*, **115**, 401–5.

Steffan W. I., Walker B. H., Ingram J. S. I. and Koch G. W. (1992). Global change and terrestrial ecosystems: the operational plan. IGBP, Stockholm, Sweden.

Stevenson Hamilton J. (1911). The relation between game and tsetse-flies. *Bulletin of Entomological Research*, **2**, 113–18.

Stevenson Hamilton J. (1947). *Wildlife in South Africa*. Cassel, London.

Stuart C. and Stuart T. (1996). *Africa's vanishing wildlife*. Swan Hill Press, Shrewsbury.

Stuart C. and Stuart T. (1997). *Field guide to the larger mammals of Africa*. Struik, Cape Town.

Swynnerton G. H. (1958). Fauna of the Serengeti National Park. *Mammalia*, **22**, 435–50.

Talbot L. M. and Talbot M. H. (1963). *The wildebeest in western Masai-land, East Africa*. Wildlife Monograph No. 12, The Wildlife Society.

Taylor R. J. (1984). *Predation*. Chapman and Hall, New York London.

Thompson J. J. (1989). A comparison of some avian census techniques in a population of lovebirds at Lake Naivasha, Kenya. *African Journal of Ecology*, **27**, 157–66.

Thoules C. R. (1994). Conflict between humans and elephants on private land in northern Kenya. *Oryx*, **28**, 119–27.

Ticker C. J. J., Townsend J. R. G. and Goff T. E. (1985). African landcover classification using satellite data. *Science*, **227**, 369–75.

Torchio P. and Manconi M. (2004). Rugged pursuit. *Swara*, **27**, 70–2.

Trinkel M. and Kastberger G. (2005). Competitive interactions between spotted hyaenas and lions in the Etosha National Park, Namibia. *African Journal of Ecology*, **43**, 220–24.

Tyson P. D. (1986). *Climatic change and variability in southern Africa*. Oxford University Press, Oxford, UK.

Vandermeer J. (1984). The evolution of mutualism. In B Shorrocks, ed. *Evolutionary Ecology*, pp. 221–32, Blackwell Scientific Publications, Oxford.

Varley G. C. and Gradwell G. R. (1970). Recent advances in insect population dynamics. *Annual Review of Entomology*, **15**, 1–24.

Vesey-Fitzgerald D. F. (1960). Grazing succession among East African game mammals. *Journal of Mammalogy*, **41**, 161–72.

Vesey-Fitzgerald D. F. (1973). Browse production and utilization in Tarangire National Park. *East African Wildlife Journal*, **11**, 291–306.

Vesey-Fitzgerald D. F. (1974). Utilization of the grazing resources by buffaloes in the Arusha National park, Tanzania. *East African Wildlife Journal*, **12**, 107–34.

Voeten M. M. and Prins H. H. T. (1999). Resource partitioning between sympatric wild and domestic herbivores in the Tarangire region of Tanzania. *Oecologia*, **120**, 287–94.

Waithaka J. M. (1996). Elephants: a keystone species. Box 11.2, p. 284, In T. R. McClanahan and T. P. Young, eds. *East African ecosystems and their conservation.* Oxford University Press, Oxford.

Walter H. (1971). *Ecology of tropical and subtropical vegetation.* Oliver and Boyd, Edinburgh, UK.

Walter, H. (1973). *Vegetation of the Earth in relation to climate and the ecophysiological conditions.* Springer-Verlag, Berlin, 237pp.

Webb W. L., Lauenroth W. K., Szarek S. R. and Kinerson R. S. (1983). Primary production and abiotic controls in forests, grasslands and desert ecosystems in the United States. *Ecology*, **64**, 134–51.

Weston D. and Ssemakula (1981). The future of the savannah ecosystem: ecological islands or faunal enclaves? *African Journal of Ecology*, **19**, 7–19.

White F. (1983). *The Vegetation of Africa.* UNESCO, Paris.

Whitehouse A. M., Hall-Martin A. J. and Knight M. H. (2001). A comparison of methods used to count the elephant population of the Addo Elephant National Park, South Africa. *African Journal of Ecology*, **39**, 140–5.

Whittaker R. H. (1975). *Communities and ecosystems.* 2nd edn. Macmillan, New York.

Whyte I. J. (1985). *The present ecological status of the blue wildebeest* (Connochaetes taurinus taurinus, *Burchell, 1823*) *in the central district of the Kruger National Park.* MSc thesis, University of Natal, Pietermaritzburg, South Africa.

Willmer P., Stone G. N. and Mafunde D. (1998). Ants, pollinators and acacias in Mkomazi. In M. Coe, N. McWilliam, G. Stone and M. Packer, eds. *Mkomazi. The ecology, biodiversity and conservation of a Tanzanian savanna*, pp. 337–60, Royal Geographic Society (with the Institute of British Geographers), London.

Wilson D. E., Cole F. R., Nichols J. D., Rudran R. and Foster M. S. (1996). *Measuring and Monitoring Biological Diversity. Standard Methods for Mammals.* Smithsonian Institution Press, Washington and London.

Wirtz P. and Lörscher J. (1983). Group sizes of antelopes in an East African park. *Behaviour*, **84**, 135–55.

Wittemyer G. (2001). The elephant population of Samburu and Buffalo Springs National Reserve, Kenya. *African Journal of Ecology*, **39**, 357–65.

Woodford M. H. and Sachs R. (1973) The incidence of cysticercosis, hydratidosis, and sparganosis in wild herbivores of the queen Elizabeth National Park, Uganda.*Bulletin of Epizooic Diseases of Africa*,21, 265–72.

Woodroffe R., Ginsburg J. and Macdonald D. (1997). *The African wild dog.* IUCN, Switzerland & Cambridge.

Wragg F. C. (2002). Biodiversity and conservation of African mammals. PhD thesis, University of Leeds, UK.

Yadava P. S. (1991) Savannas of north-east India. In P. A. Werner, ed. *Savanna ecology and management*, pp. 3–10, Blackwell Scientific Publications.

Young T. P., Stubblefield C. H. and Isbell L. A. (1997). Ants on swollen-thorn acacias: species coexistence in a simple system. *Oecologia*, **109**, 98–107.

Young T. P., Palmer T. M. and Gadd M. E. (2005). Competition and compensation among cattle, zebras, and elephants in a semi-arid savanna in Laikipia, Kenya. *Biological Conservation*, **122**, 351–9.

Index

aardvark (*Orycteropus afer*) 24, 96
aardwolf (*Proteles cristatus*) 104
absolute population estimates 114
Acacia 16, 19, 45, 46
  ants 199–201
  birds 66
  giraffes 81
  insects 65
  mutualistic interactions 196–7
  phenology 39
  seasonal flowering patterns 39
  Serengeti–Mara 62
Acacia brevispica (prickly aka wait-a-bit thorn) 46
Acacia clavigera 63
Acacia commiphora 17, 86
Acacia drepanolobium (whistling thorn) 31, 46, 62, 115, 161–3, 200–2
Acacia erubescens (blue thorn) 46–7
Acacia gerrardii (grey-haired aka Gerrard's thorn) 46–7, 51, 63
Acacia hebeclada (candle-pod) 47
Acacia hockii 51, 184–5
Acacia kirkii (Kirk's aka flood-plain) 47
Acacia laeta (black-hooked thorn) 47–8
Acacia mellifera (hook thorn aka wait-a-bit thorn) 48
Acacia nigrescens (nob-thorn) 48
Acacia nilotica (Egyptian thorn) 45, 46, 48–9
Acacia robusta (splendid) 49, 51
Acacia senegal (three-thorned aka gum arabic) 49–50, 51
Acacia seyal (white thorn) 50, 184–5, 201
Acacia sieberiana (paperbark) 38, 50
Acacia tortilis (umbrella thorn) 45, 49, 51, 62, 196–7
Acacia xanthophloea (fever tree) 35, 51
Acacia zanzibarica 200
Aculeiferum senegal 45
aerial surveys 121–4, 147, 198
African elephant (Elephantidae family) 96
African horse sickness 180
African mahogany (*Afzelia africana*) 52
Afrotheria 68
age-specificity 127–8, 135
Alcelaphini tribe (hartebeests, topi, wildebeests and impala) 89–92
Amboseli National Park 25, 239
amensalism 155
Anacardiaceae 54–5
Andropogon greenwayi (grass) 61, 159
Andropogoneae 40
Angola 20–1, 22, 23
angular-stemmed commiphora (*Commiphora karibensis*) 54
animals 64–112
  birds 66–8
  insects 64–6
  see also mammals
annuals 34, 38
antelope 18, 122, 164, 176, 177
  dwarf 93–6
  horse 86–7
  spiral-horned 82–6
  see also roan; sable
anthrax (*Bacillus anthracis*) 144, 179, 180
Antilopine tribe (gazelles and dwarf antelopes) 92–6
antipredation 100
ants 7, 65, 200
Aonyx capensis (otter) 109
Aonyx congica (otter) 109
Apis mellifera 65
aquatic birds 4
area 27
Arecaceae 43
'arid' zone 12–13
arid/eutrophic savannahs 141–3
artificial markings 116–18
artiodactyls 68–70, 75
Arundinoideae 40
Arusha National Park (Tanzania) 158
assembly rules 219–24
assimilation efficiency 210–11
attack, risk of 174
Australia 1, 6–8, 9
avian fauna *see* birds

baboons 43, 55, 110–11, 234
badgers (Mustelidae family) 108–9, 196
Baikiaea plurijuga 22
Baikiaea woodlands 22, 51

Balanitaceae 53
Bambusoideae 40
banded mongoose (*M. mungo*) 106
banded weasel (*Poecilictis libyca*) 109
baobab (*Adansonia digitata*) 55
baobab family (Bombacaceae) 55
bark thickness 35
base ramification (trees) 35
bat-eared fox (*Otocyon megalotis*) 108
bats 7
beechwood (*Faurea salgina*) 43, 44
behavioural attributes 165
beisa oryx (*Oryx gazella beisa*) 85, 86
best fit model 139, 140
Bignoniaceae 57
biomass 141, 154
  /rainfall relationship 142–3, 153
  resistance 213
birds 18, 22, 24, 43, 66–8
  aquatic 4
  Australia 7
  South America 5
birth 127
  African buffalo 129
  zebras 140
black rhinoceros (aka hook-lipped) (*Diceros bicornis*) 19, 77, 78–9, 142, 143
  conservation 231–2
  ear notching 116–17
  predator/prey interactions 164
  skulls 74
black sable (*Hippotragus niger niger*) 87
black wildebeest (*Conochaetes gnou*) 91
black-backed jackal (*Canis mesomeles*) 97, 100, 107–8, 168
black-faced impala 91–2
black-footed cat (*Felis nigripes*) 103
black-hooked thorn (*Acacia laeta*) 47–8
blade (grass leaf) 36
blowflies 180
blue thorn (*Acacia erubescens*) 46
blue wildebeest 91
body size 98, 167, 168–9, 176–7, 191–2, 221
body weight 192, 220, 222, 223
bohor reedbuck (*Redunca redunca*) 89, 142
Bombacaceae 55
Bontebok National Park (southern Africa) 187
Borassus 16
borassus palm (*Borassus aethiopum*) 43
Boscia senegalensis 44
Boswellia carteri 17, 53
Botswana 22, 238
Bovidae family:
  buffalo 82
  gazelles and dwarf antelopes 92–6
  hartebeests, topi, wildebeests and impala 89–92
  horse antelopes 86–7
  reedbucks, kob and waterbuck 87–9
  spiral-horned antelopes 82–6
bovids 157, 223
Bovini tribe (buffalo) 82
Brachystegia 16
branching (grasses) 37
brindled gnu (*Connochaetes taurinus*) 91
broad-leaved beechwood (*Faurea speciosa*) 43–4
brown hyaena (*Parahyaena brunnea*) 100, 104
browsers 24, 72
  acacia 50–1
  competitive interactions 190–3
  conservation 220, 221, 223
  gazelles and dwarf antelopes 92, 94–5
  giraffes 81
  hartebeests, topi, wildebeests and impala 91
  horse antelopes 86
  predator/prey interactions 159–61
  reedbucks, kob and waterbuck 88, 89
  rhinoceroses 78
  spines 35
  spiral-horned antelopes 82–3, 84
  zebras 76
buffalo (*Syncerus caffer*) 9, 77, 82, 126–7, 128, 141–4, 147–9
  competitive interactions 186, 192
  conservation 212, 213, 214, 216, 219, 222, 231–2, 234
  feet 69
  mutualistic interactions 197
  numbers estimation, changes in 129–35
  predator/prey interactions 158–9, 166–9, 171, 175, 177, 180
Buffon's kob (*Kobus kobus kob*) 88
burke (*Burkeae africana*) 52
burning 33, 38, 65
Burseraceae 53–4
bush pigs 234
bushbuck (*Tragelaphus scriptus*) 74, 84, 85, 115, 142, 219

C_3 plants 1
C_4 plants 1
Caesalpinioideae 51
Caessalpiniodes 44
camels' foot trees (*Bauhinia* genus) 51
CAMPFIRE 235
Canada 165
candelabra tree (*Euphorbia candelabrum*) 54
candle-pod acacia (*Acacia hebeclada*) 47

Canidae family (dogs and jackals) 107–8
canids 168
canine distemper 178, 180
Cape fox (*Vulpes chama*) 108
cape grey mongoose (*Herpestes pulverulenta*) 106
Cape griffin vulture (*Gyps coprotheres*) 203
caper family (Capparaceae) 44
Capparis spinosa 44
capybara 4, 5–6, 10
caracal (*Felis (Caracal) caracal*) 103–4, 168
carbon dioxide 1
carcass biomass 151–2
carnivores 18, 24, 98–109, 165
  badgers, weasels, polecats and otters (Mustelidae family) 108–9
  cats (Felidae family) 101–4
  conservation 229, 230
  dogs and jackals (Canidae family) 107–8
  genets, civets and mongooses (Viverridae family and Herpestidae) 105
  hyaenas (Hyaenidae family) 104–5
  rainfall, plant biomass and grass-tree mixture 29
  species richness patterns 28
carrying capacity 140
caryopsis 37
cascading indirect interactions (in communities) 213–14
cassava plant (*Manishot esculenta*) 54
catenas/catenary sequences 61, 197
cats (Felidae family) 101–4, 168
cattle 211, 236
cellulase 1
cellulose 5–6, 70–1, 111
cerrado 3, 10
chacma baboon (*Papio hamadryas ursinus*) 110, 111
chaco 4–5
chamaephytes 33, 34, 35
cheetah (*Acinonyx jubatus*) 97, 100, 102–3, 152, 153, 154
  competitive interactions 195–6
  mutualistic interactions 204
  predator/prey interactions 167, 168, 170, 172, 177
chital 9
Chloridoideae 40
Chloris gayana 37
Chloris pycnothrix 62
Chobe National Park (Botswana) 22, 195
Chrysobalanaceae 44
CITES 231–3
civet (*Civettictis civetta*) 106
climatic effects 180
climatic patterns 10–15
climatically determined savannahs 31
coconut palm (*Cocos nucifera*) 43
coexistence 182, 185, 193
Coke's hartebeest (*Alcelpahus buselaphus cokii*) 89, 127, 190
coliiformes 66
Combretaceae 55–6
Combretum genus 16, 20, 55, 81
Combretum molle 63
commensalism 155
Commiphora abyssinica 17, 53
Commiphora africana (poison-grub) 53–4
Commiphora karibensis (angular-stemmed) 54
Commiphora mollis (soft-leaved) 54
Commiphora trothae 62–3
common eland (*Taurotragus oryx*) 77, 82–3
common hartebeest (*Alcelaphus buselaphus*) 89
common hippopotamus (*Hippopotamus amphibius*) 80–1
common or plains zebra (*Equus burchelli*) 75–6, 77, 122, 148, 166–7, 168
common (small-spotted) genet (*Genetta genetta*) 106
common (southern) reedbuck (*Redunca arundinum*) 89, 187
common waterbuck (*Kobus ellipsiprymus ellipsiprymus*) 87–8, 148
Community Areas Management Project for Indigenous Resources (CITES) (Zimbabwe) 233–4
community patterns 183
comparison of major savannahs 9–10
compensatory growth 157
competition 155
  apparent 155–6
  balanced 185
  exploitation 180–1
  interference 180–1, 194–5
  past 183, 190
  two-species 113, 182
  see also competitive interactions; interspecific; intraspecific
competitive interactions 180–96
  kleptoparasitism: African wild dogs and spotted hyaenas 194–6
  resource competition: niche differentiation between herbivores 186–94
  trees and grass 183–6
complexity and stability 212–13
'condition' (of animal) 134–5
Congo 25
connectance 212

conservation 205–39
  assembly rules 219–24
  energy flow 208–12
  habitat destruction and wildlife-human conflict 234–8
  hunting and poaching 230–4
  island biogeography 224–9
  wildlife tourism 238–9
  see also food webs
consumers 208
  primary 209
  secondary 209, 210
consumption efficiency 210–11
coprophagy (reingestion of faeces) 6
correspondence analysis 193–4
counting errors 127
Crematogaster mimosae (ant) 200–1
Crematogaster nigriceps (ant) 200–1
Crematogaster sjostedi (ant) 200–1
critically endangered species 75, 78, 87, 102
crop loss 234
Croton 63
Ctenium concinnum 40
Ctenium newtoni 40
Ctenium somalense 40
Cynodon dactylon grassland 61, 158, 159–60

D. macroblephora 61, 62
date palm (*Phoenix dactylifera*) 43
death:
  rates 127
  risk of 173–4
  see also mortality
declining species 150, 151
decomposers 209
decomposition rate (dung) 124
deer 5
defassa waterbuck (*Kobus ellipsiprymus defassa*) 85, 87
defecation rate 124
Demidoff's galago (*Galagoides demidoff*) 109
density (trees) 20
density dependence 140, 141
density independence 141
desert date family (Balanitaceae) 53
desert dwarf mongoose (*Herpestes hirtula*) 106
desert warthog (*Phacochoerus aethiopicus*) 80
detection, risk of 174
detritivores 65
Diamphidia beetle 54
diastema 73, 111
Dichanthium-Cenchrus-Lasiurus savannah 9
dicotyledons 43, 72
dik dik 95, 97, 122, 177
dilution effect 173
direct effects 205
disease 133, 134, 135, 144–5, 177–8, 179–80, 204
  see also rinderpest
disturbance driven savannahs 31
dogs (Canidae family) 107–8
  see also jackals; wild dog
doka savannah 22, 25
drought 33
dry season 38
dung 71
  beetles 65
  counts 124, 235–6
  decomposition 124
dwarf antelopes (Bovidae family, Antilopine and Neotragini tribes) 92–6
dwarf mongoose (*Helogale parvula*) 106
dynamic theory of island biogeography 225–6

ear notch markings 116–17
ebony family (Ebenaceae) 56–7
ebony tree (*Diospyros ebenum*) 57
Echinochloa pyramidalis 42
ecological features of savannahs 10
ecological niches 181–2, 184
ecotone 169, 171
'effect' (classification) 155
Egyptian mongoose (*Herpestes ichneumon*) 106
Egyptian thorn (*Acacia nilotica*) 45, 46, 48
Egyptian vulture (*Neophron percnopterus*) 203
eland 20, 122, 126, 127, 142, 143, 148, 149
  common 77, 82–3
  conservation 219
  predator/prey interactions 166–7, 177
elephant (Elephantidae family) 96–8, 119–20, 122, 126, 127, 142, 143, 157
  conservation 216, 218, 219, 231–2, 234
  forest (*Loxodonta cyclotis*) 96, 98, 222
  and mango trees 54
  mutualistic interactions 197
  population estimates 125
  predator/prey interactions 161–3, 164, 165
  savannah (*Loxodonta africana*) 96, 222
  teeth 74
  transects 124
elephant grass (*Pennisetum purpureum*) 20
endangered species 75, 76, 82, 84, 86, 90, 98, 103, 106
endemism 22
energy flow 208–12
energy turnover 192

Equidae family *see* zebras
Ethiopia 17
Ethiopian wolf (*Canis simensis*) 107
Etosha National Park (Namibia) 22, 180, 195
Euarchordata (+ Glires) 68
Eucalyptus 7, 8
Euclea 63
Euclea pseudebenus 56
Euphorbia guerichiana 19
Euphorbis family (Euphorbiaceae) 54
Eurasian otter (*Lutra lutra*) 109
Europe 229
excretion 209–10
extinct species 75, 222–3
extinction curves 225–6, 228

fatty acids 71
feeding facilitation and grazing succession 197–9
feeding height stratification 193
Felidae family (cats) 101–4
fennec fox (*Vulpes zerda*) 107
fever trees (*Acacia xanthophloea*) 35, 51
fire 31, 32, 34, 65, 162–3, 164
  competitive interactions 184, 186
  conservation 209
  damage 35
  reduction of risk 159, 165
  Serengeti–Mara ecosystem 60–2
  see also burning
flamingos 239
fleeing 176
floral richness 27
florets 37
flowers 45
food chains 212
food supply 136, 141
food webs 212–19
  cascading interactions 213–14
  complexity and stability 212–13
  inside park sub model 217
  outside park sub model 217
  predator sub model 216–17
  Serengeti–Mara 214–19
  ungulate sub model 216
  vegetation sub model 216
foot and mouth disease 180
forest buffalo (aka red buffalo) (*Syncerus caffer nanus*) 82
forest elephant (*Loxodonta cyclotis*) 96, 222
forest–savannah mosaic 15, 25–6, 44, 66
fox 107, 108
frankincense 53
'freezing' 177
freshwater habitats 66
fringe-eared oryx (*Oryx gazella callotis*) 86
frogs 7
fundamental niche 182

gallery forests 26
gazelles (*Gazella*) 18, 70, 92–6, 168
  see also Grant's gazelle; Thomson's gazelle
gemsbok (*Oryx gazella gazella*) 86, 144
genets (Viverridae family and Herpestidae) 105
geophytes 33, 34, 37
Geosphere–Biosphere Programme 206
geoxyles 37
gerbil (*Gerbillus*) 17
gerenuk (giraffe-necked antelope) (*Litocranius walleri*) 92, 122
giant diospyros (*Diospyros abyssinica*) 57
giant eland (*Taurotragus derbianus*) 20
giant sable (*Hippotragus p. varianii*) 87
giraffe (*Giraffa camelopardalis*) 81–2, 122, 126–7, 142–4, 146–9, 151
  acacia 51
  competitive interactions 193
  conservation 214, 219
  predator/prey interactions 161, 166–7, 168
  reticulated 77, 115–16
  skulls 74
  western 20
Giraffidae family *see* giraffe
Global Change and Terrestrial Ecosystems 206
golden jackal (*Canis aureus*) 107, 168
Gombe Game Controlled Area (Tanzania) 24
gorilla (*Gorilla gorilla*) 109
Gran Sabana 5
Grant's gazelle (*Gazella granti*) 85, 92–4, 122, 126, 127, 142, 143
  competitive interactions 190, 192
  conservation 219
grass 3, 17, 19, 20–1, 26, 40–2, 189–90
  conservation 215
  cover 17
  greeness 189–90
  height 190
  long 61, 65
  predator–prey interactions 157
  production 65
  rainfall, plant biomass and grass–tree mixture 30
  Serengeti–Mara 66
  short 65
  and shrub savannah 15, 16–19, 28, 44
  tall 7, 61
  /tree coexistence 29–32, 184
  tufted 19
  tussock 36, 37
grasshoppers 7–8, 65–6, 218

grazers 23–4, 34, 72
Australia 8
buffalo 82
competitive interactions 190–3
conservation 212, 214, 220, 221, 222, 223
cyclic 159
damage caused by 35, 37
gazelles and dwarf antelopes 93–4
hartebeests, topi, wildebeests and impala 89, 90, 91
hippopotamuses 80
horse antelopes 86, 87
pigs 79
predator–prey interactions 157, 158, 159
reedbucks, kob and waterbuck 88
rhinoceroses 78
spiral-horned antelopes 82–3
teeth 73–4
zebras 76
grazing:
cyclic 159
lawns 158, 163, 197
succession and feeding facilitation 197–9
greater honey guide (*Indicator indicator*) 196
greater kudu (*Tragelaphus strepsiceros*) 83, 142, 148
Grevillea family (Proteaceae) 43–4
Grevy's zebra (*Equus grevyi*) 76, 77
grey-haired acacia (aka Gerrard's thorn) (*Acacia gerrardii*) 46–7
griffon vulture (*Gyps* sp.) 203
ground pangolin (*Smutsia temminckii*) 24
group living 100, 169, 171, 172–3
group size 171–2, 174–7
Guenther's dik dik (*Madoqua guentheri*) 95–6
guiding behaviour 196
Guinea 25
Guinea baboon (*Papio hamadryas papio*) 110
Guinea savannahs 34

habitat 189–90
destruction 234–8
diversity 226
hamadryas baboon (*Papio hamadryas hamadryas*) 110
hares 165, 167
hartebeests (Bovidae family, Alcelaphini tribe) 89–92, 126, 142, 216
Coke's 89, 127, 190
Hausa (Villier's) genet (*Genetta thierryi*) 105–6
heart-fruit (*Hymenocardia acida*) 54
height (trees) 20, 35
hemicryptophytes 33, 34
herb layer biomass 29–30
herbaceous species 3
herbivores 35, 141, 143, 144, 145
mutualistic interactions 196–7
niche differentiation 186–94
/plant interactions 155, 157–65
rainfall, plant biomass and grass–tree mixture 29
South America 5
tree recruitment 164–5
see also herbivores and plants
herbivores and plants 157–65
browsing 159–61
fire 162–5
grazing 157–9
grazing lawns 158–9, 163
herbivory 31, 184, 186
Herpestidae family (genets, civets and mongooses) 105–6
Heteropogon contortus 42
hiding 176, 177
hindgut fermentation 70–1, 73
hippopotamus (Hippopotamidae family) 77, 80–1, 142
competitive interactions 186
conservation 222, 234
feet 69
skulls 74
teeth 73
Hippotragini tribe (horse antelopes) 86–7
Hippotragus niger roosevelti (horse antelope) 87
Holling's disc equation 138, 215
honey badger *see* ratel
honeybees (*Apis*) 65
hooded vulture (*Necrosyrtes monachus*) 203
Horn of Africa 17, 18
horse 69
horse antelopes (Bovidae family, Hippotragini tribe) 86–7
human population growth 217
hunting 203, 230–4
behaviour 98
cursorial 98
hyaena (Hyaenidae family) 104–5, 144
brown 100, 104
conservation 217
ear notching 18
predator/prey interactions 168, 178
see also spotted hyaena
hyraxes 96

immigration 129
curves 225–6, 228
impala (*Aepyceros melampus*) 85, 89–92, 122, 126–7, 142, 144, 146–9, 151
competitive interactions 190, 192, 193
conservation 216, 219
predator/prey interactions 161, 172, 175
India 8–9
Indigofera basiflora (herb) 61

'inhibiting' 164
insectivores 24
insects 8, 64–6
inselbergs 65
inside park sub model 217
International Biological Programme 206
internodes 36
interspecific competition 180, 181, 182, 183
  conservation 220
  mutualistic interactions 198–9
  resource competition 186, 187, 188, 189
  tree and grass competition 184, 185
Intertropical Convergence Zone 11–13, 58
intraspecific competition 185, 188
intraspecific interactions 180, 182
invertebrates 7
island biogeography 224–9
  protected savannah areas 226–9
Isoberlinia doka 25
IUCN 75

jacaranda family (Bignoniaceae) 57
Jacaranda mimosifolia 57
jackals (Canidae family) 97, 100, 107–8, 168
Jackson's hartebeest (*Alcelpahus buselaphus jacksoni*) 89
jaw muscles 72

k factor analysis 131, 132, 135, 136–7, 141
Kafue National Park (Zambia) 169
Kalahari sand terminalia (*Terminalia brachystemma*) 56
kangaroo 8, 10
karoo bushes/shrubs 19
Kasungu National Park (Malawi) 234
Kenya 17, 18, 19, 171–2, 238–9
  Wildlife Service 237
  see also Laikipia
'key factor' 133
keystone species 218
'killing' 164
Kirk's acacia (aka flood-plain acacia) (*Acacia kirkii*) 47
Kirk's dik dik (*madoqua kirki*) 95, 97
kleptoparasitism 194–6, 203–4
klipspringer 219
kob (*Kobus kob*) 87–9, 120, 142
kongoni 122, 219
kopjes 65
korrigum (*Damaliscus lunatus korrigum*) 90
Kruger National Park (South Africa) 22, 143–53
  competitive interactions 186, 193
  predator/prey interactions 159–60, 161, 166–7, 179–80
kudu 143, 144, 149, 151
  competitive interactions 193
  conservation 211, 212
  predator/prey interactions 161, 180
  see also greater kudu; lesser kudu

Laikipia (Kenya) 25, 139–41, 161, 200–1
  conservation 234, 235–7
Laikipia Wildlife Forum 237
Lake Manyara National Park (Tanzania) 144, 159–60, 169
Lake Nakuru National Park 239
Lamina (grass) 36
lammergeier (bearded vulture) (*Gypaetus barbatus meridionalis*) 203
land use, changes in 217
Lannea humilis 184–5
lappet-faced vulture (*Torgus tracheliotus*) 203
large false mopane (*Guibourtia coleosperma*) 52
large sour plum (*X. caffra*) 44
large-leaved euclea (*Euclea natalensis*) 56
large-leaved munondo (*Julberlandia paniculata*) 52
large-leaved terminalia (*Terminalia mollis*) 56
large-spotted genet (*Genetta tigrina*) 106
Laurasiatheria 68
Lead-wood (*Combretum imberbe*) 55–6
leaves 35–6, 45
lechwe (*Kobus leche*) 169
Leguminosae 44–52
leopard (*Panthera pardus*) 100, 102, 144, 152, 153, 154
  mutualistic interactions 204
  predator/prey interactions 167, 168, 170, 177
lesser kudu (*Tragelaphus imberbis*) 84, 122
Lewa Wildlife Conservancy 237–8
Lichenstein's hartebeest (*Alcelpahus (Sigmoceros) lichensteinii*) 89
life forms 32–8
  annuals 38
  perennials with above ground seasonal vegetation 36–8
  perennials with woody ground structures (trees and shrubs) 35–6
ligule (grass) 36
lily family (Liliaceae) 43
Lincoln–Peterson estimate 118
lion (*Panthera leo*) 107, 100–1, 115, 144–5, 147, 150–4
  competitive interactions 195–6
  conservation 215, 239
  mutualistic interactions 204
  predator/prey interactions 167, 168, 169, 170, 171, 177, 178, 180
livestock ranches 236–7

lizards 7, 8
llanos de Moxos 4
llanos del Orinoco 5
locusts 43
Lotka Volterra model 165, 182, 184
lovebirds (*Agapornis* sp.) 119
lower risk species75, 87, 90, 91, 93, 94, 104–5, 111
Luangwa Valley 180
Lutra maculicollis (otter) 109
lynx (*Felis lynx*) 165, 167

macropods 8
mahobohobo (*Uapaca kirkiana*) 54
Malawi 234
malnutrition 150, 180
 see also under nutrition
mammals 18, 19, 23, 68–112
 forest-savannah mosaic 26
 India 9
 Primates109–11
 rodents 111–12
 Serengeti–Mara 65
 South America 5
 species richness patterns 28
 subungulates 96–8
 see also carnivores; ungulates
mango family (Anacardiaceae) 54–5
'many eyes' theory (increased vigilance) 171
marabou stork (*Leptoptilos crumeniferus*) 204
marsh deer (*Blastocerus dichotomus*) 5
marula (*Sclerocarya birrea*) 54
mathematical description/model 137
mean annual precipitation 31–2, 184, 186
meerkat (*Suricata suricata*) 106
mesquites (*Prosopis* spp.) 51
metal ring tagging 116
midgrass savannahs 7
Mimosoideae 44, 51
miombo (aka Prince-of-Wales' feathers) (*Brachystegia boehimii*) 23, 52
miombo (msasa) (*Brachystegia spiciformis*) 22–3, 27, 43–4, 52
miombo (Muuyombo) 23
miombo woodland savannah 24, 28, 35, 52, 54, 55, 143
missing species 222, 223
mixed feeders 72
Mkomazi Game Reserve (Tanzania) 65, 200
mobola family (Chrysobalanaceae) 44
mobola plum (*Parinari curatellifolia*) 44
mongalla gazelle (*Gazella thomsonii albonotata*) 93
mongooses (Viverridae family and Herpestidae) 105–6
monkeys 4, 109–10, 234
monocotyledons 43, 72
Monocymbium ceresiforms 42
monsoon tallgrass savannahs 7
mopane tree (*Colophospermum mopane*) 20–1, 22, 44, 52
morphology and life history 32–40
 life forms 32–8
 phenology 38–40
mortality 145–8
 African buffalo 129, 130, 131, 132, 133, 134, 135
 density dependent 128
 density independent 128
 wildebeest 136, 137, 139
mountain acacia (*Brachystegia glaucascens*) 52
mountain reedbuck (*Redunca fulvorufula*) 89
mountain zebra (*Equus zebra*) 19
mousebirds 66
Mozambique 238
muhutu (*Terminalia brownii*) 56
multiple mark/release/recapture methods 118–19
munondo (*Julberlandia globiflora*) 52
Mustelidae family (badgers, weasels, polecats and otters) 108–9
mutanga (*Elaeodendron buchanii*) 57
mutualistic interactions 155, 196–204
 Acacia ants 199–201
 feeding facilitation and grazing succession 197–9
 scavenging vultures 201–4
myrrh family (Burseraceae) 53–4

Nairobi National Park 171–2
Namibia 19, 21, 22
 see also Etosha
narrow-leaved mahobohobo (*Uapaca nitida*) 54
natural experiments 186–7, 188
natural markings 115–16
Nazinga Game Ranch 124
Neotragini tribe (gazelles and dwarf antelopes) 92–6
neutralism 155
niche differentiation 186–94, 220
 buffalo 186, 191
 Coke's hartebeest 190
 giraffe 191, 193
 Grant's gazelle 190, 191, 192
 hippopotamus 186
 impala 190, 191, 192, 193
 kudu 193
 oribi 186–7
 steenbok 193
 Thomson's gazelle 187–92
 topi 190, 191, 192
 warthog 190
 waterbuck 190

white rhinoceros 186
wildebeest 187–92
zebra 187–92
niches:
ecological 181–2, 184
fundamental 182
realized 182
root 184, 185
separation 185
see also niche differentiation
nitrogen 39–40, 44, 134
nob-thorn (*Acacia nigrescens*) 48
nodes 36, 37
nodulated legumes 196
nonruminants *see* ungulates
North America 229
northern quoll (*Dasyurus hallucatus*) 8
not endangered species 82, 83, 88, 91, 92
not threatened species 75, 76, 81
numbers estimation 114–36
aerial survey 121–4
African buffalo 129–35
artificial markings 116–18
multiple mark/release/recapture methods 118–19
natural markings 115–16
removal method 119–20
road transects 120–1, 224
wildebeest 135–6

oil palm (*Elaeis guineensis*) 43
Olaceae 44
olive baboon (*Papio hamadryas anubis*) 110
open tree savannah 19
open woodland savannah 63
oribi (*Ourebia ourebi*) 142, 146, 147, 168, 186–7, 219
Orthoptera 65–6
oryx (*Oryx gazella*) 17–18, 85–6, 142, 143, 144, 219
ostrich (*Struthio camelus*) 66–7, 126, 127
otters (Mustelidae family) 108–9
outside park sub model 217

palatability 39–40
pale fox (*Vulpes pallida*) 108
palm (*Borassus aethiopum*) 35
palm civet (*Nandinia binotata*) 106
palm family (Arecaceae) 43
pampas deer (*Ozotoceros bezoarticus*) 5
Paniceae 40
panicle 37, 40
Panicoideae 40
Panicum maximus 42, 62
pantanal 4
paperbark acacia (*Acacia sieberiana*) 50
Papilionoideae 44
parasite/host interactions 155, 177–80
parasites 133–4
park revenue 217
patas monkey (*Erythrocebus patas*) 109–10
Pennisetum mezianum (grass) 62, 159
Pennisetum purpureum (grass) 37
Pennisetum schimperi (grass) 76
Pennisetum ultramineum (grass) 159
perennials:
with above ground seasonal vegetation 34, 36–8
with non-lignified storage organs 36
with permanent above ground woody structures 34
with woody ground structures (trees and shrubs) 35–6
with woody underground storage organs 36, 37–8
perissodactyls 68
phanerophytes 33, 34, 35
phenological niche separation 185
phenology 38–40
photosynthesis 1, 207
photosynthetic carbon reduction pathway (Calvin-Benson Cycle) 1
Phragmites-Saccharum-Imperata savannah 9
physical attributes 165
physical dimension 181
physionomy 15–16, 22, 26–7
phytochorian 22
pigs (Suidae family) 70, 73, 79–80, 234
plant:
biomass 29–32
food species separation 190
parts eaten separation 190
poaching 164, 217, 230–4
pod-bearing family (Leguminosae) 44–52
poison-grub commiphora (*Commiphora africana*) 53–4
polecat (*Mustela putorius*) 109
polecats (Mustelidae family) 108–9
pollination 38–9, 196
Pooideae 40
population:
density 114–15, 118, 120, 130, 150
dynamics 217
estimates 166
models 136–41
wildebeest 137–9
zebra 139–41
numbers 148, 149
size 114–15, 118, 119, 137
potential evaporation 207–8
predation 133, 152
competitive interactions 186
mutualistic interactions 198, 199
regulating populations 145–6
single species populations 144–8, 150–3
wildebeest 136

predator/predators 18, 19
  attack success 217
  confusion 172–3
  forest-savannah mosaic 26
  mutualistic interactions 204
  removal 146–7
  sub model 216–17
predator–prey type interactions 155, 156–80
  parasites and hosts 177–80
  predators and prey 165–77
    dilution effect 173
    groups 169, 171–2, 174–7
    habitat changes 169
    physical size 167–8
    predator confusion 172–3
    quasi-cyclic fluctuations 166–7
    risk of death 173–4
    stable limit cycles 165–6
    vigilance 171–2
  see also herbivores and plants
pregnancy rate 137
prey handling time 217
prickly acacia (*Acacia brevispica*) (aka wait-a-bit-thorn) 46
primary production 141, 158, 206–8, 209, 210
  above ground 209
  efficiency 210–11
  of green grass 157
  gross 206
  net 206–8, 209, 211
primates 24, 99, 109–11
Prosopis africana 51
Protea caffra 44
Proteaceae 43–4
protected areas 22, 24, 26, 226–9
protection 196
protein 71, 72, 134, 135, 192, 211
pseudogall 199–200
puku (*Kobus vardani*) 89
puku (*Kobus vardonii*) 169
purple-leaved albizia (*Albizia antunesiana*) 51
purple-pod terminalia (*Terminalia prunoides*) 56
pygmy hippopotamus (*Hexaprotodon liberiensis*) 80

quadrats 124
Queen Elizabeth National Park (Uganda) 186
quiver tree 19

rabies 178
radio transmitter tracking 117–18
raid frequency of crops 234
rainfall 2–3, 9–11, 14, 16–17, 19–23, 29–32
  Australia 6–7
  and buffalo 129, 135
  competitive interactions 186
  conservation 215, 216
  dependent model 140, 141
  forest-savannah mosaic 26
  mediated density dependent model 140, 141
  predator/prey interactions 166, 167
  Serengeti–Mara 57–60
  single species populations 147–52
  and wildebeest 137, 138
  see also biomass/rainfall relationship; mean annual precipitation
rainforests 26, 66
ratel (honey badger) (*Mellivora capensis*) 109, 196
realized niche 182
red bushwillow (*Combretum apiculatum*) 55
red fox (*Vulpes vulpes*) 107
red hartebeest (*Alcelpahus buselaphus caama*) 89
red kangaroo (*Macropus rufus*) 8
red-leaved rock fig (*Ficus ingens*) 38
Reduncini tribe *see* kob; reedbucks; waterbuck
reedbucks (Bovidae family, Reduncini tribe) 87–9, 142, 187, 219
reintroductions 186–7
relative population estimates 114
relaxation 226
removal method 119–20, 186
resource competition *see* niche differentiation
resource dimensions 181
resource utilization curves 182
respiration 209–10
reticulated giraffe 77, 115–16
'reversing' 164
Rhigozum genus 57
rhinoceros (Rhinocerotidae family) 78–9
  feet 69
  perissodactyls 75
  predator/prey interactions 161
  teeth 72–3
  see also black rhinoceros; white rhinoceros
Rhodesian teak (*Baikiaea plurijuga*) 51–2
Rift Valley Fever 180
rinderpest 133, 134, 144, 178–9
road transects 120–1
roan antelope (*Hippotragus equinus*) 20, 86–7, 142–3, 148–9, 151, 219
rock wallaby 8
rodents 111–12, 157
root niches 184, 185
roots 36, 37
roughage eaters *see* grazers
Rukwa Valley (Tanzania) 197
ruminants 70–2
  digestion 196

skulls 74
see also ungulates
Ruppell's fox (*Vulpes ruppelli*) 107
Ruppel's griffon vulture (*Gyps rupellii*) 203

sable antelope (*Hippotragus niger*) 24, 85, 87, 142, 148, 149, 151
safari parks 19
Sahelian savannahs 16, 17, 34
sambar 9
sarcoptic mange 180
sausage tree (*Kigelia africana*) 57
savannah baboon (*Papio hamadryas*) 110–11
savannah buffalo (*Syncerus caffer caffer*) 82
savannah elephant (*Loxodonta africana*) 96, 106, 222
scavenging 100, 196, 201–4
Schmidtia bulbosa 42
scimitar oryx (*Oryx dammah*) 17–18
seasonal habitat changes 169
secretary bird (*Sagittarius serpentarius*) 67–8
seed dispersal 196
Sehima-Dichantium savannah 9
selective eaters *see* browsers
Selous Game Reserve (Tanzania) 24, 173–4, 175
Senegalia 46
separation by habitat 190
Serengeti–Mara 17–18, 25, 57–63, 125–7, 141, 143–4, 147, 152–3
competitive interactions 187–90, 191, 192–3, 194, 195, 196
conservation 213–19, 228, 230, 231, 232, 239
mutualistic interactions 197, 198, 199, 204
predator/prey interactions 157–9, 162–3, 164–5, 167–8, 169, 175, 178, 180
rainfall and green biomass 57–60
single species populations 144, 146–7
soils, fire and vegetation 60–3
seroprevalence 177–8
serval (*Felis (Leptailurus) serval*) 97, 103, 168
sheath 36
Sheppard's tree (*Boscia albitrunca*) 44
shrub and grass 23
shrub and woodland savannah 62
shrub savannah (bushveld) 19
sickle bush (*Dichrostachys cinerea*) 51
sickle-leaved albizia (*Albizia harveii*) 51
side-striped jackal (*Canis adustus*) 107–8
silver terminalia (*Terminalia sericea*) 56
single species populations 113–54
arid/eutrophic savannahs 141–3
biomass 141, 154
biomass/rainfall relationships 142–3, 153
disease 144–5
food supply 141
malnutrition 150
mortality 145–8
predation 144–8, 150–3
rainfall 147–52
soil nutrients 141–3
see also numbers estimation; population models
skulls of carnivores 99
slender mongoose (*Herpestes sanguineus*) 106
small false mopane (*Guibourtia conjugata*) 52
small sour plum (*X. americana*) 44
smelly boscia (*Boscia foetida*) 44
snakes 7
snowberry tree (*Securinega virosa*) 54
snowshoe hares (*Lepus americanus*) 165, 167
soft-leaved commiphora (*Commiphora mollis*) 54
soil 16, 34
competitive interactions 186
nutrients 141–3
Serengeti–Mara ecosystem 60–3
Solanum incanum (herb) 61
Solanum panduraeforme 161
solar radiation 207–8
Somalia 17
Somali-Masai dry savannah 17
sour plum family (Olaceae) 44
sourveld 40
South Africa 22
see also Kruger National Park
South America 1, 3–6, 9
South Luangwa National Park (Zambia) 24
southern reedbuck 142
species interaction 155–204
see also competitive interactions; mutualistic interactions; predator–prey type interactions
species richness patterns 26–8
species–area relationship 224–5, 226–8, 229
spikelets 37, 40
spines 35, 45
spiral-horned antelopes (Bovidae family, Strepsicerotini tribe) 82–6
splendid acacia (*Acacia robusta*) 49, 52
Sporobolus marginatus 159
Sporobolus nitens 42
spotted hyaena (*Crocuta crocuta*) 97, 100, 105, 118, 145, 152–4, 194–6
conservation 215
mutualistic interactions 204
predator/prey interactions 167, 168, 170, 177, 179

springbok (*Antidorcus marsupialis*) 94, 142, 144
stabilizing species 151
stalking and/or ambush 98
starvation 204
steenbok (steinbok) (*Raphicerus campestris*) 95, 122, 161
  competitive interactions 193
  conservation 219
  predator/prey interactions 177
Strepsicerotini tribe (spiral-horned antelopes) 82–6
striped hyaena (*Hyaena hyaena*) 104–5
striped polecat (zorilla) (*Ictonyx striatus*) 109
striped weasel (*Poecilictis libyca*) 109
Struthioniformes 66
subungulates 96–8
Sudan–Zambezian savannahs 34
Suidae family (pigs) 84–5
survival 140, 150–1
suvivorship curve 138–9
Swayne's hartebeest (*Alcelpahus buselaphus swaynei*) 90
sweetveld 40

Tanzania 17, 18, 19, 24, 39, 141–3, 197
  see also Lake Manyara; Mkomazi; Selous
Tarangire ecosystem 219
Techlea 63
teeth 72–4, 98, 99, 111
temperature 2, 9, 10–11, 14, 16, 17, 19, 20, 21, 22, 23
  Australia 6–7
  forest-savannah mosaic 26
  Serengeti–Mara 57–8
Terminalia genus 16, 20, 55, 56
Terminalia mollis 63
termites 1, 7–8, 61, 65
Tetraponera penzigi (ant) 200–1
theileriosis 180
Themeda arundinella 9
Themeda triandra 61, 62
therophytes 33, 34, 38
Thomson's gazelle (*Gazella thomsonii*) 93, 94, 126, 127, 144, 146, 147
  competitive interactions 187–9, 190, 192
  conservation 213, 214, 216, 219
  mutualistic interactions 197–9
  predator/prey interactions 172
  Serengeti–Mara 60
thorn-tree zone 62
three-thorned acacia (aka gum arabic) (*Acacia senegal*) 49–50, 52
tiang (*Damaliscus lunatus tiang*) 90
tillering 37
time foraging 172
time vigilant 172
topi (*Damaliscus lunatus*) 85, 89–92, 126, 127, 142, 143, 146
  competitive interactions 190, 192
  conservation 216, 219
  mutualistic interactions 197
topi (*Damaliscus lunatus jumela*) 147
tora (*Alcelaphus buselaphus tora*) 89, 90
tourism 217, 230–1, 235, 237, 238–9
tree and shrub savannah 3, 15, 16, 19–22, 61
trees 26, 43–57
  baobab family (Bombacaceae) 55
  caper family (Capparaceae) 44
  cover 31
  desert date family (Balanitaceae) 53
  ebony family (Ebenaceae) 56–7
  Euphorbis family (Euphorbiaceae) 54
  and grass, competition between 183–6
  Grevillea family (Proteaceae) 43–4
  jacaranda family (Bignoniaceae) 57
  lily family (Liliaceae) 43
  low-branched 35
  mango family (Anacardiaceae) 54–5
  Mimosoidea 51
  mobola family (Chrysobalanaceae) 44
  myrrh family (Burseraceae) 53–4
  palm family (Arecaceae) 43
  pod-bearing family (Leguminosae) 44–52
  recruitment 164–5, 186
  sour plum family (Olaceae) 44
  terminalia family (Combretaceae) 55–6
  see also tree and shrub savannah
trophic approach 206
tsessebe (*Damaliscus lunatus lunatus*) 90, 142, 143, 148, 149, 151
tsetse flies 118
two species competition 113, 182

Uganda 24, 186
Ugandan kob (*Kobus kobus thomasi*) 88
umbrella thorn (*Acacia tortilis*) 49
under nutrition 133, 134–5, 136, 150
unexpected effects 205
ungulates 1, 21–2, 24, 70–1, 72–3, 141
  Australia 8
  competitive interactions 191
  conservation 219, 220
  nonruminants 75–81
    hippopotamuses (Hippopotamidae family) 80–1
    pigs (Suidae family) 79–80
    rhinoceroses (Rhinocerotidae family) 78–9
    zebras (Equidae family) 75–7
  predator/prey interactions 179
  ruminants 81–96
    buffalos (Bovidae family, Bovini tribe) 82

gazelles and dwarf antelopes (Bovidae family, Antilopine and Neotragini tribes) 92–6
giraffes (Giraffidae family) 81–2
hartebeests, topi, wildebeests and impala (Bovidae family, Alcelaphini tribe) 89–92
horse antelopes (Bovidae family, Hippotragini tribe) 86–7
reedbucks, kob and waterbuck (Bovidae family, Reduncini tribe) 87–9
spiral-horned antelopes (Bovidae family, Strepsicerotini tribe) 82–6
Serengeti–Mara 65
skulls 74
South America 5
species richness patterns 28
sub model 216
teeth 73–4
United Kingdom 229
urea recycling 71
utilization function 181

Vachellia 46
variable combretum (*Combretum collinum*) 56
vegetation 29–63
grasses 40–2
patterns 15–26
forest-savannah mosaic 25–6
grass and shrub savannah 16–19
tree and shrub savannah 19–22
woodland savannah 22–5
rainfall, plant biomass and grass–tree mixture 29–32
sub model 216
see also morphology and life history; Serengeti–Mara; trees
velvet-leaved combretum (*Combretum molle*) 56
verbal description/model 137
vervet (savannah or green) monkey (*Chlorocebus aethiops*) 110, 234
Viverridae family (genets, civets and mongooses) 105–6
vulnerable species 75, 86, 91, 92, 101
vultures 203–4

wait-a-bit-thorn (aka hook thorn) (*Acacia mellifera*) 48
wallaby 8
warthog (*Phacochoerus africanus*) 77, 79–80, 122, 126–7, 142, 144, 146–9, 151
competitive interactions 190
conservation 219
predator/prey interactions 167
skulls 74
teeth 73
water 207
water buffalo 9
water storage facilities (trees) 36
waterbuck (*Kobus ellipsiprymus*) 85, 87–9, 126–7, 142–4, 148–51
competitive interactions 190
conservation 219
weasel (*Mustela nivalis*) 109
weight ratios 224
western giraffe (*Giraffa camelopardalis peralta*) 20
western hartebeest (*Alcelpahus buselaphus major*) 89
western rhigozum (*Rhigozum brevispinosum*) 57
western woody euphorbia (*Euphorbia guerichiana*) 54
wet season 38
'wet' zone 12
whistling thorn (*Acacia drepanolobium*) 46, 62, 200–1, 202
white bauhinia (*Bauhinia petersiana*) 51
white rhinoceros (aka square-lipped) (*Ceratotherium simum*) 78, 79, 186, 222
white thorn (*Acacia seyal*) 50
white-backed griffin (*Gyps africanus*) 203
white-bearded wildebeest 91
white-eared kob (*Kobus kob leucotis*) 88, 120
white-headed vulture (*Trigonoceps occipitalis*) 203
white-tailed deer (*Odocoileus virginianus*) 5
white-tailed mongoose (*Ichneumia albicauda*) 106
wild cat (*Felis silvestris*) 103, 168
wild dog (*Lycaon pictus*) 97, 100, 107–8
mutualistic interactions 204
predator/prey interactions 167, 168, 170
single species populations 128, 152–3
species interactions 173–4, 177, 178, 194–6
wild green-hair tree (*Parkinsonia africana*) 52
wildebeest (*Connochaetes taurinus*) 10, 85, 89–92
competitive interactions 187–9, 190, 192
conservation 213, 214, 215, 216, 217, 218, 219, 239
mutualistic interactions 197–9
numbers estimation, changes in 135–6
predator/prey interactions 158, 159, 164, 168, 173–4, 179, 180
Serengeti–Mara 58, 60
single species populations 128, 132–3, 141–2, 144–5, 148–51
transects 123–6
wildebeest model 137–9
wildlife tourism 238–9

wildlife–human conflict 234–8
wolves 107
wooded grassland 26
woodland savannah 15, 16, 20, 21, 22–5
  mobola family 44
  see also miombo
woody species 17
world-wide distribution of savannahs 2–10
  Africa 6
  Australia 6–8
  comparison of major savannahs 9–10
  India 8–9
  South America 3–6

Xenartha 68

yearling/adult ratio 137
yellow baboon (*Papio hamadryas cyanocephalus*) 110
yellow mongoose (*Cyntis penicillata*) 106
yellow tree bauhinia (*Bauhinia tomentosa*) 51

Zambezian region 20–1, 26
Zambia 23, 24, 169
zebra model 139–41
zebras (Equidae family) 75–7, 115, 141, 142, 143, 144, 145, 149, 151
  common (plains) 75, 76, 77, 122, 148, 166–7, 168
  competitive interactions 187–99, 190, 192
  conservation 213, 214, 216, 219, 235–6
  hindgut fermentation 70
  mountain 19
  mutualistic interactions 197–9
  perissodactyls 75
  predator/prey interactions 167
  Serengeti–Mara 60
  skulls 74
  teeth 73
Zimbabwe 22, 23, 233–4